LONDON MATHEMATICAL SOCIETY LECTURE NOTE SERIES

The books in the series listed below are available from booksellers, or, in case of difficulty, from Cambridge University Press.

34 Representation theory of Lie groups, M.F. ATIYAH *et al*
36 Homological group theory, C.T.C. WALL (ed)
39 Affine sets and affine groups, D.G. NORTHCOTT
40 Introduction to H_p spaces, P.J. KOOSIS
46 p-adic analysis: a short course on recent work, N. KOBLITZ
49 Finite geometries and designs, P. CAMERON, J.W.P. HIRSCHFELD & D.R. HUGHES (eds)
50 Commutator calculus and groups of homotopy classes, H.J. BAUES
57 Techniques of geometric topology, R.A. FENN
59 Applicable differential geometry, M. CRAMPIN & F.A.E. PIRANI
62 Economics for mathematicians, J.W.S. CASSELS
66 Several complex variables and complex manifolds II, M.J. FIELD
69 Representation theory, I.M. GELFAND *et al*
74 Symmetric designs: an algebraic approach, E.S. LANDER
76 Spectral theory of linear differential operators and comparison algebras, H.O. CORDES
77 Isolated singular points on complete intersections, E.J.N. LOOIJENGA
79 Probability, statistics and analysis, J.F.C. KINGMAN & G.E.H. REUTER (eds)
80 Introduction to the representation theory of compact and locally compact groups, A. ROBERT
81 Skew fields, P.K. DRAXL
82 Surveys in combinatorics, E.K. LLOYD (ed)
83 Homogeneous structures on Riemannian manifolds, F. TRICERRI & L. VANHECKE
86 Topological topics, I.M. JAMES (ed)
87 Surveys in set theory, A.R.D. MATHIAS (ed)
88 FPF ring theory, C. FAITH & S. PAGE
89 An F-space sampler, N.J. KALTON, N.T. PECK & J.W. ROBERTS
90 Polytopes and symmetry, S.A. ROBERTSON
91 Classgroups of group rings, M.J. TAYLOR
92 Representation of rings over skew fields, A.H. SCHOFIELD
93 Aspects of topology, I.M. JAMES & E.H. KRONHEIMER (eds)
94 Representations of general linear groups, G.D. JAMES
95 Low-dimensional topology 1982, R.A. FENN (ed)
96 Diophantine equations over function fields, R.C. MASON
97 Varieties of constructive mathematics, D.S. BRIDGES & F. RICHMAN
98 Localization in Noetherian rings, A.V. JATEGAONKAR
99 Methods of differential geometry in algebraic topology, M. KAROUBI & C. LERUSTE
100 Stopping time techniques for analysts and probabilists, L. EGGHE
101 Groups and geometry, ROGER C. LYNDON
103 Surveys in combinatorics 1985, I. ANDERSON (ed)
104 Elliptic structures on 3-manifolds, C.B. THOMAS
105 A local spectral theory for closed operators, I. ERDELYI & WANG SHENGWANG
106 Syzygies, E.G. EVANS & P. GRIFFITH
107 Compactification of Siegel moduli schemes, C-L. CHAI
108 Some topics in graph theory, H.P. YAP
109 Diophantine Analysis, J. LOXTON & A. VAN DER POORTEN (eds)
110 An introduction to surreal numbers, H. GONSHOR
111 Analytical and geometric aspects of hyperbolic space, D.B.A.EPSTEIN (ed)
113 Lectures on the asymptotic theory of ideals, D. REES

114 Lectures on Bochner-Riesz means, K.M. DAVIS & Y-C. CHANG
115 An introduction to independence for analysts, H.G. DALES & W.H. WOODIN
116 Representations of algebras, P.J. WEBB (ed)
117 Homotopy theory, E. REES & J.D.S. JONES (eds)
118 Skew linear groups, M. SHIRVANI & B. WEHRFRITZ
119 Triangulated categories in the representation theory of finite-dimensional algebras, D. HAPPEL
121 Proceedings of *Groups - St Andrews 1985*, E. ROBERTSON & C. CAMPBELL (eds)
122 Non-classical continuum mechanics, R.J. KNOPS & A.A. LACEY (eds)
124 Lie groupoids and Lie algebroids in differential geometry, K. MACKENZIE
125 Commutator theory for congruence modular varieties, R. FREESE & R. MCKENZIE
126 Van der Corput's method for exponential sums, S.W. GRAHAM & G. KOLESNIK
127 New directions in dynamical systems, T.J. BEDFORD & J.W. SWIFT (eds)
128 Descriptive set theory and the structure of sets of uniqueness, A.S. KECHRIS & A. LOUVEAU
129 The subgroup structure of the finite classical groups, P.B. KLEIDMAN & M.W.LIEBECK
130 Model theory and modules, M. PREST
131 Algebraic, extremal & metric combinatorics, M-M. DEZA, P. FRANKL & I.G. ROSENBERG (eds)
132 Whitehead groups of finite groups, ROBERT OLIVER
133 Linear algebraic monoids, MOHAN S. PUTCHA
134 Number theory and dynamical systems, M. DODSON & J. VICKERS (eds)
135 Operator algebras and applications, 1, D. EVANS & M. TAKESAKI (eds)
136 Operator algebras and applications, 2, D. EVANS & M. TAKESAKI (eds)
137 Analysis at Urbana, I, E. BERKSON, T. PECK, & J. UHL (eds)
138 Analysis at Urbana, II, E. BERKSON, T. PECK, & J. UHL (eds)
139 Advances in homotopy theory, S. SALAMON, B. STEER & W. SUTHERLAND (eds)
140 Geometric aspects of Banach spaces, E.M. PEINADOR and A. RODES (eds)
141 Surveys in combinatorics 1989, J. SIEMONS (ed)
142 The geometry of jet bundles, D.J. SAUNDERS
143 The ergodic theory of discrete groups, PETER J. NICHOLLS
144 Introduction to uniform spaces, I.M. JAMES
145 Homological questions in local algebra, JAN R. STROOKER
146 Cohen-Macaulay modules over Cohen-Macaulay rings, Y. YOSHINO
147 Continuous and discrete modules, S.H. MOHAMED & B.J. MÜLLER
148 Helices and vector bundles, A.N. RUDAKOV et al
149 Solitons, nonlinear evolution equations and inverse scattering, M.A. ABLOWITZ & P.A. CLARKSON
150 Geometry of low-dimensional manifolds 1, S. DONALDSON & C.B. THOMAS (eds)
151 Geometry of low-dimensional manifolds 2, S. DONALDSON & C.B. THOMAS (eds)
152 Oligomorphic permutation groups, P. CAMERON
153 L-functions in Arithmetic, J. COATES & M.J. TAYLOR
154 Number theory and cryptography, J. LOXTON (ed)
155 Classification theories of polarized varieties, TAKAO FUJITA
156 Twistors in mathematics and physics, T.N. BAILEY & R.J. BASTON (eds)
157 Analytic pro-p groups, J.D. DIXON, M.P.F. SU SAUTOY, A. MANN & D. SEGAL
158 Geometry of Banach spaces, P.F.X. MÜLLER & W. SCHACHERMAYER (eds)
159 Groups St Andrews 1989 Volume 1, C.M. CAMPBELL & E.F. ROBERTSON (eds)
160 Groups St Andrews 1989 Volume 2, C.M. CAMPBELL & E.F. ROBERTSON (eds)
161 Lectures on block theory, BURKHARD KÜLSHAMMER
162 Harmonic analysis and representation theory for groups acting on homogeneous trees, A. FIGA-TALAMANCA & C. NEBBIA
163 Topics in Varieties of Group representations, S.M. VOVSI
166 Surveys in Combinatorics, 1991, A.D. KEEDWELL (ed)

London Mathematical Society Lecture Note Series, 166

Surveys in Combinatorics, 1991

A.D. Keedwell
Department of Mathematical and Computer Sciences, University of Surrey

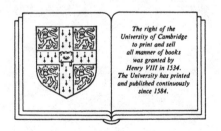

The right of the University of Cambridge to print and sell all manner of books was granted by Henry VIII in 1534. The University has printed and published continuously since 1584.

CAMBRIDGE UNIVERSITY PRESS

Cambridge

New York Port Chester Melbourne Sydney

CAMBRIDGE UNIVERSITY PRESS
Cambridge, New York, Melbourne, Madrid, Cape Town, Singapore, São Paulo

Cambridge University Press
The Edinburgh Building, Cambridge CB2 2RU, UK

Published in the United States of America by Cambridge University Press, New York

www.cambridge.org
Information on this title: www.cambridge.org/9780521407663

First published 1991

A catalogue record for this publication is available from the British Library

ISBN-13 978-0-521-40766-3 paperback
ISBN-10 0-521-40766-4 paperback

Transferred to digital printing 2005

CONTENTS

S. W. Golomb, "Radar Signal Patterns from Combinatorial Designs." 1

G. Korchmáros "Old and New Results on Ovals of Finite Projective Planes." 40

I. G. MacDonald, "Schubert Polynomials." 73

R. Mathon, "Computational Methods in Design Theory." 101

J. Nešetřil [with M. Loebl], "Unprovable Combinatorial Statements." 119

O. Pretzel, "Orientations and Edge Functions on Graphs." 161

P. Rowlinson, "Graph Perturbations." 187

J. A. Bondy, "A Graph Reconstructor's Manual." 221

Z. Füredi "Turán Type Problems." 253

PREFACE

Interest in Combinatorics continues to grow. For this, the thirteenth British Combinatorial Conference, our mailing list was almost worldwide and contained more than 950 names.

As usual, the British Combinatorial Committee has chosen nine Invited Speakers for the Conference each of whom has been requested to give a lecture of survey type at the Conference and also to prepare a paper for these Proceedings on the subject matter of their talk. Once again, the hoped-for result of a valuable collection of papers to act as a reference source over a wide spectrum of combinatorial topics has been achieved.

In order that this volume should be available for distribution to participants on the first day of the Conference, the usual tight schedule had to be adhered to. The Editor is therefore especially grateful to those who submitted their papers on time and to the referees who almost all carried out their task promptly as well as thoroughly. (At the time this Preface is being written, it is still not certain whether it will be possible to include all nine papers because one or two were either submitted in final form very late and/or were not refereed in time.)

The Rado lecture (which commemorates the late Richard Rado's contribution to British combinatorics) will, on this occasion, be that given by Z. Füredi.

Arrangements have been made for the contributed papers of the Conference to be published in a special issue of Discrete Mathematics and the organizers are grateful to D. R. Woodall for agreeing to act as guest editor for this purpose.

Finally, thanks are due to Cambridge University Press for their courteous advice and help in the preparation of this volume and to the London Mathematical Society for providing back-up financial support to the Conference.

A. D. Keedwell,
Guildford, April 1991.

CONSTRUCTION OF SIGNALS WITH FAVOURABLE CORRELATION PROPERTIES

S. W. GOLOMB
Communication Sciences Institute
University of Southern California
Los Angeles, CA 90089 USA

CHAPTER 1. GENERAL PROPERTIES OF CORRELATION

1.1. What is Correlation?

Correlation is a measure of the *similarity*, or *relatedness*, between two phenomena. When properly normalized, the correlation measure is a real number between −1 and +1, where a correlation value of "+1" indicates that the two phenomena are identical, a correlation value of "−1" means that they are diametrically opposite, and a correlation value of "0" means that they are "uncorrelated" -- i.e. that they agree exactly as much as they disagree.

In statistics, the correlation between two sets of data is called their *covariance*. In linear algebra, the correlation between two vectors is their *(normalized) dot product*. Specifically, let $\alpha = (a_1, a_2, \ldots, a_n)$ and $\beta = (b_1, b_2, \ldots, b_n)$ be two n-dimensional vectors of real numbers, which could represent two sets of experimental data. The *magnitudes* of these vectors are $|\alpha| = \left(\sum_{i=1}^{n} a_i^2\right)^{1/2}$ and $|\beta| = \left(\sum_{i=1}^{n} b_i^2\right)^{1/2}$. The *normalized vectors* are $\alpha' = \dfrac{\alpha}{|\alpha|}$ and $\beta' = \dfrac{\beta}{|\beta|}$. The *correlation* between the a_i's and the b_i's is the *covariance* of the two data sets:

$$C(\alpha,\beta) = \frac{(\alpha\cdot\beta)}{|\alpha||\beta|} = \frac{\sum_{i=1}^{n} a_i b_i}{\left(\sum_{i=1}^{n} a_i^2\right)^{1/2}\left(\sum_{i=1}^{n} b_i^2\right)^{1/2}} \tag{1.1}$$

which is also the *normalized dot product* of the two vectors, i.e. the dot product of the two normalized vectors: $(\alpha'\cdot\beta') = \left(\dfrac{\alpha}{|\alpha|}\cdot\dfrac{\beta}{|\beta|}\right) = \dfrac{(\alpha\cdot\beta)}{|\alpha||\beta|} = C(\alpha,\beta)$.

Geometrically, $(\alpha\cdot\beta) = |\alpha||\beta|\cos\theta$, so that

$$C(\alpha,\beta) = \frac{(\alpha\cdot\beta)}{|\alpha||\beta|} = \cos\theta \tag{1.2}$$

where θ is the angle between the vectors α and β. When the vectors are *orthogonal* (i.e. *perpendicular*), we have

$$C(\alpha,\beta) = \cos 90° = 0, \tag{1.3}$$

so that *uncorrelated data sets* correspond to *orthogonal vectors*. Note that for all angles θ, $-1 \le \cos\theta \le +1$.

The only real vector which cannot be normalized is the zero vector, $\vec{0} = (0,0, \ldots ,0)$. However, since the (unnormalized) dot product of this vector with any other has the (scalar) value *zero*, the zero vector is usually regarded as *uncorrelated* with all vectors.

1.2. Continuous Correlation

Suppose that $f(x)$ and $g(x)$ are real-valued functions which are "square-integrable" on the interval $[0,T]$. That is, the definite integrals $\int_0^T f^2(x)dx$ and $\int_0^T g^2(x)dx$ are well-defined. If $\phi = \left[\int_0^T f^2(x)dx\right]^{1/2}$ and $\gamma = \left[\int_0^T g^2(x)dx\right]^{1/2}$ are non-zero, then the *correlation* between $f(x)$ and $g(x)$ on $[0,T]$ is defined to be

$$C(f,g) = \frac{\int_0^T f(x)g(x)dx}{\phi\gamma} . \tag{1.4}$$

This is the natural extension of *covariance*, or of *(normalized) dot product*, to the continuous case. If we regard the vector $\alpha = (a_1,a_2, \ldots ,a_n)$ as a step function $\alpha(x)$ on the interval $[0,n]$, where $\alpha(x) \equiv a_i$ on the subinterval $(i-1,i]$, and similarly regard $\beta = (b_1,b_2, \ldots ,b_n)$ as the step function $\beta(x)$ on $[0,n]$ where $\beta(x) \equiv b_i$ for $i-1 < x \le i$, then clearly, $\int_0^n \alpha(x)\beta(x)dx = \sum_{i=1}^n a_i b_i$, so that $C(\alpha,\beta)$ has the same value whether computed by (1.1) or (1.4).

1.3. Binary Correlation

Suppose that $\alpha = (a_1,a_2, \ldots ,a_n)$ and $\beta = (b_1,b_2, \ldots ,b_n)$ are both *binary* vectors, and specifically that the a_i's and b_i's are restricted to the two values $+1$ and -1. Then $|\alpha| = \left[\sum_{i=1}^n a_i^2\right]^{1/2} = \sqrt{n} = \left[\sum_{i=1}^n b_i^2\right]^{1/2} = |\beta|$, from which

$$C(\alpha,\beta) = \frac{1}{n}\sum_{i=1}^n a_i b_i = \frac{1}{n}(A-D) = \frac{A-D}{A+D} \tag{1.5}$$

where A is the number of times, for i from 1 to n, that a_i and b_i *agree*, and D is the number of times that a_i and b_i *disagree*. Clearly $A+D = n$, since at each value of i, a_i and b_i either agree or disagree.

Since $(+1)\cdot(+1) = (-1)\cdot(-1) = +1$, while $(+1)\cdot(-1) = (-1)\cdot(+1) = -1$, each *agreement* between a_i and b_i contributes +1, and each *disagreement* contributes −1, to the sum $\sum_{i=1}^{n} a_i b_i$. If a_i and b_i agree completely, then $A=n$, $D=0$, and $C(\alpha,\beta) = \frac{A-0}{A+D} = \frac{n}{n} = +1$. If a_i and b_i disagree everywhere, then $A=0$, $D=n$, and $C(\alpha,\beta) = \frac{0-D}{A+D} = \frac{-n}{n} = -1$. If agreements and disagreements occur equally often, then $A = D = \frac{n}{2}$, $A-D = 0$, and $C(\alpha,\beta) = \frac{A-D}{A+D} = \frac{0}{n} = 0$.

When binary vectors are used where the two values taken on are r and s, with $r>s$, the correlation defined as a normalized dot product is in general less useful than the *modified* correlation obtained from the comparison of agreements and disagreements: $\bar{C}(\alpha,\beta) = \frac{A-D}{A+D}$.

1.4. Complex Correlation

When working with complex-valued vectors or functions, the appropriate notion of dot product is the *Hermitian dot product*, for which, if $\alpha = (a_1,a_2,\ldots,a_n)$ and $\beta = (b_1,b_2,\ldots,b_n)$, we have $|\alpha| = \sqrt{(\alpha\cdot\alpha)} = \left[\sum_{i=1}^{n} a_i a_i^*\right]^{1/2} = \left[\sum_{i=1}^{n} |a_i|^2\right]^{1/2}$, $|\beta| = \sqrt{(\beta\cdot\beta)} = \left[\sum_{i=1}^{n} \beta_i \beta_i^*\right]^{1/2} = \left[\sum_{i=1}^{n} |\beta_i|^2\right]^{1/2}$, $(\alpha\cdot\beta) = \sum_{i=1}^{n} a_i b_i^*$, and the correlation $C(\alpha,\beta)$ is given by

$$C(\alpha,\beta) = \frac{(\alpha\cdot\beta)}{|\alpha||\beta|} = \frac{\sum_{i=1}^{n} a_i b_i^*}{\left[\sum_{i=1}^{n} |a_i|^2\right]^{1/2} \left[\sum_{i=1}^{n} |b_i|^2\right]^{1/2}} \tag{1.6}$$

where z^* denotes the *complex conjugate* of z. That is, if $z = x+y\sqrt{-1}$ then $z^* = x-y\sqrt{-1}$, so that $zz^* = x^2+y^2 = |z|^2$.

If the complex numbers happen to be real numbers, these definitions reduce to the previous ones for real-valued vectors/data/functions, since $z=z^*$ if (and only if!) z is real. However, in order for a complex number to have a correlation of +1 *with itself* (which is clearly what we desire), it is necessary to use the Hermitian dot product. Note that in the simple case $z = \sqrt{-1}$, if we used $(z\cdot z) = z^2$ we would get $(z\cdot z) = -1$, whereas with $(z\cdot z) = zz^* = |z|^2$ we get $(z\cdot z) = +1$.

1.5. Problems

1. Let a,b,x,y be four real numbers, with $a>b$ and $x>y$. Show that $ax+by > ay+bx$.

2. Let $\alpha = (a_1, a_2, \ldots, a_n)$ be a vector of real numbers, and eqnlet $\beta = (b_1, b_2, \ldots, b_n)$ be a vector whose components are a permutation of the components of α. Among all $n!$ such permutations, prove that $(\alpha \cdot \beta)$ is maximized when $\beta = \alpha$.
3. Suppose $\alpha = (a_1, a_2, \ldots, a_n)$ and $\beta = (b_1, b_2, \ldots, b_n)$ are any two vectors with components restricted to the values 0 and 1. Find a linear relation between the correlation $C(\alpha,\beta) = (\alpha \cdot \beta)/|\alpha||\beta|$ and the modified correlation $\bar{C}(\alpha,\beta) = (A-D)/(A+D)$.
4. In the case of the Hermitian dot product, show that $(\beta \cdot \alpha) = (\alpha \cdot \beta)^*$.

1.6. Mutual Orthogonality

It is well-known that there are many possible infinite sets of pairwise "orthogonal functions" on a given interval $[0,T]$. For example, the Fourier functions $\{1, \cos nx, \sin nx\}$, $1 \le n < \infty$, are mutually orthogonal on the interval $[0,2\pi]$. The Walsh functions provide an infinite set of mutually orthogonal *binary-valued* functions on $[0,1]$. Since *orthogonal* means "a correlation value of 0," it is possible to have an arbitrarily large number of mutually uncorrelated vectors, if we do not limit the dimensionality (i.e. the number of components) of the vectors.

The Walsh Functions are square-wave functions on $[0,1]$, which jump between the values -1 and $+1$, and there are 2^n such functions when the unit interval is divided into 2^n sub-intervals. That is, there are 2^n orthogonal vectors of $+1$'s and -1's with 2^n components. (More generally, with d-component real vectors, there are *at most d* mutually orthogonal vectors, since mutually orthogonal vectors are linearly independent, and the maximum number of linearly independent vectors it contains is the *dimension* of a vector space.) The Walsh functions will be properly defined, and further discussed, in Chapter 2.

In a similar way, the number of mutually orthogonal Fourier sine and cosine functions on $[0,2\pi]$ is finite if one imposes a "bandwidth constraint" -- i.e. an upper limit on the *frequency* of the sine and cosine functions which can be used.

1.7. The Simplex Bound on Mutual Negative Correlation

It is possible for any finite number n of vectors (data sets, functions) to be mutually negatively correlated, if there is no dimensionality limitation on the vectors.

Two vectors can point in opposite directions, in which case the angle between them is 180°, and $\cos 180° = -1$. If α is any "unit vector," i.e. a vector for which $|\alpha| = 1$, and $\beta = -\alpha$, then $C(\alpha,\beta) = (\alpha \cdot -\alpha) = -|\alpha|^2 = -1$, the lowest possible value of correlation. The vectors α and β lie on a single line through the origin, so that one dimension suffices to provide this example.

Three vectors in the plane can be picked with a mutual separation of 120° between each pair of them. Since cos 120° = −½, we can find three unit vectors $\alpha_1,\alpha_2,\alpha_3$ for which $(\alpha_1\cdot\alpha_2) = (\alpha_1\cdot\alpha_3) = (\alpha_2\cdot\alpha_3) = -\frac{1}{2}$. For example, the vectors $\alpha_1 = (1,0)$, $\alpha_2 = (-\frac{1}{2},\frac{1}{2}\sqrt{3})$, $\alpha_3 = (-\frac{1}{2},-\frac{1}{2}\sqrt{3})$ have this property. These three vectors define the vertices of an equilateral triangle inscribed in the unit circle of the xy-plane.

Similarly, four vectors in three-dimensional space can be located at the vertices of a regular tetrahedron inscribed in the unit sphere. The angle between each pair of these vectors is 109°28′16.394 · · · ″, the cosine of which is $-\frac{1}{3}$.

In general, n unit vectors can be placed at the n vertices of the $(n-1)$-dimensional "regular simplex" inscribed in the $(n-1)$-dimensional "unit hypersphere," achieving a mutual negative correlation of $-1/(n-1)$ for all $n \ge 2$. Fortunately, the fact that no better set of vectors exists can be proved by simple arithmetic, with no need of geometric knowledge or intuition.

Theorem 1. Let $A = \{\alpha_1,\alpha_2,\ldots,\alpha_n\}$ be a set of n unit vectors (i.e. $(\alpha_i\cdot\alpha_i) = 1$ for all i, $1 \le i \le n$) and let $c_{ij} = (\alpha_i\cdot\alpha_j)$ for all i,j. Then

$$\min_{\text{all } A}\ \max_{i\ne j}\ c_{ij} \ge \frac{-1}{n-1}\,. \tag{1.7}$$

(Note that $\max_{i\ne j} c_{ij}$ is the *worst* -- *largest* -- correlation value between any two vectors in the set A, and then applying "$\min_{\text{all } A}$" gives us the result for the *best* choice of a set of n vectors.)

Proof. For the members of any finite set S, the *maximum* value of some parameter must be at least as big as the *average* value. Hence

$$\max_{i\ne j} c_{ij} \ge \operatorname*{average}_{i\ne j} c_{ij} = \frac{1}{n(n-1)}\sum\sum_{i\ne j} c_{ij}$$

$$= \frac{1}{n(n-1)}\left\{\sum_{i=1}^{n}\sum_{j=1}^{n} c_{ij} - \sum_{i=1}^{n} c_{ii}\right\} = \frac{1}{n(n-1)}\left\{\sum_{i=1}^{n}\sum_{j=1}^{n}(\alpha_i\cdot\alpha_j) - \sum_{i=1}^{n}|\alpha_i|^2\right\}$$

$$= \frac{1}{n(n-1)}\left\{\left(\left[\sum_{i=1}^{n}\alpha_i\right]\cdot\left[\sum_{j=1}^{n}\alpha_j\right]\right) - \sum_{i=1}^{n} 1\right\}$$

$$= \frac{1}{n(n-1)}\left\{\left|\sum_{i=1}^{n}\alpha_i\right|^2 - n\right\} \ge \frac{1}{n(n-1)}(0-n) = -\frac{1}{n-1}\,.$$

Thus, $\min_{\text{all } A}\ \max_{i\ne j}\ c_{ij} \ge \frac{-1}{n-1}$, since $\left|\sum_{i=1}^{n}\alpha_i\right|^2 \ge 0$ because the sum vector $\sum_{i=1}^{n}\alpha_i$ has a non-negative magnitude. ■

Selecting n vectors which correspond to the vertices of the $(n-1)$-dimensional simplex is the basic way to achieve this bound. If the m components of the vectors are restricted to the binary values $\frac{+1}{\sqrt{m}}$ and $\frac{-1}{\sqrt{m}}$, the same proof idea yields a slightly stronger result. (This is equally true if the "vectors" are in fact step functions on $[0,T]$ which take on only the values $\frac{+1}{\sqrt{T}}$ and $\frac{-1}{\sqrt{T}}$.)

Theorem 2. Let $A = \{\alpha_1, \alpha_2, \ldots, \alpha_n\}$ be a set of n binary unit vectors (i.e. the m components of α_i are restricted to the values $\pm\frac{1}{\sqrt{m}}$ for each i, $1 \le i \le n$) and let $c_{ij} = (\alpha_i \cdot \alpha_j)$ for all i,j. Then

$$\min_{\text{all } A} \max_{i \neq j} c_{ij} \ge \begin{cases} \dfrac{-1}{n-1}, & n \text{ even} \\ \dfrac{-1}{n}, & n \text{ odd} \end{cases} \tag{1.8}$$

Proof. We proceed as in the proof of Theorem 1, up to $\max_{i \neq j} c_{ij} \ge \frac{1}{n(n-1)}\left\{\left|\sum_{i=1}^{n} \alpha_i\right|^2 - n\right\}$. We now observe that each of the m components of $\sum_{i=1}^{n} \alpha_i$ is a sum of n terms, each of which is either $+\frac{1}{\sqrt{m}}$ or $-\frac{1}{\sqrt{m}}$. If n is *odd*, the sum of n such terms cannot be 0, and must have an absolute value of at least $\frac{1}{\sqrt{m}}$. (For the sum to be 0, half the n terms would have to be $+\frac{1}{\sqrt{m}}$ and half $-\frac{1}{\sqrt{m}}$, which cannot happen for odd n.) Hence we have $\left|\sum_{i=1}^{n} \alpha_i\right|^2 \ge \left|\left(\frac{1}{\sqrt{m}}, \frac{1}{\sqrt{m}}, \ldots, \frac{1}{\sqrt{m}}\right)\right|^2 = m \cdot \left(\frac{1}{\sqrt{m}}\right)^2 = 1$, from which

$$\max_{i \neq j} c_{ij} \ge \frac{1}{n(n-1)}(1-n) = -\frac{1}{n}.$$

Hence in this case,

$$\min_{\text{all } A} \max_{i \neq j} c_{ij} \ge \begin{cases} -\dfrac{1}{n-1}, & n \text{ even} \\ -\dfrac{1}{n}, & n \text{ odd} \end{cases}$$

■

Note. When $n=4$, the vectors $\alpha_1 = \frac{1}{\sqrt{3}}(1,1,1)$, $\alpha_2 = \frac{1}{\sqrt{3}}(1,-1,-1)$, $\alpha_3 = \frac{1}{\sqrt{3}}(-1,1,-1)$, $\alpha_4 = \frac{1}{\sqrt{3}}(-1,-1,1)$ have the property

$$(\alpha_i \cdot \alpha_j) = -\tfrac{1}{3} \quad \text{for all } i \neq j.$$

1.8. Problems

1. Find a set of *six* binary vectors which achieves the bound of Theorems 1 and 2. (Do not assume that $m \leq n$.)
2. Suppose the vectors of the set $A = \{\alpha_1, \alpha_2, \ldots, \alpha_n\}$ have m components each, all restricted to odd integer values. Show that the bound of Theorem 2 applies.
3. Find a set of five real vectors which achieves the bound of Theorem 1.
4. Where would you place 6 vectors on the unit sphere (the ordinary three-dimensional kind) so as to minimize the maximum correlation between any two of them?
5. Where would you place 8 vectors on the unit sphere so as to minimize the maximum correlation between any two of them? What about 12 vectors?
6. In n dimensions, consider the set B of $2n$ unit vectors, $B = \{e_1, e_2, \ldots, e_n, -e_1, -e_2, \ldots, -e_n\}$, where $e_1 = (1,0,0,\ldots,0)$, $e_2 = (0,1,0,\ldots,0)$, ..., $e_n = (0,0,0,\ldots,1)$. What are the correlation values c_{ij} which occur among this set of vectors? Calculate $\max_{i \neq j} c_{ij}$ and $\operatorname{average}_{i \neq j} c_{ij}$. How does this relate to Theorem 1? How does it relate to Problem 4 above? (**Note.** B is an example of what is called a *biorthogonal code.*)

1.9. Autocorrelation

Def. The *autocorrelation function* $C(\tau)$ of a function $f(t)$ is the correlation between $f(t)$ and $f(t+\tau)$, regarded as a function of τ.

It is customary to distinguish three cases:

(i) *Finite* autocorrelation is computed on the assumption that $f(t)$ is identically 0 outside some interval $[0,T]$. This leads to:

$$C_f^F(\tau) = \frac{\int_0^{T-\tau} f(t) f(t+\tau)\, dt}{\int_0^T |f(t)|^2 dt}. \tag{1.9}$$

(ii) *Infinite* autocorrelation is computed on the assumption that $f(t)$ is defined for all t, $-\infty < t < \infty$, with $\int_{-\infty}^{\infty} |f(t)|^2 dt < \infty$, and we calculate

$$C_f^I(\tau) = \frac{\int_{-\infty}^{\infty} f(t) f(t+\tau)\, dt}{\int_{-\infty}^{\infty} |f(t)|^2 dt} \tag{1.10}$$

(iii) *Periodic* autocorrelation is computed on the assumption that $f(t)$ is periodic with some period P, so that $f(t+P) = f(t)$ for all t. In this case, we compute

$$C_f^P(\tau) = \frac{\int_0^P f(t) f(t+\tau) dt}{\int_0^P |f(t)|^2 dt} . \tag{1.11}$$

The same concept of autocorrelation applies to *sequences*. If $S = \{s_i\}$ is a sequence defined for $1 \le i \le n$, the *finite* autocorrelation of S is given by

$$C_S^F(\tau) = \frac{\sum_{i=1}^{n-\tau} s_i s_{i+\tau}}{\sum_{i=1}^{n} |s_i|^2} . \tag{1.12}$$

If $S = \{s_i\}$ is defined for all i, $-\infty < i < \infty$, and $\sum_{i=-\infty}^{\infty} |s_i|^2 < \infty$ then the *infinite* autocorrelation of S is given by

$$C_S^I(\tau) = \frac{\sum_{i=-\infty}^{\infty} s_i s_{i+\tau}}{\sum_{i=-\infty}^{\infty} |s_i|^2} . \tag{1.13}$$

If $S = \{s_i\}$ is periodic with period P, so that $s_i = s_{i+P}$ for all i, then the *periodic* autocorrelation of S is defined as

$$C_S^P(\tau) = \frac{\sum_{i=1}^{P} s_i s_{i+\tau}}{\sum_{i=1}^{P} |s_i|^2} . \tag{1.14}$$

Note that in all the cases considered, $C(0) = 1$, and $C(-\tau) = C(\tau)$. However, if we are working with complex values, and therefore use the Hermitian dot product, we find that $C(-\tau) = C^*(\tau)$.

1.10. Crosscorrelation

Suppose that $f(t)$ and $g(t)$ are two functions of the continuous variable t. The *crosscorrelation function* $R_{f,g}(\tau)$ between f and g is defined, analogous to the autocorrelation function, on one of the following three assumptions:

(i) *Finite* crosscorrelation is computed on the assumptions that $f(t)$ and $g(t)$ are identically 0 outside of some interval $[0,T]$, leading to

$$R_{f,g}^{F}(\tau) = \frac{\int_{0}^{T-\tau} f(t)g(t+\tau)dt}{\left[\int_{0}^{T} |f(t)|^2 dt\right]^{1/2} \left[\int_{0}^{T} |g(t)|^2 dt\right]^{1/2}} . \tag{1.15}$$

(ii) *Infinite* crosscorrelation is computed on the assumption that $f(t)$ and $g(t)$ are defined and square-integrable on $(-\infty,\infty)$, leading to

$$R_{f,g}^{I}(\tau) = \frac{\int_{-\infty}^{\infty} f(t)g(t+\tau)dt}{\left[\int_{-\infty}^{\infty} |f(t)|^2 dt\right]^{1/2} \left[\int_{-\infty}^{\infty} |g(t)|^2 dt\right]^{1/2}} . \tag{1.16}$$

(iii) *Periodic* crosscorrelation assumes that both $f(t)$ and $g(t)$ are periodic with a common periodicity P, so that $f(t+P) \equiv f(t)$ and $g(t+P) \equiv g(t)$ for all t. In this case, we have

$$R_{f,g}^{P}(\tau) = \frac{\int_{0}^{P} f(t)g(t+\tau)dt}{\left[\int_{0}^{P} |f(t)|^2 dt\right]^{1/2} \left[\int_{0}^{P} |g(t)|^2 dt\right]^{1/2}} . \tag{1.17}$$

Similarly, if we have two *sequences* $S = \{s_i\}$ and $T = \{t_i\}$, both defined for $1 \le i \le n$, then the *finite* crosscorrelation between S and T is defined by

$$R_{S,T}^{F}(\tau) = \frac{\sum_{i=1}^{n-\tau} s_i t_{i+\tau}}{\left[\sum_{i=1}^{n} |s_i|^2\right]^{1/2} \left[\sum_{i=1}^{n} |t_i|^2\right]^{1/2}} . \tag{1.18}$$

If $S = \{s_i\}$ and $T = \{t_i\}$ are defined for all i, $-\infty < i < \infty$, and if both $\sum_{i=-\infty}^{\infty} |s_i|^2 < \infty$ and $\sum_{i=-\infty}^{\infty} |t_i|^2 < \infty$, then the *infinite* crosscorrelation between S and T is given by

$$R_{S,T}^{I}(\tau) = \frac{\sum_{i=-\infty}^{\infty} s_i t_{i+\tau}}{\left[\sum_{i=-\infty}^{\infty} |s_i|^2\right]^{1/2} \left[\sum_{i=-\infty}^{\infty} |t_i|^2\right]^{1/2}} . \tag{1.19}$$

Finally, if both $S = \{s_i\}$ and $T = \{t_i\}$ are periodic with period P, so that $s_{i+P} = s_i$ and $t_{i+P} = t_i$ for all i, then the *periodic* crosscorrelation of S and T is defined to be

$$R^P_{S,T}(\tau) = \frac{\sum_{i=1}^{P} s_i t_{i+\tau}}{\left(\sum_{i=1}^{P} |s_i|^2\right)^{1/2} \left(\sum_{i=1}^{P} |t_i|^2\right)^{1/2}} . \tag{1.20}$$

We note that in all cases, the crosscorrelation reduces to the autocorrelation if the two functions, or the two sequences, being correlated are in fact the same. That is:

$$R^X_{f,f}(\tau) \equiv C^X_f(\tau), \quad \text{where } X \text{ is any of } F, I, \text{ or } P, \tag{1.21}$$

and

$$R^X_{s,s}(\tau) = C^X_s(\tau), \quad \text{where } X \text{ is any of } F, I, \text{ or } P. \tag{1.22}$$

When we correlate $f(t)$ against $g(t+\tau)$, the effect is basically the same as correlating $g(t)$ against $f(t-\tau)$, except in the complex case, where we must remember the effect of the Hermitian dot product. Similarly, s_i correlated with $t_{i+\tau}$ is essentially the same as t_i correlated with $s_{i-\tau}$. In the most general case (i.e. the complex case), we therefore have

$$R^X_{f,g}(-\tau) = \left[R^X_{g,f}(\tau)\right]^*, \quad \text{where } X \text{ is any of } F, I, \text{ or } P, \tag{1.23}$$

and

$$R^X_{S,T}(-\tau) = \left[R^X_{T,S}(\tau)\right]^*, \quad \text{where } X \text{ is any of } F, I, \text{ or } P. \tag{1.24}$$

We should also note that the *convolution* V between $f(t)$ and $g(t)$, (or between $\{s_i\}$ and $\{t_i\}$), is in fact the unnormalized crosscorrelation between $f(t)$ and $g(-t)$, (or between $\{s_i\}$ and $\{t_{-i}\}$). Thus

$$V^F_{f,g}(\tau) = \int_0^{\tau} f(t)g(\tau-t)dt \tag{1.25}$$

$$V^I_{f,g}(\tau) = \int_{-\infty}^{\infty} f(t)g(\tau-t)dt \tag{1.26}$$

$$V^P_{f,g}(\tau) = \int_0^{P} f(t)g(\tau-t)dt \tag{1.27}$$

and

$$V^F_{S,T}(\tau) = \sum_{i=0}^{\tau} s_i t_{\tau-i} \tag{1.28}$$

$$V_{S,T}^{I}(\tau) = \sum_{i=-\infty}^{\infty} s_i t_{\tau-i} \tag{1.29}$$

$$V_{S,T}^{P}(\tau) = \sum_{i=1}^{P} s_i t_{\tau-i} \ . \tag{1.30}$$

<u>Note</u>: For $V_{S,T}^{F}(\tau)$, we let $S = \{s_0, s_1, s_2, \ldots, s_n\}$ and $T = \{t_0, t_1, t_2, \ldots, t_n\}$, with $0 \le \tau \le n$. This leads to simpler notation than one gets with n-term sequences.

1.11. Problems

1. Find a closed-form expression for $C_f^I(\tau)$ when $f(t) = e^{-t^2/2}$.
2. Find a closed-form expression for $C_f^P(\tau)$ when $f(t) = \cos t$ and $P = 2\pi$.
3. Let $S = \{+1,+1,+1,-1,-1,+1,-1\}$ repeating periodically with period $P=7$. Calculate $C_S^P(\tau)$ for this sequence.
4. Taking $S = \{+1,+1,+1,-1,-1,+1,-1\}$ as a finite sequence of length 7, calculate $C_S^F(\tau)$ for this sequence.
5. Find a closed-form expression for $R_{f,g}^P(\tau)$ when $f(t) = \cos t$, $g(t) = \sin t$, and $P = 2\pi$.
6. Let $A(x) = \sum_{i=0}^{m} a_i x^i$ and $B(x) = \sum_{j=0}^{n} b_j x^j$ be two polynomials in x, and write their product as $A(x)B(x) = C(x) = \sum_{k=0}^{m+n} c_k x^k$. In what sense is $\{c_k\}$ the *convolution* of $\{a_i\}$ and $\{b_j\}$?

CHAPTER 2. APPLICATION OF CORRELATION IN THE COMMUNICATION OF INFORMATION

2.1. The Maximum Likelihood Detector

In order to communicate information from a sender to a receiver, there must be more than one possible *message* which the sender is able to transmit and which the receiver is able to detect. If only one message were possible, its receipt would be a foregone conclusion, and it would convey no information.

It is important to distinguish here between a *signal* and a *message*. If the sender is capable of transmitting only one *signal*, but has the choice of whether or not to send it at a given time, then "signal" and "no signal" are two distinct *messages*, and the decision of which one to send does indeed convey information.

According to Claude Shannon's *Theory of Information* [A], if there are N possible messages, $m_1, m_2, \ldots, m_N$, which might be sent, and the a priori probabilities of these N messages are $p_1, p_2, \ldots, p_N$, respectively, then the amount of information conveyed by knowing which one of these messages was actually sent is given by the expression

$$H(p_1, p_2, \ldots, p_N) = -\sum_{i=1}^{N} p_i \log_2 p_i \tag{2.1}$$

where the information measure H is called the *entropy* of the probability distribution $\{p_1, p_2, \ldots, p_N\}$, and measures both the a priori *uncertainty* concerning what will occur, and the a posteriori *information gained* as a result of removing this uncertainty. In expression (2.1), the use of logarithms to the base 2 has the effect of measuring information in *bits*, where one *bit* of information is the amount of uncertainty removed when one learns the outcome of an experiment which, a priori, had two equally likely possible outcomes.

Most real communication channels are *noisy*. That is, the signals which are received do not look identical to the signals which were sent. As a result, the receiver must make a decision as to which signal was actually sent, given the actual signal which was received. As a result of noise in the channel, there is some probability, hopefully small, that an incorrect decision will have been made. The remaining uncertainty as to what was sent, *given* what was received, is called the *equivocation* in the channel. If the set of possible transmitted signals is represented by X, and the set of possible received signals is denoted by Y, then the a priori uncertainty, as calculated in expression (2.1), is written as $H(X)$, and the a posteriori uncertainty, or *equivocation*, is denoted $H_Y(X)$, the uncertainty regarding X given Y. The amount of information actually communicated in such a case is

$$I(X, Y) = H(X) - H_Y(X). \tag{2.2}$$

(For a fuller treatment of these basic concepts of Information Theory, see [A], [B], or [C].)

The basic problem which serves as a model of *detection theory* concerns the situation where there are two possible transmitted signals, represented by the real numbers 0 and 1, and these are similarly corrupted by "gaussian noise". That is, the receiver does not receive 0 or 1, but instead receives a sample from a gaussian distribution having mean M and standard deviation σ, where M is either 0 or 1. The larger the value of σ, the noisier the channel, and the greater the probability that the receiver will make an incorrect decision as to what was sent. A picture of this idealized detection theory situation (where both distributions have the same standard deviation σ) is shown in Figure 2.1.

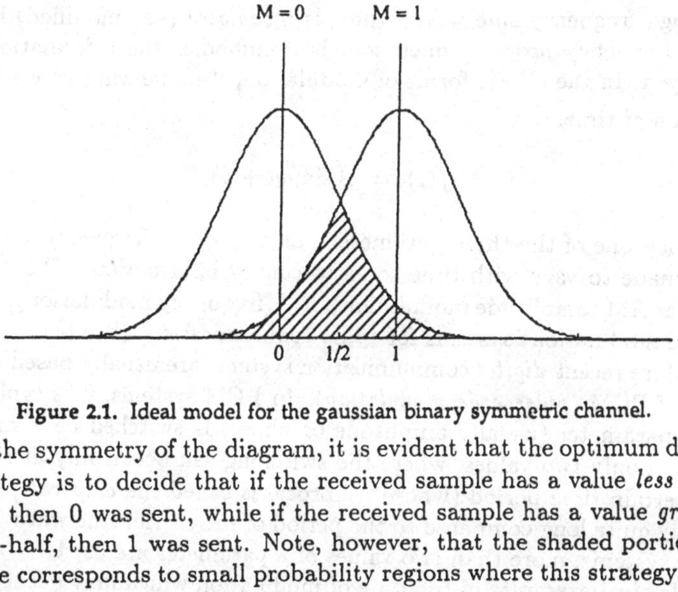

Figure 2.1. Ideal model for the gaussian binary symmetric channel.

By the symmetry of the diagram, it is evident that the optimum detection strategy is to decide that if the received sample has a value *less than* one-half, then 0 was sent, while if the received sample has a value *greater than* one-half, then 1 was sent. Note, however, that the shaded portion of the figure corresponds to small probability regions where this strategy will, unavoidably, lead to an incorrect decision. (For further discussion of this model, see [D].)

A considerably more general result is that when the receiver is trying to decide which one of a set of N signals was actually sent, over a channel corrupted by gaussian noise, the optimum decision process is to perform "correlation detection" — i.e. to calculate the correlation between the actual received signal and ideal models of each of the possible transmitted signals, and to decide that the highest value of the correlation corresponds to the signal that was actually sent. (For a proof of this theorem, see [E].) The optimum detector for a given channel is known as the *matched filter* for that channel, and the result we have just mentioned is frequenctly stated as "the matched filter for the gaussian channel is a *correlation detector*".

2.2. Problems

1. Evaluate Shannon's entropy expression (2.1) when $p_1 = p_2 = \ldots = p_N$.

2. In the "binary symmetric channel", the symbols 0 and 1 may be sent (each in unit time) and the same two symbols may be received. However, because of noise in the channel, the *correct* symbol is received with probability p, while the *other* symbol is received with probability $q = 1 - p$. If 0 and 1 are equally likely to be sent, and occur at random, what information rate (in bits per unit time) can be achieved over this channel?

2.3. Coherent vs. Incoherent Detection

Most forms of electronic communication involve a *carrier signal*, which is a high-frequency sine wave, which is *modulated* (i.e. modified) by some lower frequency process which somehow embodies the information to be conveyed. In the classic forms of modulation, the sine wave is written as a function of time:

$$f(t) = A\sin(\omega t + \phi), \tag{2.3}$$

and then one of the three parameters, amplitude A, frequency ω or phase ϕ, is made to vary with time so as to convey information. These are the familiar AM (amplitude modulation), FM (frequency modulation), and PM (phase modulation) systems for radio communication. (See [F].)

More recent *digital* communication systems are usually based on some form of PCM (*pulse code modulation*). In PCM systems, it is typical that some parameter (usually amplitude or phase) is switched back and forth between only two values, where the switching can occur only at multiples of a certain time period (whose reciprocal is called the *chip rate*) which is usually quite long compared to the period of the carrier sine wave. In some PCM systems, more than two values of a parameter are used.

Typical examples of the type of modulation which may occur are:

i) The amplitude A may be switched between the values 1 and 0. (This is equivalent to having the signal turned *on* and *off*.)

ii) The amplitude A may be switched between the values $+1$ and -1.

iii) The phase ϕ may be switched between the values 0 and π (i.e. 0° and 180°). Note that this system of *phase reversal* is indistinguishable from ii) above, where the amplitude undergoes sign reversal, since $\sin(x + \pi) = -\sin x$.

iv) The phase ϕ may be switched between two values ϕ_1 and ϕ_2 which are not necessarily 180° apart. (Any such system is referred to as *biphase modulation*.)

v) The phase ϕ may be switched among a finite set of values, $\phi_1, \phi_2, \ldots, \phi_n$, which are usually equally spaced modulo 360°. For example, when $n = 4$ (called *quadriphase modulation*), it is customary to use $\phi_1 = 0°, \phi_2 = 90°, \phi_3 = 180°, \phi_4 = 270°$.

In order to obtain the maximum information rate from the kinds of phase modulation enumerated above, it is necessary to maintain a fixed rational relationship between the chip rate (the frequency with which phase shifts are allowed to occur) and the frequency of the sine wave carrier. In such systems, which are called *coherent* communication systems, a phase change is allowed to occur every M cycles of the carrier sine wave, for some fixed integer M, and at no other times. In coherent systems, the clocking of the phase changes must be rigidly related to timing of the underlying carrier sine wave.

Where the distinction between coherent and non-coherent communication clearly matters is in the context of correlation detection. When phase coherence has been maintained, the distinction between a correlation value (between the incoming signal and a locally generated model of it at the receiver) of $+\delta$ (where $\delta > 0$) and a correlation value of $-\delta$ is meaningful, and can be used to convey information. When phase coherence is not maintained, it is impossible to distinguish between correlation values of $+\delta$ and $-\delta$. This is illustrated in Figure 2.2.

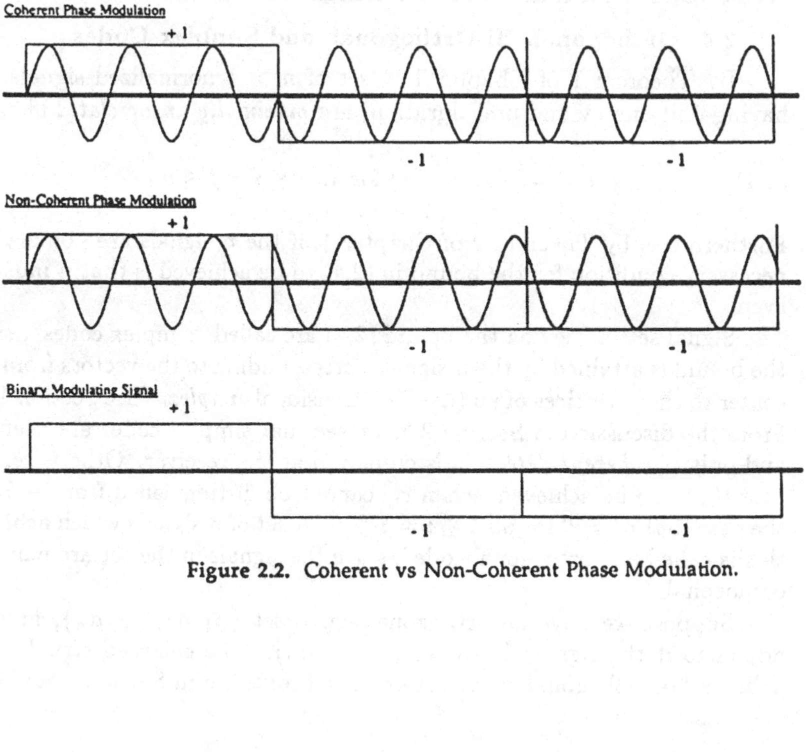

Figure 2.2. Coherent vs Non-Coherent Phase Modulation.

In the coherent example, when a reference sine wave of one chip duration is compared with one chip of the received signal (at least in the ideal, noiseless case), the sine waves will line up exactly (when we have selected the correct reference sine wave), and the correlation over one chip duration between them will be +1. However, in the non-coherent case, the assumption is that the receiver has not kept track of any "absolute phase reference" for the carrier sine wave, and the phase transitions occur somewhat randomly relative to the timing of the carrier cycles. In such a situation, the reference sine wave is as likely to be 180° out of phase as it is to be exactly in phase with the received signal. (In fact, all phase relationships from 0° to 360° are equally likely.) In such a case, the *sign* of correlation contains no useful information. In this non-coherent condition, it is common for the receiver to perform an "envelope detection" of the pattern of phase modulation (the bottom illustration in Figure 2.2.) and totally discard the "fine structure" information contained in the carrier sine wave.

In the case of coherent communication, signals can be negatively correlated, and this makes them more distinguishable from each other than merely being uncorrelated. The greatest distinguishability occurs between diametrically opposite signals, $s(t)$ and $-s(t)$, which have a normalized crosscorrelation of -1. On the other hand, in the case of non-coherent communications, maximum distinguishability occurs when two signals are *orthogonal*, i.e. when their crosscorrelation has the value 0.

2.4. Orthogonal, Bi-Orthogonal, and Simplex Codes

By Theorem 1 of Chapter 1, a set of $n > 1$ normalized signals (all having unit energy and unit duration) are *maximally uncorrelated* iff

$$c_{ij} = -1/(n-1) \text{ for all } 1 \leq i \neq j \leq n. \tag{2.4}$$

Furthermore, by Theorem 2 of Chapter 1, if the n signals are "binary", a necessary condition for the bound in (2.4) to be achieved is that n must be even.

Signal sets achieving the bound (2.4) are called "simplex codes", since the bound is attained by the n signals corresponding to the vectors from the center to the n vertices of an $(n-1)$-dimensional *simplex*. (See Section 1.7.) From the discussion in Section 2.3, we see that simplex codes are useful if and only if *coherent detection* is employed at the receiver. Otherwise, the best that can be achieved (when c_{ij} cannot be distinguished from $-c_{ij}$) is the case that $c_{ij} = 0$ for all $1 \leq i \neq j \leq n$. A set of n signals which achieves this is called an "orthogonal code", since the signals in the set are pairwise orthogonal.

Suppose we have an orthogonal signal set $\{\alpha_1, \alpha_2, \ldots, \alpha_n\}$, and we adjoin to it the signals $\{-\alpha_1, -\alpha_2, \ldots, -\alpha_n\}$. The enlarged signal set is called a "bi-orthogonal code". (Compare Problem 6 in Section 1.8.) If we let $\{\beta_1, \beta_2, \ldots, \beta_{2n}\} = \{\alpha_1, \alpha_2, \ldots, \alpha_n, -\alpha_1, -\alpha_2, \ldots, -\alpha_n\}$, we have

$$(2.5) \qquad (\beta_i \cdot \beta_j) = \left\{ \begin{array}{rll} 1 & \text{if} & i = j \\ -1 & \text{if} & |i-j| = n \\ 0 & \text{otherwise} & \end{array} \right\}.$$

(Obviously, in order to distinguish α_i from $-\alpha_i$, this is meaningful only in the case of coherent detection.) For the set of $2n$ signals β_i we see that

$$(2.6) \qquad \underset{i \neq j}{\text{Average}}\ c_{ij} = \frac{1}{(2n)^2 - (2n)}(-2n) = \frac{-1}{2n-1},$$

since for each of the $2n$ choices of β_i there is exactly one choice of β_j with $(\beta_i \cdot \beta_j) = -1$, and all the other choices of $j \neq i$ give $(\beta_i \cdot \beta_j) = 0$. Since the size of the signal set $\{\beta_i\}$ is $2n$, (2.6) achieves the "simplex bound" of Theorem 1 (Chapter 1) for the *average* value of crosscorrelation, though not for the *maximum* value.

2.5 Hadamard Matrices and Code Construction

A *Hadamard matrix* H of order n is an $n \times n$ matrix whose entries are restricted to the values $+1$ and -1, with the property

$$(2.7) \qquad HH^T = nI,$$

where H^T is the transpose of H, and I is the $n \times n$ identity matrix. (If the rows of H are $\alpha_1, \alpha_2, \ldots, \alpha_n$, this equation requires $(\alpha_i \cdot \alpha_j) = 0$ for all $i \neq j$.)

Theorem 3. The order n of a Hadamard matrix H is a member of the set $\{1, 2, 4t\}$ where t runs through the positive integers.

Proof. $[+1]$ and $\begin{bmatrix} +1 & +1 \\ +1 & -1 \end{bmatrix}$ are Hadamard matrices of orders 1 and 2. For $n > 2$, let the rows of H be $\alpha_1, \alpha_2, \ldots, \alpha_n$. Multiplying any of the columns of H by -1 has no effect on the pairwise orthogonality of the rows, nor does any permutation of the columns of H. For convenience, we multiply those columns of H by -1 where α_1 had a -1, so that in the normalized matrix H', $\alpha_1' = (+1, +1, \ldots, +1)$. We also *permute* the columns so that $\alpha_2' = (+1, +1, \ldots, +1, -1, -1, \ldots, -1)$. Since $(\alpha_1' \cdot \alpha_2') = 0$, α_2' must consist of equally many $+1$'s and -1's, hence $n/2$ of each, and n must be *even*. We further permute the columns, without affecting the appearance of α_1' or α_2', to get $\alpha_3' = (+1, \ldots, +1, -1, \ldots, -1, +1, \ldots, +1, -1, \ldots, -1)$ where we have r $+1$'s followed by $(\frac{n}{2} - r)$ -1's, and then s $+1$'s followed by $(\frac{n}{2} - s)$ -1's. Since $(\alpha_1' \cdot \alpha_3') = 0$, we have $(\alpha_1' \cdot \alpha_3') = (r+s) - (\frac{n}{2} - r) - (\frac{n}{2} - s) = 0$, from which $r + s = \frac{n}{2}$. Since $(\alpha_2' \cdot \alpha_3') = 0$, we have $(\alpha_2' \cdot \alpha_3') = r - (\frac{n}{2} - r) - s + (\frac{n}{2} - s) = 0$, from which $r - s = 0$. Hence $r = s = \frac{n}{4}$, and n must be a multiple of 4. ■

It has long been *conjectured* that Hadamard matrices exist for all $n = 4t$, but this is still far from proved. However, the smallest value of $4t$ for which no Hadamard matrix is known has been steadily increasing, and now stands at 428. For a description of many of the systematic methods for constructing Hadamard matrices, see [H].

Theorem 4. If there is a Hadamard matrix H of order $n > 1$, then there is a binary orthogonal code consisting of n vectors each of length n; there is a binary bi-orthogonal code consisting of $2n$ vectors each of length n; and a binary simplex code consisting of n vectors each of length $n-1$.

Proof. The row vectors of H need only to be normalized by multiplication of each row by $\frac{1}{\sqrt{n}}$ to obtain the n vectors of an orthogonal code. If to these n vectors their negatives are also adjoined, the resulting set of $2n$ vectors is a bi-orthogonal code. Finally, we may transform H to a new Hadamard matrix H' in which the first column of H' consists entirely of $+1$'s, by multiplying those rows of H that begin with -1 by the scalar -1. (Since $(\alpha_i \cdot \alpha_j) = 0$ we will still have $(a\alpha_i \cdot b\alpha_j) = 0$ for any scalars a and b.) Then we remove the first column of H' to get n row vectors each of length $n-1$. Call these row vectors $\sigma_1, \sigma_2, \ldots, \sigma_n$. Then $(\sigma_i \cdot \sigma_j) = \frac{-1}{n-1}$ for all $i \neq j$, because σ_i and σ_j differ from α_i and α_j by dropping a coordinate in which α_i and α_j were equal. (For $(\alpha_i \cdot \alpha_j)$ we had $\frac{A-D}{A+D} = \frac{0}{n}$, so for $(\sigma_i \cdot \sigma_j)$ we have $\frac{A-D}{A+D} = \frac{-1}{n-1}$.) ■

The converse of Theorem 4 is obviously also true: If any one of the binary $n \times n$ orthogonal code, or the $2n \times n$ bi-orthogonal code, or the $n \times (n-1)$ simplex code, exists, then all three exist, as does the Hadamard matrix of order n (which really *is* the $n \times n$ orthogonal code).

Slightly less obvious is

Theorem 5. If there is a Hadamard matrix H of order $2n > 2$, then there is a simplex code with n binary codewords, each of length $2n-2$.

Proof. Since $HH^T = 2nI$, $\frac{1}{\sqrt{2n}}H$ is an *orthogonal matrix* (in the usual sense of matrix theory), so the columns of H (as well as the rows) are mutually orthogonal. As in Theorem 4, we first transform H to H', where all the rows of H' begins with $+1$, and then we drop this "all $+1$'s" column to get H''. Since the second column of H' is orthogonal to the first column of H', it consists of equally many $+1$'s and -1's. This second column of H' is the first column of H''. Thus H'' has n rows beginning with $+1$ and n rows beginning with -1. Keep only the n rows beginning with $+1$'s, and from these drop the initial $+1$, leaving a set S of n vectors of length $2n-2$. Since both of the dropped positions were "agreements" between any two of the corresponding rows of H', for any two distinct vectors σ_i and σ_j in S we have $(\sigma_i \cdot \sigma_j) = \frac{A-D}{A+D} = \frac{-2}{2n-2} = \frac{-1}{n-1}$, the "simplex bound". ■

2.6. Exercises

1. If $A = (a_{ij})$ is an $n \times n$ matrix and $B = (b_{k\ell})$ is an $m \times m$ matrix,

the *Kronecker Product* $A * B$ is defined as

$$A * B = \begin{bmatrix} a_{11}B & a_{12}B & \dots & a_{1n}B \\ a_{21}B & a_{22}B & \dots & a_{2n}B \\ \vdots & & & \vdots \\ a_{n1}B & a_{n2}B & \dots & a_{nn}B \end{bmatrix}$$

regarded as an $mn \times mn$ matrix. Show that if A and B are both Hadamard matrices, then so too are $A * B$ and $B * A$.

2. Show that Hadamard matrices of order 2^k exist for all $k \geq 0$.

3. Show that if there is a Hadamard matrix of order n, then there is a Hadamard matrix of order $2n$.

4. Suppose that A, B, C, D are $t \times t$ symmetric circulant matrices of $+1$'s and -1's with $A^2 + B^2 + C^2 + D^2 = 4tI$. ($A = (a_{ij})$ is a symmetric circulant matrix iff

$$A = \begin{bmatrix} a_{11} & a_{12} & a_{13} & \dots & a_{1t} \\ a_{12} & a_{13} & a_{14} & \dots & a_{11} \\ \vdots & \vdots & \vdots & & \vdots \\ a_{1t} & a_{11} & a_{12} & \dots & a_{1,t-1} \end{bmatrix}.)$$

Show that $$H = \begin{bmatrix} A & B & C & D \\ -B & A & -D & C \\ -C & D & A & -B \\ -D & -C & B & A \end{bmatrix}$$

is a Hadamard matrix of order $4t$. [This result is "Williamson's Theorem", and the construction for H is known as "Williamson's Construction".]

5. Use Williamson's Construction to obtain Hadamard matrices of order 12 ($t = 3$) and 20 ($t = 5$).

2.7. Cyclic Hadamard Matrices

For $n > 1$, the only known example of a Hadamard matrix which is also a circulant matrix has order $n = 4$:

$$H = \begin{bmatrix} -1 & +1 & +1 & +1 \\ +1 & -1 & +1 & +1 \\ +1 & +1 & -1 & +1 \\ +1 & +1 & +1 & -1 \end{bmatrix}.$$

(There is a significant literature of partial results that no such example exists with $n > 4$. See, for example, [I].)

However, there are many examples of $n \times n$ Hadamard matrices H which consist of an $(n-1) \times (n-1)$ circulant matrix with a "border" added (top-most row and left-most column of H) consisting entirely of $+1$'s. Some examples of these are

$$H_4 = \begin{bmatrix} +1 & +1 & +1 & +1 \\ +1 & +1 & -1 & -1 \\ +1 & -1 & +1 & -1 \\ +1 & -1 & -1 & +1 \end{bmatrix}, H_8 = \begin{bmatrix} + & + & + & + & + & + & + & + \\ + & - & - & - & + & - & + & + \\ + & + & - & - & - & + & - & + \\ + & + & + & - & - & - & + & - \\ + & - & + & + & - & - & - & + \\ + & + & - & + & + & - & - & - \\ + & - & + & - & + & + & - & - \\ + & - & - & + & - & + & + & - \end{bmatrix},$$

$$\text{and } H_{12} = \begin{bmatrix} + & + & + & + & + & + & + & + & + & + & + & + \\ + & - & + & - & + & + & + & - & - & - & + & - \\ + & - & - & + & - & + & + & + & - & - & - & + \\ + & + & - & - & + & - & + & + & + & - & - & - \\ + & - & + & - & - & + & - & + & + & + & - & - \\ + & - & - & + & - & - & + & - & + & + & + & - \\ + & - & - & - & + & - & - & + & - & + & + & + \\ + & + & - & - & - & + & - & - & + & - & + & + \\ + & + & + & - & - & - & + & - & - & + & - & + \\ + & + & + & + & - & - & - & + & - & - & + & - \\ + & - & + & + & + & - & - & - & + & - & - & + \\ + & + & - & + & + & + & - & - & - & + & - & - \end{bmatrix}.$$

Examples of this type are called "cyclic Hadamard matrices", and are in one-to-one correspondence with "cyclic Hadamard difference sets". (For an extensive treatment of these, see [G].)

All known examples of cyclic Hadamard matrices of order $n = 4t$ have $n - 1$ belonging to one of three sequences:

a) $4t - 1 = 2^k - 1$, $k \geq 1$.

b) $4t - 1 = p$, p a prime.

c) $4t - 1 = p(p+2)$, where p and $p+2$ form a twin prime.

Examples of type a) can be obtained for all $k \geq 1$ by taking the top row of the circulant to be an "m-sequence", i.e. a maximum-length linear shift register sequence, of period $2^k - 1$, and replacing the 0's and 1's of the m-sequence by $+1$'s and -1's, respectively. (For the theory of m-sequences, see [K].) Additional examples of type a), but only for certain composite values of k, are obtained by the Gordon-Mills-Welch (GMW) construction (see [J]).

Examples of type b) can be obtained for all primes $p = 4t - 1$ by the "Legendre sequence" construction, taking the top row of the circulant to be

$$-1, \left(\frac{1}{p}\right), \left(\frac{2}{p}\right), \ldots, \left(\frac{p-1}{p}\right),$$

where $(\frac{a}{p})$ is the "Legendre symbol", or "quadratic character", modulo p, defined for $1 \leq a \leq p-1$ by

$$\begin{aligned}\left(\tfrac{a}{p}\right) &= +1 \text{ if for some } x, x^2 \equiv a \pmod{p},\\ \left(\tfrac{a}{p}\right) &= -1 \text{ otherwise.}\end{aligned}$$

Additional examples of type b) are obtained by Hall's "sextic residue sequence" construction when $p = 4t - 1 = 4u^2 + 27$ (see [H]).

If $a \equiv 0 \pmod{p}$, the Legendre symbol $(\frac{a}{p})$ is defined to be 0. If p and q are distinct odd primes, the *Jacobi symbol* $(\frac{a}{pq})$ is defined to be $(\frac{a}{p})(\frac{a}{q})$, the product of the Legendre symbols. In the "Legendre sequence construction" for examples of type b), we replaced $(\frac{0}{p}) = 0$ with the value –1. To get the "twin prime construction", for cyclic Hadamard matrices of type c) with $4t - 1 = p(p+2) = pq$, we take the top row of the circulant matrix to be a modification of the sequence of Jacobi symbols $\{(\frac{0}{pq}), (\frac{1}{pq}), (\frac{2}{pq}), \ldots, (\frac{pq-1}{pq})\}$ where we use the Jacobi symbol $(\frac{a}{pq})$ whenever this is non-zero; we replace it by +1 for $a = \{0, q, 2q, \ldots, (p-1)q\}$; and we replace it by –1 for $a = \{p, 2p, 3p, \ldots, (q-1)p\}$.

Sequence lengths of types b) and c) are obviously disjoint sets, since no prime is a product of twin primes. The only overlap of lengths of types a) and c) occurs with $2^4 - 1 = 3 \cdot 5$, and here the matrix examples are in fact the same. The overlaps of the sequence lengths of types a) and b) are precisely the Mersenne primes, $p = 4t - 1 = 2^k - 1$. However, the matrix examples are the same only for $p = 2^2 - 1 = 3$ and $p = 2^3 - 1 = 7$, if we are considering m-sequences and Legendre sequences. At $p = 2^5 - 1 = 31$, the m-sequence gives the same result as Hall's "sextic residue sequence" construction. At $p = 2^7 - 1 = 127$, the m-sequence, the Legendre sequence, and the sextic residue sequence constructions all give inequivalent examples. Moreover, at $4t - 1 = 127$, Baumert [G], by complete search, found *three more* inequivalent examples (thus six inequivalent constructions altogether) which do not form part of any known families.

2.8. Exercises

1. For all $t \leq 100$, make a table of the values of $4t - 1$, showing which correspond to the constructions of type a), of type b), of type c), or to no known constructions.

2. Use the Legendre sequence construction to obtain examples of cyclic Hadamard matrices of orders $n = 8, n = 12$, and $n = 20$.

3. Use the twin prime construction to obtain examples of cyclic Hadamard matrices of orders $n = 16$ and $n = 36$.

4. Asymptotically, how many sequence lengths $4t - 1 \leq x$ will you find, as $x \to \infty$, of each of the three types a), b) and c) ? [It is known that $\pi(x)$, the number of primes $\leq x$, is asymptotic to $x/\ell n x$, and that half of these (asymptotically) are of the form $4t - 1$. It is not *known* that

the number of twin primes is infinite, but it is *conjectured* (see[L]) that the number of such pairs up to x, $T(x)$, is asymptotic to $cx/\ell n^2 x$, where $c = 2\prod_{\text{all } p>2}(1 - \frac{1}{(p-1)^2}) = 1.320\ldots$, where the product is extended over all primes $p > 2$.]

CHAPTER 3. APPLICATIONS TO RADAR, SONAR, AND SYNCHRONIZATION

3.1. Overview

In all three of the applications mentioned in the title of this chapter, one of the objectives (often the major objective) is to determine a point in time with great accuracy. In radar and sonar, we want to determine the round trip time from transmitter to target to receiver very accurately, because the one-way time (half of the round-trip time) is a measure of the distance to the target (called the *range* of the target).

The simplest approach would be to send out a pure "impulse" of energy, and measure the time until it returns. The ideal "impulse" would be virtually instantaneous in duration, but with such high amplitude that the total energy contained in the pulse would be significant, much like a Dirac "delta function". However, the Dirac "delta function" not only fails to exist as a mathematical *function*, but it is also unrealizable as a physical *signal*. Close approximations to it — very brief signals with very large amplitudes — may be valid mathematically, but are impractical to generate physically. Any actual transmitter will have an upper limit on "peak power" output, and hence a short pulse will have a very restricted amount of total energy : at most, the peak power times the pulse duration. More total energy can be transmitted if we extend the duration; but if we transmit at uniform power over an extended duration, we do not get a sharp determination of the round trip time. This dilemma is illustrated in Figure 3.1.

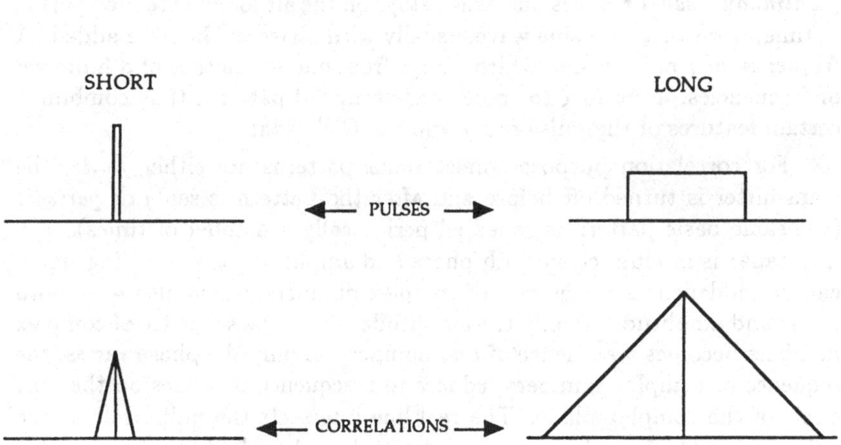

Figure 3.1. The shorter the pulse, the sharper the autocorrelation function.

In the presence of noise, the time of the peak of the autocorrelation function of the extended pulse becomes difficult to determine with high precision.

It is here that clever combinatorial mathematics comes to the rescue. By using a suitably coded pulse pattern, or coded signal pattern, we extend the duration, thereby increasing the total transmitted energy as much as desired, while still maintaining a sharp spike in the autocorrelation function to mark the round trip propagation time with great precision. Such a technique was successfully used in 1960 by the Jet Propulsion Laboratory to bounce a radar signal off the surface of Venus and detect it back on earth. Not only was that the first successful radar ranging of another planet in the solar system, but the distance thus measured made it possible to improve the accuracy of the "Astronomical Unit" (the mean radius of the earth's orbit around the sun, and the basic yardstick for distances within the solar system) by three orders of magnitude.

3.2. Types of Signals and Correlations

A *monostatic radar* is one which has transmitter and receiver at the same location, which requires that the transmitter be turned off while the signal is being received. A *bistatic radar* is one which has two separate antennas, often at a considerable distance from one another, for transmitting and receiving. With a bistatic radar, it is possible, in principle, to be transmitting all the time.

A *pulse radar* is one which transmits a succession of pulses, not necessarily uniformly spaced, but which is turned off between pulses. A *CW* (*continuous wave*) *radar* is one which stays on the air for an extended period of time, transmitting a sine wave, usually with phase modulation added. A *frequency hop radar* is one which jumps from one to another of a finite set of frequencies, according to some predetermined pattern, thus combining certain features of the pulse radar and the CW radar.

For correlation purposes, most radar patterns are either *finite* (the transmitter is turned off before and after the pattern is sent) or *periodic* (the same basic pattern is repeated periodically a number of times). If a CW radar is modulated in both phase and amplitude, the resulting signal can be modelled as a sequence of complex numbers, which also have both phase and amplitude. If only the amplitude varies, the sequence of complex numbers becomes a sequence of real numbers. If only the phase varies, the sequence of complex numbers reduces to a sequence of values on the unit circle of the complex plane. The real line intersects the unit circle at the points $+1$ and -1, and binary modulation by $+1$ and -1 can be regarded either as phase modulation or as amplitude modulation.

The returning signal can be correlated against either an ideal model of itself, or against some other signal (often called a "complementary" signal or sequence), especially designed to highlight some specific feature of the returning signal. In all cases, the ratio of the value of correlation for $\tau = 0$, i.e. when the signal is aligned with itself, versus the maximum value of the correlation for $\tau \neq 0$, is a measure of the clarity with which range can be measured in a noisy environment.

If there is a relative motion (either toward or away) between the transmitter and the target, then the returning radar or sonar signal is shifted not only in time, but also in frequency. The frequency shift ("Doppler shift") is proportional to the time derivative of range — i.e. to the velocity of approach or separation, between communicator and target. The two-dimensional autocorrelation function of the signal, in the time and frequency domains, is called its *ambiguity function*, and the ideal shape of such a function is a spike, or an inverted *thumb-tack* (or *drawing pin*, in the U.K.).

3.3. Barker Sequences

R. Barker [M] asked: for what lengths L do binary sequences of $+1$'s and -1's, $\{a_j\}_{j=1}^{L}$, exist, with finite, unnormalized auto-correlation $K(\tau)$, defined by

$$K(\tau) = \sum_{j=1}^{L-\tau} a_j a_{j+\tau}, \tag{3.1}$$

bounded by 1 in absolute value for $\tau \neq 0$? That is, the requirement is

$$|K(\tau)| \leq 1 \text{ for } 1 \leq |\tau| \leq L-1. \tag{3.2}$$

Barker gave examples having the following lengths:

L	*Sequence*
1	$+1$
2	$+1,+1$
3	$+1,+1,-1$
4	$+1,+1,-1,+1$
5	$+1,+1,+1,-1,+1$
7	$+1,+1,+1,-1,-1,+1,-1$
11	$+1,+1,+1,-1,-1,-1,+1,-1,-1,+1,-1$
13	$+1,+1,+1,+1,+1,-1,-1,+1,+1,-1,+1,-1,+1$

Turyn and Storer showed [N] that there are no other "Barker Sequences" for *odd* lengths $L > 13$. It is still unproven that even length Barker Sequences with $L > 4$ do not exist, though this is generally believed. It has been shown (see[I]) that even-length Barker sequences for $L \geq 4$ correspond to circulant Hadamard matrices of order L, as described in Section 2.7.

3.4 Generalized Barker Sequences

In [O], Golomb and Scholtz defined a *Generalized Barker Sequence* of length L to be a sequence $\{a_j\}_{j=1}^{L}$ of complex numbers on the unit circle (that is, $|a_j| = 1$ for $1 \le j \le L$) with finite unnormalized autocorrelation $K(\tau)$, defined by

$$(3.3) \qquad K(\tau) = \sum_{j=1}^{L-\tau} a_j a_{j+\tau}^*,$$

satisfying

$$(3.4) \qquad |K(\tau)| \le 1 \text{ for } 1 \le |\tau| \le L-1.$$

In [O], a group of $4L^2$ transformations on the sequence $\{a_j\}_{j=1}^{L}$ is identified which preserves the "Barker property" (3.4), and examples of Generalized Barker Sequences with $L \le 16$ are given. Subsequently, examples have been found for all lengths $L \le 19$. (See [P]. The example given in [P] for $L = 19$ contains an error, but other examples of this length have been found.)

With $K(\tau)$ defined as in (3.3), $K(0) = L$, and hence the *normalized* correlation $C(\tau)$ for a Generalized Barker Sequence must satisfy

$$(3.5) \qquad |C(\tau)| = \frac{1}{L}|K(\tau)| \le \frac{1}{L}, \text{ all } 1 \le |\tau| \le L-1.$$

This condition becomes increasingly difficult to satisfy as L increases, and it is extremely unlikely that any Generalized Barker Sequences exist for $L > 30$ (if indeed they exist for any $L > 19$). However, for applications to radar, conditions considerably weaker than (3.5) are still quite useful. For example, if $\{a_j\}_{j=1}^{L}$ is a sequence for which $K(\tau)$, as defined in (3.3), satisfies

$$(3.6) \qquad |K(\tau)| \le cL^{1/2}, \quad 1 \le |\tau| \le L-1,$$

then the normalized correlation $C(\tau)$ satisfies

$$(3.7) \qquad |C(\tau)| = \frac{1}{L}|K(\tau)| \le cL^{-1/2}, \quad 1 \le |\tau| \le L-1.$$

Such a family of complex sequences, existing for every length $L = n^2$, using n^{th} roots of unity as the terms of the sequence, was described by R. Frank [Q]. Another family of sequences satisfying (3.7), for every integer length $L \ge 1$, using L^{th} roots of unity as the terms, is the subject of a forthcoming paper by N. Zhang and S. W. Golomb.

3.5. Huffman's "Impulse-Equivalent Pulse Trains"

The Generalized Barker Sequences of the previous section correspond to phase modulation on a sinusoidal radar carrier signal, where the modulation lasts for L time intervals, changing from each interval to the next, and where the radar transmitter is turned off before and after the signal consisting of the L phase-modulated intervals.

In [R], D. Huffman considered the corresponding problem for an amplitude modulated radar signal, and imposed an even stronger restriction than Barker's on the out-of-phase values of the autocorrelation:

$$K(\tau) = \sum_{j=1}^{L-\tau} a_j a_{j+\tau} = 0 \text{ for } 1 \leq |\tau| \leq L-2, \tag{3.8}$$

where all the terms a_j of the sequence $\{a_j\}_{j=1}^{L}$ are *real*. As usual,

$$K(0) = \sum_{j=1}^{L} a_j^2 = \sum_{j=1}^{L} |a_j|^2 \tag{3.9}$$

is the *total energy* of the signal; and

$$K(L-1) = a_1 a_L = K(-(L-1)) \tag{3.10}$$

cannot be 0 if the sequence truly has length L.

As a general method of studying the finite autocorrelation function of the sequence $\{a_0, a_1, a_2, \ldots, a_n\}$ of complex numbers of length $L = n+1$, we consider the two associated polynomials

$$P(x) = a_0 x^n + a_1 x^{n-1} + \ldots + a_{n-1} x + a_n = a_0 \prod_{i=1}^{n} (x - r_i) \tag{3.11}$$

and

$$Q(x) = x^n P\left(\frac{1}{x}\right) = a_0 + a_1 x + a_2 x^2 + \ldots + a_n x^n = a_0 \prod_{i=1}^{n} (1 - r_i x), \tag{3.12}$$

where the r_i, in general, are complex numbers.

It is readily seen that

$$P(x) Q^*(x) = \sum_{\tau=-n}^{n} K(\tau) x^{n+\tau}, \tag{3.13}$$

where $K(\tau) = \sum_{j=0}^{n-\tau} a_j a_{j+\tau}^*$, from which $K(-\tau) = K^*(\tau)$. Here $K(0) = \sum_{j=0}^{n} |a_j|^2 = E$ is the "total energy" in the signal; and $K(n) = a_0 a_n^*$, $K(-n) = a_0^* a_n = K^*(n)$. We normalize the sequence by requiring $|K(n)| = |K(-n)| = 1$. If $K(n) = \eta$ with $|\eta| = 1$, then $K(-n) = \eta^*$. We will be particularly interested in the case that η is *real*, and hence either

$K(n) = K(-n) = 1$ or $K(n) = K(-n) = -1$. With Huffman's additional restriction, namely that $K(\tau) = 0$ for $1 \le |\tau| \le n-1$, (3.13) becomes

$$(3.14) \qquad P(x)Q^*(x) = \sum_{\tau=-n}^{n} K(\tau)x^{n+\tau} = \eta x^{2n} + Ex^n + \eta^*, \quad |\eta| = 1,$$

and we will focus on the two cases

$$(3.15) \qquad x^{2n} + Ex^n + 1 = P(x)Q^*(x)$$

or

$$(3.16) \qquad -x^{2n} + Ex^n - 1 = P(x)Q^*(x).$$

In the former case, (3.15), we can write

$$(3.17) \qquad x^{2n} + Ex^n + 1 = (x^n + R^n)(x^n + \frac{1}{R^n}), \quad R > 0,$$

where $2 < E = R^n + \frac{1}{R^n} < \infty$. In the latter case, (3.16), we can write

$$(3.18) \qquad -x^{2n} + Ex^n - 1 = -(x^n - R^n)(x^n - \frac{1}{R^n}), \quad R > 0,$$

where again $2 < E = R^n + \frac{1}{R^n} < \infty$.

Hence, for (3.15), the roots of $P(x)Q^*(x) = 0$ are the complex roots of $(x^n + R^n)(x^n + \frac{1}{R^n}) = 0$, which are $\{R\alpha^t\}$ and $\{\frac{1}{R\alpha^t}\}$, where $\alpha = e^{\pi i/n}$, and $t = 1, 3, 5, \ldots, 2n-1$. Similarly, for (3.16), the roots of $P(x)Q^*(x) = 0$ are the complex roots of $(x^n - R^n)(x^n - \frac{1}{R^n}) = 0$, which are $\{R\beta^u\}$ and $\{\frac{1}{R}\beta^u\}$ where $\beta = e^{2\pi i/n}$ and $u = 1, 2, \ldots, n$.

The sequence $\{a_j\}_{j=0}^{n}$ is a sequence of *real* numbers if and only if all coefficients of $P(x)$ (and hence also of $Q(x)$) are real, which occurs if and only if the subset of the $2n$ roots of $P(x)Q^*(x) = 0$ which are roots of $P(x)$ is closed with respect to complex conjugation.

An example of such a sequence $\{a_j\}$ of length $L = n + 1 = 5$ is $\{1, -1, \frac{1}{2}, 1, 1\}$, with $P(x) = x^4 - x^3 + \frac{1}{2}x^2 + x + 1$, $Q(x) = x^n P(\frac{1}{x}) = 1 - x + \frac{1}{2}x^2 + x^3 + x^4$, and $P(x)Q^*(x) = x^8 + 4\frac{1}{4}x^4 + 1 = (x^4 + 4)(x^4 + \frac{1}{4})$. The roots of $x^4 + 4 = 0$ are $\{1+i, 1-i, -1+i, -1-i\}$, and the roots of $x^4 + \frac{1}{4} = 0$ are $\{\frac{1+i}{2}, \frac{1-i}{2}, \frac{-1+i}{2}, \frac{-1-i}{2}\}$. The roots of $P(x) = 0$ are $\{1+i, 1-i, \frac{-1+i}{2}, \frac{-1-i}{2}\}$, verified by $(x-(1+i))(x-(1-i))(x-\frac{-1+i}{2})(x-\frac{-1-i}{2}) = (x^2-2x+2)(x^2 + x + \frac{1}{2}) = x^4 - x^3 + \frac{1}{2}x^2 + x + 1$. Also, for the sequence $\{1, -1, \frac{1}{2}, 1, 1\}$, the autocorrelation $K(\tau)$ is given by $K(0) = 4\frac{1}{4}, K(1) = K(2) = K(3) = 0$, and $K(4) = 1$.

Long Huffman sequences are useful to the extent that the magnitudes of the terms are approximately equal, so that the transmitted energy is distributed fairly uniformly throughout the broadcast interval. Methods for achieving this type of uniformity are not adequately understood for the general case.

3.6 Exercises

1. Find a Generalized Barker Sequence of Length 6, using only sixth roots of unity as terms.

2. Using only sixth roots of unity as terms, find at least one Generalized Barker Sequence of length L for $2 \leq L \leq 12$.

3. Let $\alpha = e^{2\pi i/n}$, and define the sequence $S_n = \{a_j\}_{j=1}^{n}$ where $a_j = \alpha^{\frac{j(j-1)}{2}}$. What bounds can you obtain on $\max_{\tau \neq 0} |K(\tau)|$, where $K(\tau)$ is defined in (3.3)?

4. Show that the sequence $\{1, -1, -1, \frac{5}{2}, -\frac{3}{4}, -\frac{1}{2}, -2, -1, -1\}$ is an "impulse-equivalent pulse train" in the sense of Huffman, and explain the factorization of $P(x)Q^*(x) = 0$ corresponding to this sequence.

3.7. Pulse Patterns and "Optimal Rulers"

A pulse radar is able to send one or more pulses of radio-frequency energy toward a target. It is a convenient and rather realistic assumption to require all the pulses to be the same in both duration and amplitude. The signal design problem then reduces to devising *patterns* of these identical pulses, so that the autocorrelation function of the pattern is as impulse-like as possible.

The problem is usually restated as follows: For each positive integer n, what is the shortest length $L = L(n)$ for which there is a sequence $\{a_1, a_2, \ldots, a_n\}$ with $0 = a_1 < a_2 < \ldots < a_n = L$, such that the set of $\binom{n}{2}$ differences $\{a_j - a_i\}$, with $1 \leq i < j \leq n$, are all distinct?

The model underlying this restatement is the following. At each integer a_i in the sequence, there is a pulse of brief duration and unit amplitude. Thus, the sequence corresponds to a pulse pattern with n pulses, spread out over a total duration of L (or, more precisely L plus one pulse duration) with the unnormalized autocorrelation function $K(\tau)$ satisfying $K(0) = n, 0 \leq K(\tau) \leq 1$ for $1 \leq \tau \leq L$, and $K(\tau) = 0$ for $\tau > L$. (The distinctness of the differences $\{a_j - a_i\}$ guarantees $K(\tau) \leq 1$ for all $\tau, |\tau| \geq 1$.) Finding the *shortest* length $L = L(n)$ for the sequence achieves the desired signal parameters with the shortest duration for the pulse pattern.

The sequence model is also described in terms of a certain class of *rulers* (the measuring devices, not the monarchs or autocrats).

A ruler of length L has only n marks on it, at integer positions

$a_1, a_2, \ldots, a_n$, where $a_1 = 0$ and $a_n = L$ are the two endpoints of the ruler. If every integer distance $d, 1 \leq d \leq L$, can be measured in one and only one way as a distance between two of the n marks, then the ruler is called a *perfect ruler*. For a perfect ruler, $L = \binom{n}{2}$, since there are exactly $\binom{n}{2}$ distances between the n marks, and these must be some permutation of $\{1, 2, 3, \ldots, L\}$. In Figure 3.2, we see perfect rulers for $n = 2, 3$, and 4, with the corresponding radar pulse patterns and their autocorrelation functions $K(\tau)$.

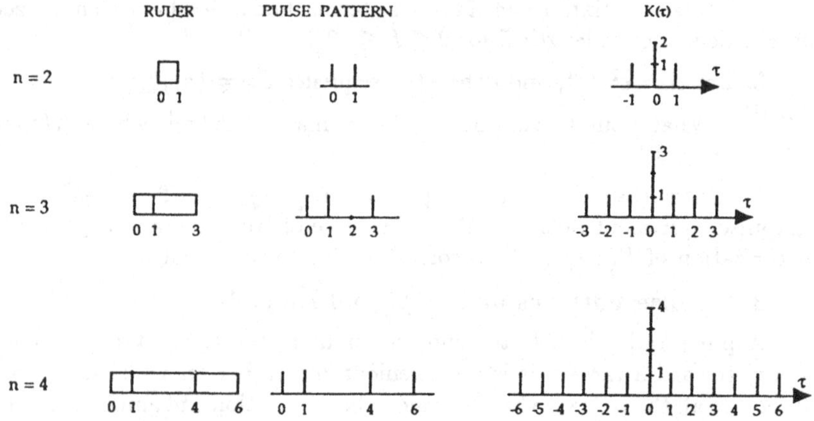

Figure 3.2. Perfect rulers for n=2, n=3, and n=4, with the corresponding pulse radar patterns and their unnormalized autocorrelation functions.

Unfortunately, for $n > 4$, there are no perfect rulers.

Theorem 4. For $n > 4$, no perfect rulers exist.

Proof. A ruler with n marks has $n - 1$ *intervals* between marks. For a *perfect* ruler, these $n - 1$ intervals must all have distinct positive integer lengths, and the sum of these lengths must be $L = \binom{n}{2}$. Hence these intervals must be (in some order) $1, 2, 3, \ldots, n - 1$, since any *other* set of $n - 1$ distinct positive integers will have a larger sum than $\binom{n}{2}$. Since *all* distances between marks on the ruler must be distinct, the interval of length 1 cannot be next to an interval of length $\leq n - 2$ (since the sum of two consecutive intervals is a measured distance of the ruler, and every length from 1 to $n - 1$ is *already* measured, as a single interval). This can only be achieved if the interval of length 1 is at one end of the ruler, and is immediately followed by the interval of length $n - 1$. Similarly, the interval of length 2 cannot be next to any interval of length $\leq n - 3$ (to avoid two consecutive intervals with a total length $\leq n-1$, equalling the length of a single interval), nor can it be next to the interval of length $n-2$ (since $2+(n-2)=n=1+(n-1)$, and the distance n would be measured in more than one way). This requires that the interval of length 2 must also be at an end of the ruler, and must also be next to the interval of length $n-1$. But then the entire ruler consists of only three intervals: $1, n-1, 2$, which means there are only four marks altogether, and $n=4$. (Note that with $n=4$, there really is a perfect ruler with the consecutive intervals of lengths 1, 3, 2.) ■

There are two obvious ways to relax the requirements on a perfect ruler to get objects which exist for all n. A *covering ruler* with n marks and length L measures every distance from 1 to L, as a distance between two marks on the ruler, in *at least one way*; while a *spanning ruler* with n marks and length L measures every distance from 1 to L, as a distance between two marks on the ruler, in *at most one way*. The interesting combinatorial problems are to determine the *longest covering ruler* with n marks, and the *shortest spanning ruler* with n marks, for each positive integer n. Both of these problems have long histories in the combinatorial literature. However, the application to pulse radar involves only finding the *shortest spanning ruler* for each n. (Martin Gardner [S] termed these objects "Golomb rulers", a name which seems subsequently to have been widely adopted.)

The behavior of $L(n)$ as a function of n, for the shortest spanning ruler, is quite erratic in detail, although it is known that asymptotically $L(n) \sim n^2$ as $n \to \infty$. The value of $L(n)$ has been determined by exhaustive computer search for all $n \leq 16$. (For the larger values of n, this search was first performed by J. Shearer of IBM.) In addition to left-right reversal of the ruler, these rulers are not unique for several of the smaller values of n. One example of a spanning ruler of length $L(n)$, for each $n \leq 16$, is shown in Table 3.1.

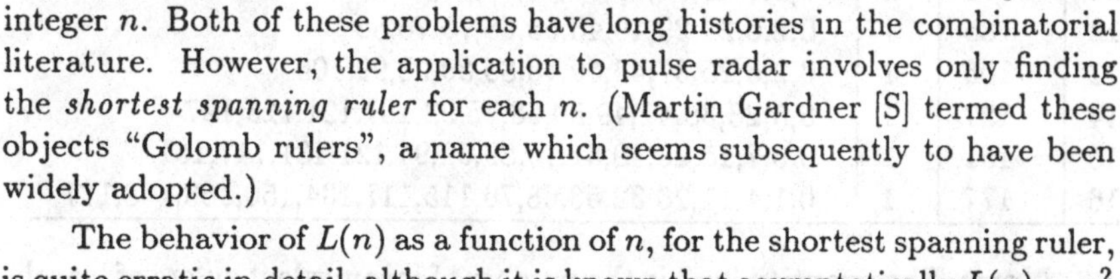

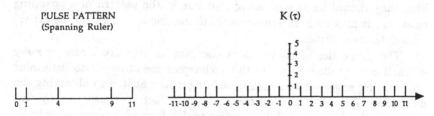

Figure 3.3. Radar pulse pattern, and its autocorrelation, for n = 5 pulses.

In Figure 3.3., we see the pulse pattern and correlation corresponding to the sequence listed for $n = 5$.

TABLE 3.1

TABLE OF THE SHORTEST SPANNING RULERS

n	L(n)	m	Sequence of Marks
2	1	1	0,1
3	3	1	0,1,3
4	6	1	0,1,4,6
5	11	2	0,1,4,9,11
6	17	4	0,1,4,10,12,17
7	25	5	0,1,4,10,18,23,25
8	34	1	0,1,4,9,15,22,32,34
9	44	1	0,1,5,12,25,27,35,41,44
10	55	1	0,1,6,10,23,26,34,41,53,55
11	72	2	0,1,4,13,28,33,47,54,64,70,72
12	85	1	0,2,6,24,29,40,43,55,68,75,76,85
13	106	1	0,2,5,25,37,43,59,70,85,89,98,99,106
14	127	1	0,5,28,38,41,49,50,68,75,92,107,121,123,127
15	151	1	0,6,7,15,28,40,51,75,89,92,94,121,131,147,151
16	177	1	0,1,4,11,26,32,56,68,76,115,117,134,150,163,168,177

Note: The quantity "m" is the number of inequivalent rulers of length $L(n)$ which are shortest spanning rulers with n marks. Only one of each set of m rulers is listed explicitly in this table.

These minimum spanning rulers have another, very different application in radio astronomy. In radio astronomy, only receiving antennas are used, and the spacing between two antennas generates a difference in the time a signal is received, which is used in making interferometry measurements on the signal. If several antennas are constructed along a straight line, they should be spaced along that line in the pattern of a spanning ruler, to get maximum interferometry information by having all the arrival time differences distinct.

The properties of these rulers also play an important role in x-ray diffraction crystallography. In this technique, one attempts to determine the bonding angles of a crystal by shining x rays at it, and observing the diffraction patterns which emerge. What in fact is measurable by this method is the *differences* of the bonding angles, from which one wishes to reconstruct the bonding angle themselves. This leads to an "inverse problem":

given the autocorrelation function $K(\tau)$, what is the set of possible signals which might have produced it ? For several decades, crystallographers relied on a "theorem" of S. Piccard which asserted (in our terminology): "If two spanning rulers have the same autocorrelation function $K(\tau)$ the two rulers are either identical or mirror images of each other." A counter-example was found by G.S. Bloom [T] for 6-mark rulers: $\{0,1,4,10,12,17\}$ and $\{0,1,8,11,13,17\}$, which are in fact *shortest* 6-mark spanning rulers. This generalizes to a two-parameter family of counter-examples, all involving 6-mark spanning rulers. No counter-examples with fewer than 6 marks are possible, and none with more than 6 marks are known. (There are partial results which suggest that counter-examples may occur only in the case $n = 6$.)

3.8. Two-Dimensional Pulse Patterns

J.P. Costas (see[U]) proposed the following problem: We wish to design an $n \times n$ frequency hop pattern, for radar or sonar, using n consecutive time intervals $t_1, t_2, \ldots, t_n$ and n consecutive frequencies $f_1, f_2, \ldots, f_n$, where some permutation of the n frequencies is assigned to the n consecutive time slots. Moreover, this should be done in such a way that, if two frequencies f_i and f_{i+r} occur at the two times t_j and t_{j+s}, then there is no i', $i' \neq i$, where the two frequencies $f_{i'}$ and $f_{i'+r}$ occur at times $t_{j'}$ and $t_{j'+s}$. This constraint corresponds to an ideal, or "thumb-tack" ambiguity function for the frequency hop pattern.

We may represent the frequency hop pattern by an $n \times n$ permutation matrix (a_{ij}), where $a_{ij} = 1$ if and only if frequency f_i is used at time t_j. (Otherwise $a_{ij} = 0$.) The extra condition is that the $\binom{n}{2}$ "vectors" connecting the n positions in the matrix where 1's are located are all distinct *as vectors* : no two vectors are the same in both magnitude and slope. One may visualize a *dot* at each position where $a_{ij} = 1$. When the pattern is shifted in both time (horizontally) and frequency(vertically), any dot can be brought into coincidence with any other dot. However, the extra "Costas" condition is that no such shift (other than the identity, which is no shift at all) will bring *two* dots into coincidence with two other dots.

Costas succeeded, initially, in finding examples, by exhaustive computer search, only for $n \leq 12$. However, several systematic constructions for these "Costas Arrays" are now known, giving examples for arbitrarily large values of n. All of these systematic constructions are based on the existence of primitive roots in finite fields. Three such constructions are the following:

1) *The Welch Construction*, for $n = p - 1$ and $p - 2$, p prime.

Let g be a primitive root modulo p. The "dots" of the permutation matrix occur at the locations (i, g^i) for $1 \leq i \leq p-1$, giving a Costas Array of order $n = p - 1$.

Since $g^{p-1} \equiv 1 \pmod{p}$, there is a dot at $(p-1, g^{p-1})$, which is at a corner of the matrix. Removing the row and column of this dot leaves a Costas Array of order $n = p - 2$.

2. *The Lempel Construction*, for $n = q - 2$, q a prime power.

Let α be a primitive element in $GF(q)$. The "dots" of the permutation matrix occur at the locations (i, j) whenever $\alpha^i + \alpha^j = 1$ in $GF(q), 1 \leq i, j \leq n - 2$. (This always produces a *symmetric* matrix.)

3. *The Golomb Construction*, for $n = q - 2$ and $q - 3$, q a prime power.

Let α and β be any two primitive elements in $GF(q)$. The "dots" of the permutation matrix occur at the locations (i, j) whenever $\alpha^i + \beta^j = 1$ in $GF(q)$, $1 \leq i, j \leq n - 2$. (The special case when $\alpha = \beta$ is the Lempel Construction.)

It has been shown that for all $q > 2$, the field $GF(q)$ contains primitive elements α and β (not necessarily distinct) with $\alpha + \beta = 1$. Using such α and β in the Golomb Construction, since $\alpha^1 + \beta^1 = 1$, we have (1,1) as the location of a "dot" in the construction. Removing the top row and left column of the matrix leaves a $(q - 3) \times (q - 3)$ Costas Array.

For proofs that these three constructions must yield Costas Arrays, see [V]. For additional variants, and the way they yield examples of Costas Arrays for many values of $n \leq 360$, see [W]. Since 1984, the smallest values of n for which no examples of Costas Arrays are known are $n = 32$ and $n = 33$.

Table 3.2

n	$C(n)$	$c(n)$	$s(n)$
2	2	1	1
3	4	1	1
4	12	2	1
5	40	6	2
6	116	17	5
7	200	30	10
8	444	60	9
9	760	100	10
10	2160	277	14
11	4368	555	18
12	7852	990	17
13	12828	1616	25
14	17252	2168	23
15	19612	2467	31
16	21104	2648	20
17	18276	2294	19
18	15096	1892	10
19	10240	1283	6
20	6464	810	4

Table of the number of Costas Arrays for $n \leq 20$; where $C(n)$ is the total number, $c(n)$ is the reduced number, and $s(n)$ is the number of symmetric Costas Arrays of order n.

The complete enumeration of Costas Arrays through $n = 13$ was reported in [W]. Subsequent values have been found by J. Silverman ($n \leq 16$) and O. Moreno ($n \leq 20$), leading to the tabulation of $C(n)$, the total number of $n \times n$ permutation matrices which are Costas Arrays; the reduced number $c(n)$, where two Costas Arrays which differ only by one of the 8 "dihedral symmetries" of the square are not considered distinct; and $s(n)$, the number of inequivalent arrays which have diagonal symmetry; all of which are shown in Table 3.2. For $n > 2$, these three quantities are linearly related by $C(n) = 8c(n) - 4s(n)$.

3.9. Exercises

1. Find *all* the inequivalent (with respect to mirror reversal) shortest spanning rulers of order n, for $n = 5, 6, 7$, and 11.

2. Extend the proof of Theorem 4 to show that $L(n) - \binom{n}{2} \geq 2$ for $n > 5$. Can you show that $L(n) - \binom{n}{2} > k$ for all $n > n_0(k)$, for every positive integer k ?

3. Generalize G. Bloom's counter-example to S. Piccard's "theorem" to give an infinite family of *pairs* of 6-mark spanning rulers having the same autocorrelation function.

4. Construct Costas Arrays of each of the following sizes.

a. $n = 10$, using the Welch Construction with $p = 11, g = 2$. Show that this can be successively reduced to give examples of sizes $n = 9$ and $n = 8$.

b. $n = 7$, using the Lempel Construction with $q = 9, \alpha = 1 + i$, where $i^2 = -1$, and $GF(9)$ consists of $\{0, \pm 1, \pm i, \pm 1 \pm i\}$.

c. $n = 14$, using the Golomb Construction with $q = 16$, where $GF(16)$ is obtained by adjoining the roots of $x^4 + x + 1 = 0$ to $GF(2)$. Let α be one such root, and let $\beta = \alpha^4$, so that $\beta + \alpha = 1$. Show that the Costas Array so constructed can be successively reduced to the sizes $n = 13$ and $n = 12$.

5. Plot the ambiguity function ($-7 \leq \Delta t \leq +7, -7 \leq \Delta f \leq +7$) for the Costas Array of order $n = 7$ from Problem 4.b. above.

3.10 Periodic Modulation

For CW radar, biphase modulation based on a periodic binary sequence can be used to obtain a "two-valued correlation" :

$$(3.19) \qquad C(\tau) = \frac{1}{P}\sum_{i=1}^{P} a_i a_{i+\tau} = \begin{Bmatrix} 1, & \tau \equiv 0 \pmod{P} \\ \alpha, & \tau \not\equiv 0 \pmod{P} \end{Bmatrix}$$

where P is the period of the binary sequence $\{a_i\}$, whose terms are either $+1$ or -1.

While *m-sequences* (maximum-length linear binary shift register sequences, mentioned earlier) are the most popular choice to generate the sequence $\{a_i\}$, the other "cyclic Hadamard sequences" (corresponding to the "cyclic Hadamard difference sets" described in Section 2.7 above) give the same behavior for $C(\tau)$, namely

$$C(\tau) = \left\{ \begin{array}{ll} 1, & \tau \equiv 0 \pmod{P} \\ -\frac{1}{P}, & \tau \not\equiv 0 \pmod{P} \end{array} \right\}. \tag{3.20}$$

More generally, one may start with any "cyclic difference set" with parameters v, k, and λ (see [H] for standard terminology), and generate a binary sequence of period v, containing k +1's and $v-k$ −1's. Then, if this sequence is compared to a phase shift of itself with $\tau \not\equiv 0 \pmod{v}$, $a_i a_{i+\tau} = (+1)(+1) = +1$ a total of λ times per period; $a_i a_{i+\tau} = (+1)(-1) = -1$ a total of $k-\lambda$ times per period; $a_i a_{i+\tau} = (-1)(+1) = -1$ a total of $k-\lambda$ times per period; and therefore $a_i a_{i+\tau} = (-1)(-1) = +1$ a total of $v-\lambda-2(k-\lambda) = v-2k+\lambda$ times per period. Hence, for such a sequence

$$C(\tau) = \frac{1}{v}\sum_{i=1}^{v} a_i a_{i+\tau} = \left\{ \begin{array}{ll} 1, & \text{if } \tau \equiv 0 \pmod{v} \\ 1 - \frac{4(k-\lambda)}{v}, & \text{if } \tau \not\equiv 0 \pmod{v} \end{array} \right\}. \tag{3.21}$$

It is especially favourable to configure the system in such a way that $C(\tau) = 0$ for all $\tau \not\equiv 0 \pmod{v}$. However, this will not happen in (3.21) for $v > 4$, since v will be odd, and cannot then equal $4(k-\lambda)$. There are several modifications which can be made to get $C(\tau) = 0$ for $\tau \neq 0$:

i) Instead of using the values +1 and −1 , the sequence $\{a_i\}$ could consist of +1 and $-b$, for some real $b \geq 0$.

ii) Instead of using the values +1 and −1, the sequence $\{a_i\}$ could consist of +1 and $e^{i\phi}$, for some phase angle $\phi \neq 0$.

iii) The *transmitted* sequence $\{a_i\}$ can consist of +1's and −1's, but the *reference* sequence against which it is correlated can consist of +1 and $-b$, for some real $b \geq 0$.

The mathematical consequences of each of these three possible modifications are explored in [X]. Possibility ii) seems especially promising.

3.11. Other Applications

A number of additional topics would have been discussed here but for space limitations. One particularly interesting set of topics involves "code division multiple access" (CDMA) communications. These techniques were originally developed for military communications (where they were usually referred to as "spread spectrum" communications), but are being increasingly used in such civilian contexts as cellular telephony and mobile radio.

In one subcategory of CDMA, sometimes called "direct sequence spread spectrum", an m-sequence or similar random-looking binary sequence is used to modulate a sine wave (the *carrier*) with phase reversal or non-reversal occurring at rather high frequency (the "chip rate"), and with information being added to the signal at a lower rate (the "bit rate"). The other major subcategory of CDMA is "frequency hop spread spectrum". The principle of frequency hopping was described in connection with radar in Section 3.8.; but it is also possible to convey *information* with a frequency hop system.

The "multiple access" aspect of CDMA refers to the possibility of a number of users operating in the same signalling environment at the same time, without drowning out each others' signals. A family of combinatorial designs which we have named "Tuscan Squares" [Y], originally introduced for frequency-hop multiple access applications, have already become the subject of an extensive combinatorial literature.

Signal design problems in communications almost invariably correspond to interesting combinatorial problems. Conversely, almost every major family of combinatorial designs can be interpreted as the solution to a family of signal design problems.

REFERENCES

[A] C. Shannon, A mathematical theory of communication, *Bell System Technical Journal*, vol. 27, 379-423 (Part I), 623-656 (Part II).

[B] R. Gallager, *Information Theory and Reliable Communication*, John Wiley and Sons, New York, 1968.

[C] S.W. Golomb, R.E. Peile, and R.A. Scholtz, *Coding and Information Theory*, Plenum Press, New York, to appear, 1991/92.

[D] I. Selin, *Detection Theory*, Princeton University Press, 1965.

[E] R. Fano, *Transmission of Information*, M.I.T. Press, 1961.

[F] H.S. Black, *Modulation Theory*, Van Nostrand, New York, 1953.

[G] L.D. Baumet, *Cyclic Difference Sets*, Lecture Notes in Mathematics 182, Springer-Verlag, 1971.

[H] M. Hall, Jr., *Combinatorial Theory, Second Edition*, Wiley - Interscience, New York, 1986.

[I] R. Turyn, Sequences with small correlation, in *Error Correcting Codes*, Henry B. Mann, editor, John Wiley & Sons, New York, 1968, 195-228.

[J] B. Gordon, W.H. Mills, and L.R. Welch, Some new difference sets, *Canadian Journal of Mathematics*, vol.14, no.4, 1962, 614-625.

[K] S.W. Golomb, *Shift Register Sequences*, Holden-Day, Inc, San Francisco, 1967. Revised edition, Aegean Park Press, Laguna Hills, CA, 1982.

[L] S.W. Golomb, The twin prime constant, *American Mathematical Monthly*, vol.67, no.8,October, 1960, 767-769.

[M] R.H. Barker, Group synchronization of binary digital systems, *Communication Theory* (Proceedings of the Second London Symposium on Information Theory), London, Butterworth, 1953, 273-287.

[N] R. Turyn and J. Storer, On binary sequences, *Proceedings of the American Mathematical Society*, vol.12, no.3, June,1961, 394-399.

[O] S.W. Golomb and R.A. Scholtz, Generalized Barker Sequences, *IEEE Transactions on Information Theory*, vol.IT-11, no.4, October, 1965, 533-537.

[P] N. Zhang and S.W. Golomb, Sixty-phase generalized Barker sequences, *IEEE Transactions on Information Theory*, vol.IT-35, no.4, July,1989, 911-912.

[Q] R.L. Frank, Polyphase codes with good nonperiodic correlation properties, *IEEE Transactions on Information Theory*, vol.IT-9, no.1, January,1963,43-45.

[R] D.A. Huffman, The generation of impulse-equivalent pulse trains, *IRE Transactions on Information Theory*,vol.IT-8, no.5,September,1962, S10-S16.

[S] M. Gardner, *Wheels, Life and Other Mathematical Amusements*, W.H. Freeman and Co., New York, 1983; Chapter 15, 152-165.

[T] G.S. Bloom and S.W. Golomb, Applications of Numbered, Undirected Graphs, *Proceedings of the IEEE*,vol.65,no.4,April,1977,562-571.

[U] J.P. Costas, A study of a class of detection waveforms having nearly ideal range-doppler ambiguity properties, *Proceedings of the IEEE*,vol.72, no.8,August,1984, 996-1009.

[V] S.W. Golomb, Algebraic Constructions for Costas Arrays, *Journal of Combinatorial Theory (A)*,vol.37,no.1,July,1984,13-21.

[W] S.W. Golomb and H. Taylor, Constructions and Properties of Costas Arrays, *Proceedings of the IEEE*,vol.72,no.9,September,1984,1143-1163.

[X] S.W. Golomb, Two-valued sequences with perfect periodic autocorrelation, *IEEE Transactions on Aerospace and Electronic Systems*, to appear, 1991.

[Y] S.W. Golomb and H. Taylor, Tuscan Squares - A New Family of Combinatorial Designs, *Ars Combinatorica*,vol.20-B,December, 1985, 115-132.

Old and New Results on Ovals in Finite Projective Planes

G. KORCHMÁROS

Dipartimento di Matematica
Università della Basilicata
via N. Sauro 85
85100 Potenza (Italy)

The classical concept of a conic leads in a natural way to the concept of an oval in an arbitrary projective plane: An *oval* is a subset Ω of points satisfying both of the following properties: *i*) no three points of Ω are collinear; *ii*) Ω has exactly one 1-secant (also called a tangent) at each one of its points. If the plane is finite and has order n, then an oval consists of n+1 points.

Ovals of *finite* projective planes have been intensively studied since 1954. The starting point was the famous theorem of B. Segre [94], [95]: In a Desarguesian plane of odd order, the ovals are exactly the irreducible conics.

This paper is a survey of known results in the following areas:

1) The classification problem for ovals in a desarguesian plane of even order.

2) Ovals in finite non desarguesian planes.

3) Pascal's theorem for ovals and abstract ovals.

4) Collineation groups fixing an oval; some characterizations of the finite desarguesian planes.

1. THE CLASSIFICATION PROBLEM FOR OVALS IN A DESARGUESIAN PLANE OF EVEN ORDER

In 1956 Segre pointed out that his result on the characterization of conics cannot be extended to desarguesian planes of even order. The classification of ovals in these planes is still an open problem and seems to be very complex.

We give a brief account of the known ovals in desarguesian planes of even order, but for detailed information concerning the extensive theory of ovals developed by Segre and his school the reader is referred in particular to the books [53], [98]. Quite recently, some new investigations have been carried out. Details will be found in the survey papers [15], [80].

In order to investigate ovals in a Desarguesian plane of even order it is usual to adopt the following canonical form:

An oval of PG(2,q), q *even and* $q>2$, *can be written in the form*

$$\mathcal{D}(F) = \{(F(t),t,1): t\in GF(q)\} \cup \{(1,0,0)\},$$

where $F(t)$ *is a polynomial over* GF(q) *such that*

$F(t)$ *is a permutation polynomial of* GF(q) *with degree* $(F)<q$ *and* $F(0)=0$, $F(1)=0$;

for each $s\in GF(q)$, $G_s(x)=[F(x+s)+F(s)]/x$ *is a permutation polynomial with* $G_s(0)=0$.

For q even, there is one class of ovals in PG(2,q) that has already been characterized. If $\mathcal{D}(F)$ is an oval and F induces an endomorphism of the additive group of GF(q), then $\mathcal{D}(F)$ is called a *translation* oval. This is because $\mathcal{D}(F)$ remains fixed under the elation e_c, $c\in GF(q)$, with equations $\rho x'_1=x_1+cx_3$, $\rho x'_2=x_2+F(c)x_3$, $\rho x'_3=x_3$: in the affine plane whose line at infinity has equation $x_3=0$, the mapping e_c is a translation. Setting

I) $F(x)=x^{2^n}$ *where* gcd(n,d)=1,

$\mathcal{D}(F)$ is a translation oval in PG(2,2^d). In fact every translation oval is of this form by a result due to Bartocci, Segre [100] and Payne [88]. An alternative proof can be found in [54].

In 1958, using a computer, Lunelli and Sce [75] found a first example of a non-translation oval in PG(2,16). A canonical form of this oval is given by $\mathcal{D}(F)$ where

II) $F(x)=x^{12}+x^{10}+\eta x^8+x^6+\eta^2x^4+\eta^9x^2$ *where* η *is a primitive element of* GF(16) *satisfying* $\eta^4+\eta+1=0$.

Several authors considered ovals $\mathcal{D}(F)$ with $F(t)=x^h$ where h is an integer. All the known examples have been found in PG(2,q), $q=2^d$ and d odd, and they are listed here below, see [99], [44]:

III) $F(x)=x^6$;

IV) $F(x)=x^{3(2^n+1)+1}$ *where* $n=2^{(h+1)/2}$;

V) $F(x)=x^{2^n+2^m}$

where $n=2^{(h+1)/2}$ *and* $\begin{cases} m=(d+1)/4 & \text{if } d\equiv 3 \pmod 4 \\ m=(3d+1)/4 & \text{if } d\equiv 1 \pmod 4 \end{cases}$;

We also notice that Glynn, who constructed the latter two examples made an exhaustive computer search for ovals of the same type, but no new examples have been found in PG(2,q), $q\le 2^{19}$. On the other hand, some necessary conditions on h are given in [74], and [94]; they depend on the Hasse-Weil theorem concerning the number of points lying on an irreducible algebraic curve in PG(2,q).

More recently, some new ovals have been found in PG(2,2^d) by Payne [89], and Cherowitzo [15]; see also [106]:

For d≥5 *odd:*

VI) $F(x)=x^{1/6}+x^{3/6}+x^{5/6}$ *where the exponents are taken mod* 2^d-1;

For d=5,7,9:

VII) $F(x)=x^{2^m}+x^{2^m+2}+x^{3(2^m+1)+1}$

$$\text{where} \begin{cases} m=(d+1)/4 & \text{if } d\equiv 3 \pmod 4 \\ m=(3d+1)/4 & \text{if } d\equiv 1 \pmod 4 \end{cases};$$

We remark that permutation polynomials $F(x)$ arising from ovals sometimes are called *o-polynomials*. General results about such polynomials can be found in some recent papers [15], [45], [84].

We point out that the connection between ovals and hyperovals provides a systematic way of obtaining other ovals starting from a given oval. This connection depends on the fact that in a projective plane of even order, the tangents to an oval Ω are concurrent. As for a conic, the point of concurrency is called the *nucleus* of Ω. Note that if Ω is written in the above canonical form, then its nucleus coincides with the point (1,0,0). The oval Ω together with its nucleus *N* forms a set of n+2 points, no three of which are collinear. Such a set is called a *hyperoval*. Now a new oval Σ can be obtained from the hyperoval by deleting one of the n+1 points which are distinct from *N*. This deleted point *P* is the nucleus of Σ. There are up to n+1 ovals which can be obtained in this way, but Ω and Σ may be equivalent under the collineation group of the plane. This occurs if and only if some collineation fixes the hyperoval and sends *N* to *P*. In order to study ovals, it is therefore useful first of all to find hyperovals and then to determine the stabilizer of each hyperoval, thus obtaining all possible pairwise non-equivalent ovals arising from it. So far as the author is aware, this has been done only for the hyperovals arising from the ovals above listed. The interested reader is referred to the papers [82], [86].

We end this section with a few remarks about small desarguesian planes. In PG(2,q), with q≤8, each hyperoval is regular, that is it arises from a conic. In PG(2,16), there is only one class of irregular ovals. This result, first found with the aid of a computer [47], has recently been proved without a computer [83]. Very recently, some new hyperovals in PG(2,32) have been construced; see [85], [86]. On the contrary, in PG(2,64) all known hyperovals are regular.

2. Ovals in Finite Non-Desarguesian Planes

The construction of ovals and their classification in finite non-desarguesian planes is an unsolved problem of considerable difficulty. One of the fascinating things about this problem is that ovals have been found by various different methods based on polarities, collineations, quasifield properties as well as on extensive computer searchs.

In this chapter we list the known ovals in finite non-desarguesian planes. We will not give the details as to how these planes are determined, although methods for constructing ovals often require a detailed knowledge of the structure of the given plane.

2.1 Inherited ovals in some known non-desarguesian planes

Let $\Sigma=\Sigma(q)$, $q=p^h$ and p prime, be the family of all affine planes of order q. Of course Σ is not empty, as it contains the affine plane AG(2,q) over the Galois field GF(q). Let σ be any plane belonging to Σ. Then, σ can be looked at as a modification of AG(2,q); the points of σ can be identified with those of AG(2,q), and the lines of σ can be regarded as point-sets in AG(2,q). If σ is a translation plane, then we assume that the translation group of AG(2,q) coincides with that of σ. This makes it possible for certain lines of AG(2,q) still to be lines in σ.

In this context, an irreducible conic C of AG(2,q) is called *an inherited arc of* σ if no three points of C are collinear in σ. Now, look at an inherited arc C in the projective closure Π of σ. If C is an ellipse in AG(2,q), then C is itself an oval in Π. Otherwise, we have to complete C to an oval by adding one or two improper points, according as C is a parabola or a hyperbola. An oval of Π which may be obtained in such a way is called an *inherited oval*. We will also use the term *elliptic, parabolic,* or *hyperbolic inherited oval* according to the type of C in AG(2,q). Examples of inherited ovals have been constructed in various planes, namely:

Hyperbolic inherited ovals: in a non-associative Moulton plane of order $q=p^h$, $h=2u$, $p^u\equiv 1 \pmod 4$ [62], in the Hughes planes [93], in the Hall planes [22], [60], in the associative André plane of odd order [12], [93].

Parabolic inherited ovals: in the associative Moulton plane of odd order q [6], in the Hall plane of even square order q, [24], [68], [81].

Elliptic inherited ovals: in a non-associative Moulton plane of order $q=p^h$, $h=2u$, $p^u\equiv 3 \pmod 4$, [62].

It seems plausible that many other planes have inherited arcs. However, there are planes which cannot have inherited ovals for all the three types. In fact, Korchmáros [68] showed that no non-desarguesian translation plane of odd order has a parabolic inherited oval. Another negative result is that if each secant and each tangent of C is a line in σ, then σ coincides with AG(2,q). Some further results concerning inherited arcs can be found in Szönyi [103].

2.2 Ovals arising from polarities

Baer [5] showed that the absolute points of a polarity in a finite projective plane of non-square odd order always form an oval. Known examples of such planes are the commutative semifield planes coordinatized by Albert's "twisted fields"; see [42], [43]. Also, the absolute points of a polarity in planes of square odd order form an oval in some cases. The cyclic planes admit such polarities, but it should be noticed that all known cyclic planes are Galois planes. Room [95] showed that in the Hughes plane there is a polarity such that the set of absolute points contains an oval properly.

2.3 Ovals arising from collineation groups

Put $q=2^{2r+1}$. Let $Sz(q)$ denote the simple group of Suzuki. Lüneburg showed that, if a projective plane Π of order q^2 has a collineation group G isomorphic to $Sz(q)$ such that G fixes no non-incident point-line pair in Π, then G fixes an oval Ω either in Π or in the dual plane Π' of Π. Furthermore, G acts on Ω as $Sz(q)$ in its natural doubly transitive permutation representation of degree q^2+1. This situation occurs in the dual Lüneburg plane; see [76] page 139, [57], [66]. It seems very likely that no other example exists, but so far this has been proved only for the smallest case q=8.

Let Π denote a projective plane of odd q^2 order having a collineation group G isomorphic to PGL(2,q). Assume that the involutorial elements of G are homologies such that no two of them have the same centre nor the same axis. It is not hard to see that then Π contains a G-invariant Baer subplane $\bar{\Pi}$. Now assume further that a cyclic subgroup of order q+1 of G has a fixed point on $\Pi-\bar{\Pi}$. Biliotti and Korchmáros have proved that Π contains an oval Ω which splits into two orbits under G, one of size q+1 on which G acts as PGL(2,q) in its triply transitive permutation reprsentation. We point out that this method works in every Hughes plane and in the generalized Hughes planes of order 25 and 49. We should notice however that the resulting ovals in the Hughes planes are not new; they are in fact the above-mentioned ovals given by Room.

2.4 Ovals constructed by ad hoc methods

Kantor [57] pointed out that the dual translation plane coordinatized by a commutative non-associative division algebra of even order contains an oval. Cherowitzo [16] constructed a very interesting oval, called *ovale di Roma*, in the Figueroa plane of odd order. Very recently, J.M.N. Brown constructed a polarity whose set of absolute points is the set of points of the *ovale di Roma* of Cherowitzo.

2.5 Ovals obtained by computer search

Cherowitzo, Kiel, and Killgrove [19] discussed the enumeration and classification of ovals in each one of the known non-desarguesian planes of order 9. Later on, Cherowitzo [18] carried out a complete enumeration of hyperovals of the non-desarguesian translation planes of order 16. He also tabulated many significant properties of particular hyperovals in such planes. Some further interesting results connected with the enumeration of ovals and hyperovals in small planes can be found in [25], [26], [27].

Let us end this chapter by mentioning an interesting connection between ovals and linear codes.

2.6 Ovals from the point of view of coding theory

It can be proved by an easy geometric argument that a projective plane of *even* order n contains an oval (and hence a hyperoval) if and only if the row space of the incidence matrix of the plane has a vector of weight n+2. As a consequence of this result, it is natural to expect that coding theory could be used directly in the attempt to solve the existence problem for ovals in some projective planes of even order. In fact, this method of attack has received little attention so far. The only use known to the author goes back to 1983 in connection with the attempt to determine whether a putative projective plane of order 10 could contain an oval; see [72], [105]. The very few known general results on this subject can be summarized as follows.

Given a projective plane Π of even order n, let A denote the row space of the incidence matrix of Π. Let $A^{\perp}$ denote the dual code of A, and B the even-weight subcode of A. If $n \equiv 2 \pmod 4$ then each vector of weight n+2 is in $A^{\perp}$. If $n \equiv 0 \pmod 4$, B has codimension 1; or, more precisely, the weight of each vector in B is congruent to 0 mod 4. It follows that no vector of weight n+2 belongs to B. Hence, should one find a projective plane of order n, $n \equiv 0 \pmod 4$, with A of dimension (n^2+n+1) one would have $B=A^{\perp}$ and the plane would have no ovals. Unfortunately, no such plane is known. For the proofs the reader is referred to [4]; see also [104].

3. PASCAL'S THEOREM FOR OVALS AND ABSTRACT OVALS

The classical Pascal theorem states that the three points of intersection of opposite sides of any hexagon inscribed in an irreducible conic are collinear. In 1966 Buekenhout [13] proved that if the Pascal theorem holds for an oval of a projective plane (finite or infinite), then the plane is Pappian and the oval is a conic. Buekenhout's theorem is of considerable interest as it shows that a geometrical property of a single oval can determine the type of the plane.

Two different proofs are known: the original one depends on permutation group theory; the other one is due essential to Artzy [3], and uses co-ordinates which are linked to operations defined on the points of the oval as given in Veblen and Young [107] for the classical case where the oval is a conic.

In the first part of this chapter we shall give a brief account of these proofs.

Before doing that we need to make more precise the definition of a Pascal oval, because both proofs deal with degenerate Pascal hexagons as well. Let Ω be an oval in a projective plane Π of order n. If $P\in\Omega$, we shall denote the tangent at P by PP; since we shall be dealing with only one oval, this notation is unambiguous. A hexagon $ABCDEF$ inscribed in Ω is a cyclically ordered set of six points on Ω, not necessarily distinct. The *sides* of this *hexagon* are AB, BC, CD, DE, EF, and FA. (If $A=B$, then the side AB is AA, the tangent at A, etc.). A hexagon inscribed in Ω is *degenerate* if two non-adjacent vertices coincide.

If a hexagon $ABCDEF$ inscribed in Ω is non-degenerate, then its six sides are distinct. The three points of intersections of pairs of opposite sides, *viz.* $P=AB\cap DE$, $Q=BC\cap EF$, $R=CD\cap FA$, are called the *diagonal* points of $ABCDEF$. If they are collinear, then the hexagon is a *Pascal hexagon.* Note that a degenerate hexagon inscribed in Ω has always collinear diagonal points which may not be pairwise distinct. The oval Ω is called a *Pascal oval* if every non-degenerate hegaxon inscribed in Ω is a Pascal hexagon. It is not difficult to show that, if every hexagon with *six distinct* vertices inscribed in Ω is a Pascal hexagon, then Ω is a Pascal oval.

A line ℓ is called a *Pascal line* of Ω, if all hexagons inscribed in Ω have the following property: if two diagonal points lie on ℓ then the third one also lies on ℓ. Clearly, Ω is a Pascal oval if and only if each line is a Pascal line.

3.1 A brief outline of Buekenhout's proof

A powerful tool in the proof is the concept of involution on an oval. To each point P not on Ω, we may associate the transposition (AB), A,B∈Ω, for each secant AB which passes through P; we define the *involution* (*with centre* P) *on* Ω, and denote it by **P**, the product of all transpositions associated to P. This concept leads to an algebraic criterion for an oval to have a Pascal line.

Proposition 3.1. *If Ω is an oval and ℓ is a line in a projective plane, then the following statements are equivalent:*

3.1 *ℓ is a Pascal line with respect to Ω;*

3.2 *the product of any three involutions* **P**,**Q**, *and* **R** *with centers on r is still an involution with centre on r.*

Proposition 3.1 allowed Buekenhout to prove that each involution on Ω may be extended to an involutorial perspectivity of the plane. It follows from 3.2 that the products **PQ** with centers P,Q on a Pascal line ℓ form an abelian group acting on the set Ω-ℓ, as a sharply transitive permutation group. Buekenhout used this consequence of 3.2 to prove for a Pascal oval Ω that the set of all products **PQ** is a group *G* when P and Q run over all points off Ω. *G* turns out to be a collineation group which is isomorphic to PGL(2,K) and acts on the oval Ω as PGL(2,K) in its 3-transitive permutation representation on the projective line over the field

K. This allows the reconstruction of the plane within G: the points not on Ω (and also different from the nucleus of Ω) are the involutions of G, the lines correspond to certain maximal subgroups of G. Let us remark that in the case when the plane is finite, K is a Galois field, and each secant consists of all non-central involutions of a dihedral subgroup of order 2(q-1) together with the two fixed points on Ω of its maximal cyclic subgroup; each tangent consists of all involutions of a subgroup of order 2q and its unique fixed point on Ω; each exterior line consists of all noncentral involutions of a dihedral subgroup of order 2(q+1). We see that to each field K there corresponds, up to isomorphism, only one projective plane containing a Pascal oval. Since the Pascal theorem holds for an irreducible conic of the projective plane over K, Buekenhout's theorem follows.

3.2 A brief outline of the alternative proof

Let Ω be an oval in an arbitrary projective plane. We use $L(a,b)$ to denote the unique line through distinct points a and b on Ω, and $L(a,a)$ to denote the tangent at a. Take any be three distinct points X, Y, and I of Ω, and put $0=X, \infty=Y, 1=I$, for future convenience. Let S denote the set of all points of Ω other than Y. We define an *oval addition on* S and an *oval multiplication* on S as follows: $L(a,b)$, $L(\infty,\infty)$ and $L(0,a\oplus b)$ are concurrent for all a and b in S; $L(a,b), L(0,\infty)$ and $L(1,a\odot b)$ are concurrent for all a and b in S. The set S together with these operations is a double loop $S(0,\infty,1,\oplus,\odot)$, which may in turn be used to provide *oval co-ordinates* for points of the plane not on $L(0,\infty)$ by analogy with the Euclidean situation for the hyperbola xy=1; see [21], Chapter 11. The existence of certain Pascal hexagons inscribed in Ω is the condition for the double loop $S(0,\infty,1,\oplus,\odot)$ to be a field as well as for the above co-ordinatization to coincide with the standard one whose fundamental quadrangle is XYOI with origin $O=L(0,0)\cap L(\infty,\infty)$. This is all that is necessary to get an almost synthetic proof of the Buekenhout's theorem. It is worth mentioning that this line of investigation works in a finite projective plane even under a weaker assumption that the oval Ω has the five-point Pascalian property, that is each non-degenerate hexagon inscribed in Ω with at most five distinct vertices is a Pascal hexagon. In fact this is a condition under which each double loop associated to Ω is an alternative division ring, and hence a Galois field by a result of Pickert. Full details can be found in [3], [28], [55], [58], [78], [92].

3.3. Configurations of Pascal lines of an oval

In connection with the Pascal theorem, Buekenhout [14] raised the general problem of determining the configuration $p(\Omega)$ of the Pascal lines of an oval Ω in a finite projective plane.

In PG(2,q), q even, the Pascal lines of an oval form one of the following sets:

the empty set;
one tangent and the oval is a translation oval;
all lines and the oval is a conic.

This was first stated in [63], and a different proof was given in [37]. In a non-desarguesian plane, $p(\Omega)$ can have a different configuration, see [32]. For instance, if Ω is the Rodriguez oval of the non-desarguesian translation plane of order 9, then $p(\Omega)$ consists of the four secants and the two tangents through a point. Another result in this direction is a generalization of Buekenhout's theorem; see [67]: if $p(\Omega)$ contains all secant and tangent lines, then the plane is a Galois plane and the oval is a conic.

Cherowitzo [17] pointed out that the following two properties are equivalent for an oval Ω of a projective plane of *even* order: (i) every nondegenerate hexagon inscribed in Ω has the four point Pascalian property; (ii) every quadrangle inscribed in Ω has collinear diagonal points lying on a tangent of Ω. He also showed that such a projective plane has order a power of 2.

Fernandes [39] proved that a translation plane of order $n \equiv 3 \pmod 4$ (or its dual plane) which contains an oval with the four-point Pascalian property can be coordinatized by a commutative semifield.

An investigation which has some connections with the Pascal theorem was suggested by Ostrom [87] in 1955, who was interested in generalizing the classical notion of harmonic sets with respect to a conic. Let Ω be an oval in a projective plane of odd order. A point P which is not on Ω is called a *harmonic point* with respect to Ω if there exists a line p such that, whenever AA' and BB' are lines through P which are not both tangents to Ω, and A,A',B,B' are points on Ω, then the diagonal points $AB\cap A'B'$ and $AB'\cap A'B$ are on p. An oval is said to be *harmonic* if every exterior point is harmonic. Ostrom proved that, if $AA'BB'CC'$ is a hexagon inscribed in a harmonic oval Ω such that the lines AA', BB', and CC' are concurrent at an exterior point to Ω, then $AA'BB'CC'$ is a Pascal hexagon. It is still an open question as to whether harmonic ovals exist in a non-desarguesian plane. For full details, the reader is referred to the survey paper [86]; see also [41].

3.4 B-ovals and closed arcs

In the remaining part of this section we present some ideas and results on *B*-ovals. This concept arose from an attempt to extend the Pascal theorem to a more general setting.

Let us recall that a *B-oval* (or *abstract oval*) is a pair (M,F) consisting of a (finite or infinite) set M and a family F of permutations of order ≤ 2 (called *involutions*) on M such that F is *quasi sharply 2-transitive* on M in the sense that

3.3 *for any* $(a_1,a_2),(b_1,b_2)\in M\times M$ *with* $a_i\neq b_j$ $(i,j=1,2)$, *there is a unique* $f\in F$ *such that* $f(a_1)=a_2$ *and* $f(b_1)=b_2$.

If M is a finite set of size n+1, then n is called the order of the *B*-oval. To avoid trivial cases we always assume that M has more than two points. For finite *B*-ovals we have that the identity permutation belongs to F if and only if n is even. If this is the case, every involution has one fixed point on M.

To each *B-oval*, $B=(M,F)$, there is a canonically associated geometric structure, $\pi(B)$, called the *ambient of B*. The points of $\pi(B)$ are the elements of $M \cup F$, the secant lines of $\pi(B)$ are the sets consisting of a pair of points of M and all the involutions of F which permute them, and the tangent lines of $\pi(B)$ are the sets consisting of a point of M and all the involutions of F which fix it. Condition 3.3 garantees that each pair of lines will intersect in a unique point.

There is a natural way of deriving a B-oval from an oval: if Ω is an oval in the projective plane Π, then from each point P not on Ω we may project Ω onto itself. The resulting involutorial permutation of the points of Ω is called the *involution* P *with centre* P. Now, if we put $M=\Omega$ and define F as the set of all involutions with center, we get a B-oval. If Ω is a conic, the corresponding B-oval is called a B-conic.

Conversely, a B-oval (M,F) is said to be *embeddable* if it is possible to find sufficiently many sets consisting of involutions with the right intersection properties so that when these sets are added to $\pi(B)$ the result is a projective plane. These new sets will be the exterior lines to the oval. We point out that a finite B-conic can be embedded only into a Galois plane, according to [101]. A useful general embeddability condition has been given in [18]. On the other hand, in 1977 Krier [73] constructed a non-embeddable B-oval of countable size. More recently, examples of non-embeddable B-ovals of order eight have been discovered by Faina & Cecconi [35], Faina [30], [31], and Cherowitzo [17]. We limit ourselve to describing that given in [31].

Let $\mathbf{V}$ be the two-dimensional vector space over GF(3), and let J denote the following permutation of the vectors of $\mathbf{V}$: $(v_1,v_2)(v_3,v_4)(v_5,v_6)(v_7,v_8)$ with $v_1=(0,1)$, $v_2=(1,0)$, $v_3=(2,2)$, $v_4=(1,2)$, $v_5=(0,2)$, $v_6=(2,0)$, $v_7=(1,1)$, and $v_8=(2,1)$. It is straightforward to check that J and its conjugates under the group $3^2{\cdot}SL(2,3)$ together with the elements of order ≤ 2 of this group form a quasi sharply 2-transitive *F* set on **V**. In other words, the pair (**V**,*F*) is a *B*-oval of order eight.

We remark that the smallest order of a non-embeddable *B*-oval is eight. It is worth mentioning that no *B*-oval of order ten exists; see [33]. The proof depends on coding theory and uses some previous results obtained by extensive computer search.

The notion of embeddable *B*-oval can be generalized to that of a closed arc. Let *K* be a k-arc in a projective plane, that is a set of k points, no three of which are collinear. A k-arc *K* is called a *closed arc* if $k \geq 4$ and there exists a set *I* of $(k-1)^2$ points in the plane such that

3.4 K and I are disjoint;

3.5 each secant of K meets I in exactly k-2 points;

3.6 each tangent of K meets I in 0 or k-1 points.

As for an oval, to each point of $P \in I$ we may associate an involution. Now, if we define *F* as the set of all these involutions, the pair (*K*,*F*) is a *B*-oval. Ovals of the plane and ovals of a proper subplane are clearly closed arcs. These are called *regular closed arcs.* In [36] it has been proved that in a Galois plane of odd order each closed arc is regular, and hence is an irreducible conic of a subplane. For the even order case, the same conclusion holds under the condition that *K* is a subset of a conic; see [1]. Cherowitzo gave an example of a non-regular closed arc in the Hughes plane of order 9. For an interesting generalization of the notion of closed arc; see [108].

Proposition 3.1 shows how the notion of Pascal line can be extended to a *B*-oval. A line ℓ of a *B*-oval (*M*,*F*) is said to be a *Pascal line* if the product of any three involutions on ℓ is still an involution on this line. If all lines of *B*-oval are Pascal lines, then it is a *Pascal B-oval.*

The main result is that finite Pascal *B*-ovals are *B*-conics. It seems plausible that the finiteness condition is unnecessary, but so far this has been proved only under additional restrictions. Both known methods of proving Buekenhout's theorem may be generalized to Pascal B-ovals, and we digress here to explain how this has been done.

In his proof, Buekenhout made use of collineations fixing the oval. Since the corresponding tool in the study of B-ovals is the concept of automorphism, we need some results about automorphisms of B-ovals.

We recall that an automorphism g of a B-oval is a permutation of M which induces a map g^* of F onto itself, that is $g^*(f)=gfg^{-1}\in F$ for each $f\in F$. It is interesting to note that an involution of F may be an automorphism; if this is the case then the involution is said to be *regular*. Let R denote the set of all regular involutions of a B-oval. Buekenhout [14] proved that R is one of the following sets:

3.7 *the empty set;*

3.8 *one secant;*

3.9 *one tangent;*

3.10 *all tangents (and the B-oval has odd order);*

3.11 *all tangents and secants.*

In the latter case, all involutions are regular, and the B-oval itself is called regular. The automorphism group generated by all involutions of a finite regular B-oval (M,F) is isomorphic to PGL(2,q) and acts on M as PGL(2,q) in its 3-transitive permutation representation, and hence a finite regular B-oval is necessarily a B-conic. This was proved by Buekenhout [14] for the even order case, and by Korchmáros [61] for the odd order case. In [61] it has also been proved that if there exists a B-oval of class 3.11, then its order is at most 59.

The full automorphism group of a B-conic of order q is PΓL(2,q), and this property characterizes it. The following stronger result is due to Buekenhout [14] and Korchmáros [68]: the B-conic of order q is the only B-oval (M,F) with an automorphism group acting on M as PSL(2,q) in its usual 2-transitive permutation representation. This characterization leads directly to the problem of determining when a B-oval (M,F) is 2-transitive, that is the automorphism group acts on M as a 2-transitive permutation group. Among B-ovals of odd order, there exist no examples different from B-conics. In the even case, some further examples have been found, and an almost complete classification has been given which depends on the automorphism group generated by R. Details will not given here, and the reader should refer to [38], [69]. The methods of proof are comparable with those used for ovals of projective planes, see section 4.

We are now in a position to explain how Buekenhout's proof may be fitted to finite Pascal B-ovals. The crucial step consists of showing that each involution f is regular, that is $gfg^{-1}\in F$ for each $f\in F$. If f is on a (secant or tangent) line through g, this property follows directly from the definiton. On the other hand, there is no reason why gfg^{-1} should always be in F, except for the finite case which may be treated by a standard counting argument. Also, to conclude the proof as Buekenhout did, we need the above-mentioned characterization of regular B-ovals which has been proved so far only for the finite case.

The investigation of Pascal B-ovals by co-ordinate methods may be carried out in much the same way as we have described before for Pascal ovals: both oval addition and multiplication are consistent with the concept of a B-oval, and the corresponding double loops may be used to translate Pascal's condition on a B-oval into a condition characterizing the B-conics. It should be pointed out, however, that the manipulation of the double loops associated with a Pascal B-oval turns out to be considerably more laborious than in the case of a B-oval because it cannot, of course, involve considerations about external lines. The required supplementary work is still to be finished, although a complete treatment is available for a wide class of Pascal B-ovals including the finite ones. A key step in this context is to show that every double loop associated with a Pascal B-oval has the distributive property (and hence is a field). This has been done so far only under some additional hypotheses such as that the B-oval is either finite, or that it has a suitable set of regular involutions, or that it has the following "weak polar" property: for any three points $a, b, c \in M$, if I is the involution of F fixing both a and b, and J is the involution fixing both c and I(c), then J sends a to b. Another crucial step is to show that if every double loop associated with a B-oval is a field then the B-oval is a B-conic; see [29]. The case when all these fields have zero or odd characteristic has been already settled. It remains, however, to prove that the condition on the characteristic is unnecessary; in order to do so, one may assume that F contains the identity permutation on M. The case when each non-trivial involution has one fixed point on M has also been settled. This includes, of course, the finite case, but it is not clear whether an infinite Pascal B-oval (M,F) such that $Id_M \in F$ can contain some fixed-point-free involutions.

We end this section by mentioning the following very recent result due to Polster [89]: for a finite B-oval (M,F), the permutation group G generated by F is either a sharply 3-transitive group on M, or the symmetric group on M, or the alternating group on M. In the first case $G \cong PGL(2,q)$, and the B-oval is a B-conic.

4. Collineation Groups Fixing an Oval; Some Characterizations of the Finite Desarguesian Planes.

In this chapter we give a fairly complete account of the known results on collineation groups of finite projective planes which fix (setwise) an oval. Almost all of the results require deep theorems on finite groups. We will mention these and, in some cases, also outline how they come into play in the proofs. Our terminology is standard; see [23].

4.1 Some characterizations of PG(2,q) with q odd in terms of collineation groups fixing an oval

Let G denote a collineation group of a projective plane Π of *odd* order such that G fixes an oval Ω.

In the classical case (Π=PG(2,q), and Ω is a conic according to Segre's theorem), G is a subgroup of PΓL(2,q). In particular, if G induces a 2-transitive permutation group on Ω, then G contains PSL(2,q) regarded in its natural doubly transitive permutation representation.

In 1964 Lüneburg [76] showed that the converse also holds: if Π has order $q=p^h$, with p prime, and G has a subgroup H isomorphic to PSL(2,q), then only the classical case can occur. Lüneburg's theorem was the first deep characterization of PG(2,q) in terms of the abstract structure of a collineation group fixing an oval.

In 1967 Cofman [20] showed that if G satisfies both the conditions

4.1 *G acts on Ω as a 2-transitive permutation group,*

4.2 *all involutions of G are homologies,*

then Π is a desarguesian plane PG(2,q), and so Ω is a conic and G contains PSL(2,q). The proof uses the Gorenstein-Walter theorem on finite groups with dihedral Sylow 2-subgroups. Kantor [18] later showed that 4.2 can be weakened to the condition that G contains some involutorial homologies. Finally, Korchmáros [64] showed that 4.2 can be entirely eliminated. Both proofs involve rather deep group theory, namely the Bender-Hering classification of 2-transitive permutation groups whose involutions have zero or two fixed points and the Hering-Kantor-Seitz theorem on 2-transitive permutation groups with the property that the onepoint stabilizer contains a normal subgroup sharply transitive on the remaining points.

These results are the early characterizations of PG(2,q) with q odd in terms of the action of collineation groups on an invariant oval.

In a recent paper [10] 4.1 is weakened to the condition that

4.3 *G acts on Ω as a primitive permutation group.*

The conclusion is that Ω is a desarguesian plane PG(2,q), q odd, Ω is a conic, and either G is doubly transitive or Π has order 9 and G acts on Ω as A_5 or S_5 in its primitive permutation representation of degree 10. Amongst other group theoretical results, the proof uses the classification of the finite simple groups of 2-rank at most 3; see section 4.2.

Now, if we weaken 4.3 further to the condition that

4.4 *G acts on Ω as a transitive permutation group,*

then both the abstract structure and the action of G become much more involved. Indeed, G may be soluble as well as fixing a non-incident point-line pair. However, it seems probable that even condition 4.4 can occur only in the classical case. In [10], the authors have investigated the case where the underlying plane Π has order $n\equiv 1 \pmod 4$. The main result is the following theorem.

Assume Π has order $n\equiv 1 \pmod 4$ and 4 divides the order of G. If G is a minimal transitive collineation group on Ω, i.e. G but not any proper subgroup of G satisfies 4.4, then one of the following holds:

4.5 *$G=O(G)\langle\tau\rangle$ for a suitable 2-element τ, and G fixes exactly one non-incident point-line pair (P,r);*

4.6 *$\Pi=PG(2,5)$, $G\cong A_4$ and G has an invariant triangle;*

4.7 *$G\cong PSL(2,q)$ and $n=q(q+1)-1$ or $n=q(q-1)+1$ according as $q\equiv 1$ (mod 4) or $q\equiv 3$ (mod 4). Further, through each point on Ω there passes exactly one homology axis;*

4.8 *G acts on Ω as a primitive permutation group.*

Notice that each of the above situations actually occurs in desarguesian planes of suitable odd order.

4.2 Non-abelian simple groups acting on a projective plane of odd order as collineation groups fixing an oval

The main result is the following theorem stated in [10]:

If G *is a non-abelian simple group then* G≅PSL(2,q), q≥5 *odd.*

The proof does not depend on the classification of simple groups. An important tool in the proof is a careful investigation of the 2-groups which can act on a projective plane and fix an oval. The main result is that the 2-rank of such groups (i.e. the maximal rank of their elementary abelian subgroups) is at most three. It is also shown that, if G is a non-abelian simple group, then all its involutions are homologies forming a unique conjugacy class. These results, together with the classification theorem of finite simple groups with 2-rank at most 3 due to Stroh [102] (see also [46] Theorems 1.86 and 2.168), show that the only possibilities for G are: PSL(2,q) with q≥5, PSL(3,q), PSU(3,q^2) with q odd and PSU(3,4^2), A_7, M_{11}, M_{12}. Each of these possibilities was investigated by a laborious case by case analysis. The conclusion is that only PSL(2,q) can actually occur.

We have already noticed that the Hughes plane of order q^2 has a collineation group isomorphic to PSL(2,q) which fixes an oval.

For useful results on Baer involutions fixing an oval in a projective plane of odd order, see [8], [40], and [64].

4.3 Collineation groups strongly irreducible on an oval in a projective plane of even order

In [7], the authors investigated projective planes Π of even order having a collineation group G which fixes an oval Ω but neither a point on Ω, nor a secant or a suboval of Ω. Such a group G is said to be *strongly irreducible on the oval* Ω. This notion can be considered as a "local" version of the concept of a strongly irreducible collineation group, due to Hering [50], [51], which played a fundamental role in the study of the

composition series of collineation groups of a finite projective plane.
The main result in [7] states that if G has even order then G contains some involutorial elations. Both the abstract structure and the action of the subgroup $<\Delta>$ generated by all involutorial elations of G are determined. If G has a fixed line then $<\Delta>$ is the semidirect product of $O(<\Delta>)$ with a subgroup of order two generated by an elation. In particular G contains no Baer involution. If G fixes no line then it acts as a "bewegend group" [52] on the dual affine plane of Π with respect to the improper line corresponding to the nucleus of Ω. From Hering's results on bewegend groups containing involutorial elations, it then follows that $<\Delta>$ is isomorphic to one of the following simple groups: PSL(2,q), Sz(q), or PSU(3,q^2), where $q \geq 4$ is a power of 2.

4.4 Collineation groups transitive on an invariant oval in a projective plane of even order

It is immediate from the definition that a collineation group G fixing an oval Ω is strongly irreducible on Ω provided that 4.4 holds. From the results quoted in section 4.3, we infer that if G has even order and fixes no line then G acts on Ω as one of the following simple groups in its natural doubly transitive representation: PSL(2,q), or $Sz(q)$, or PSU(3,q^2), where q is a suitable power of 2. If $G \cong$PSL(2,q), then only the classical situation occurs. For the case $G \cong Sz(q)$, the reader is referred to section 2. The case $G \cong$PSU(3,q^2) cannot actually occur, this has been proved in [7]. Therefore, we obtain an almost complete classification of collineation groups transitive on an invariant oval:

Let G *be a collineation group of even order of a projective plane* Π *of even order* n *which acts transitively on the points of an invariant oval* Ω*. If* Δ *is the set of all involutorial elations of* G *and* $<\Delta>$ *denotes the subgroup of* G *generated by* Δ*, one of the following cases holds:*

4.9 *$\langle\Delta\rangle$ has order twice an odd integer, $O(\langle\Delta\rangle)$ acts still transively on Ω, $|\Delta|=q+1$ and G has no Baer-involutions;*

4.10 $n=q$, $q=2^h$, $h\geq 2$, $\langle\Delta\rangle\cong PSL(2,q)$, *$\pi$ is a desarguesian plane of order q, Ω is a conic and $\langle\Delta\rangle$ acts on Ω as* $PSL(2,q)$ *in its 3-transitive representation;*

4.11 $n=q^2$ *with* $q=2^{2m+1}$, $m\geq 1$, $\langle\Delta\rangle\cong Sz(q)$ *and $\langle\Delta\rangle$ acts on Ω as* $Sz(q)$ *in its natural 2-transitive representation.*

It seems plausible that if 4.1 holds then 4.9 cannot actually occur. This has been verified so far only for the case where the underlying plane has order 2^v, $v\geq 3$, see section 4.8.

Since each desarguesian plane of even order $n\geq 8$ contains ovals other than the conics, it is of interest to state the above classification theorem for desarguesian planes:

Let G *be a collineation group of* $PG(2,q)$, $q=2^h$ *which fixes an oval Ω. If 4.9 holds then Ω is a conic and either* G *contains* $PSL(2,q)$ *or* G *is a dihedral group of order $2(q+1)$.*

Notice that the latter result was originally stated in [11].

4.5 Non-abelian simple groups acting on a projective plane of even order as collineation groups fixing an oval

The main theorem is the following:

If G is a non-abelian simple group acting on a projective plane of even n order leaving an oval invariant, then either $G\cong PSL(2,q)$, $q\geq 4$, *or* $G\cong Sz(q)$, *or* $G\cong PSU(3,q^2)$, $q\geq 4$, *where* q *is a suitable power of 2.*

We briefly sketch the proof. The crucial key is the following result on 2-groups of collineations:

If a collineation group S *of order* 2^k, $k\geq 2$, *of a projective plane of even order n fixes an oval, then either* S *contains some involutorial elations, or* S *is a cyclic group.*

Assume that G is a non-abelian simple group acting on Π as a collineation group which fixes an oval Ω. Since a Sylow 2-subgroup of G is not cyclic, G contains some involutorial elations. The involutorial elations with a common axis together with the identity collineation form a subgroup Q of G with trivial normalizer intersection. By a theorem of Hering [49], G is either PSL(2,q), or *Sz*(q), or PSU(3,q^2) where $q \geq 4$ is a power of 2.

4.6 Collineation groups transitive on an invariant hyperoval in a projective plane of even order

As we noticed in section 4.1, PG(2,16) admits only one class of irregular hyperovals. Hall also proved that an irregular hyperoval of PG(2,16) is *transitive*, in the sense that the collineation group of PG(2,16) which fixes the hyperoval acts transitively on its points. Two other planes containing transitive hyperovals are known: PG(2,2) and PG(2,4). The question arises of whether or not these three small desarguesian planes are the only projective planes containing a transitive hyperoval. This is the transitive hyperoval problem.

The desarguesian case is of special interest because of the known relationship between hyperovals and other geometric structures such as generalized quandrangles and Howell Designs. In [65] it was proved that no other desarguesian plane admits transitive hyperovals; see [65].

The first deep result concerning the transitive hyperoval problem is due to V. Abatangelo [2]. He showed that the only finite planes with a 2-transitive hyperoval are PG(2,2) and PG(2,4). Notice that Abatangelo's proof does not depend on the classification of all doubly transitive permutation groups (which is a consequence of the classification theorem of the simple groups). However, his proof uses both Aschbacher's classification theorem of doubly transitive permutation groups of degree $r\equiv 2 \pmod 4$ and Shult's theorem on the doubly transitive groups M containing an invariant set T of involutions such that the stabilizer $T\cap M_\alpha$ for every α is a triangular set.

Recently, the transitive hyperovals problem has been almost completely solved in [9]:

Let Π be a projective plane of even order n admitting a hyperoval Ω and a collineation group G such that G fixes Ω and acts transitively on the points of Ω. If 4 divides the order of G then either Π=PG(2,2) or Π=PG(2,4), or n=16.

As regards the case n=16, we know that the involutorial elations of G generate a Frobenius group of order 18 but it is not even clear if Π must be desarguesian.

Some results concerning the transitive hyperoval problem have been obtained without the use of deep theorems on finite groups. The alternative arguments are mostly of a rather combinatorial nature and work only under additional strong hypotheses. One result [71], however, does not follow from the theorems mentioned so far and so we quote it. For the basic idea underlying the proof, see section 4.7.

Let Π be a projective plane of even order n admitting a hyperoval Ω and a collineation group G such that G leaves Ω invariant and acts on Ω as a sharply transitive permutation group. If G is cyclic and its unique involution is an elation then n=2 holds.

4.7 Miscellaneus results on collineation groups fixing an oval or a hyperoval

Following Ostrom [87], an oval is called a pseudo-conic if it consists of all absolute points of an orthogonal polarity of the underlying plane. In [59], sharply-transitive pseudo-conics have been investigated. The main result is the following theorem:

Let G *be a collineation group of a projective plane* π *of odd order* $n \neq 3, 11, 23, 59$ *with an invariant pseudo-conic* Ω. *If* G *acts on* Ω *as a sharply-transitive permutation group then one of three possibilities holds:*

4.12 G *contains only one involution,* G *has two point and line orbits of length* $\frac{1}{2}(n+1)$ *and* $n-1$ *of length* $n+1$;

4.13 $n \equiv 1 \pmod 4$, G *contains* $\frac{1}{2}(n+1)$ *conjugate involutions, and* G *has* $n+1$ *point and line orbits of length* $\frac{1}{2}(n+1)$ *and* $\frac{1}{2}(n-1)$ *of length* $n+1$;

4.14 $n \equiv 3 \pmod 4$, G *contains* $\frac{1}{2}(n+3)$ *involutions in three classes, of sizes* 1, $\frac{1}{4}(n+1)$ *and* $\frac{1}{4}(n+1)$. *Further,* G *has two point and line orbits of length* $\frac{1}{4}(n+1)$, n *of length* $\frac{1}{4}(n+1)$ *and* $\frac{1}{2}(n-1)$ *of length* $n+1$.

Let us end with a short discussion of a new approach to certain problems on collineation groups fixing an oval.
Assume that Π is a projective plane of even order n containing an oval Ω and let Ω' denote the hyperoval arising from Ω. Each line r of Π leads to a one-factorization of the complete graph whose vertices are the points of Ω not lying on r. In fact, the lines through a point P outside Ω' partition Ω' into 2-subsets. Now, if r is an external line of Ω, the set of such partitions, for all P on r, is a one-factorization of the complete graph on Ω'. If r is a tangent or a secant of Ω (i.e. r meets Ω'), the partition induced by P ($\notin \Omega'$) on r contains $\Omega' \cap r$ as one part; so the set of such partitions, for all P ($\notin \Omega'$) on r, is a one-factorization of the complete graph on $\Omega' - r$. Clearly, any collineation group G of π which leaves Ω invariant and maps r onto itself induces an automorphism group of the associated one-factorization. In the case where G is cyclic and acts on $\Omega' - r$ as a sharply transitive permutation group, the resulting one-factorization is cyclic. Notice that such one-factorizations have been studied by Hartman, Mendelsohn, Rosa and others; see [79]. In [71], a generalized form of the main theorem of [48] has been used to obtain both the

final statement in section 4.6 and the following theorem:

Let G *be a cyclic group of collineations of a projective plane of even order* n *which fixes an oval* Ω. *If* G *fixes a point of* Ω, *and acts on the remaining points as a transitive permutation group, then* $n\equiv 4$ *(mod* 8*) holds.*

This result can be combined with some results on sharply doubly transitive permutation groups to give:

Let G *be a collineation group of a projective plane of even order* n *which fixes an oval* Ω *and acts on the points of* Ω *as a sharply doubly transitive group, and let* Q *denote the quasifield such that* G *is isomorphic to the group of all linear permutations* $x \to ax+b$ *with* $a(\neq 0), b\in Q$. *If* Q *is a Galois field* $GF(p^u)$, *then* $p^u\equiv 5$ (mod 8) *holds.*

Since every known finite projective plane has prime power order, it is of interest to specify the latter statement for the case where $n=2^v$. It turns out that $2^v+1=p^u$ holds for some integers u, v and a prime p. This implies that either u=1, or v=3, p=3 and u=2. Actually, the latter case cannot occur because no collineation group of PG(2,8) leaving an oval invariant acts on its points as a sharply doubly transitive permutation group. For u=1, the quasifield Q is necessarily a Galois field, so that $2^v\equiv 4$ (mod 8) follows. Hence u=1 implies n=4. Thus, no projective plane of order $n=2^v$, $v\geq 3$, admits a collineation group which leaves an oval invariant and acts on its points as a sharply doubly transitive permutation group. This result was first stated in [70]. It should be noted, however, that the final step of the proof was fallacious. Finally, we point out that PG(2,4) verifies the hypothesis of the last statement. The collineation group of PG(2,4) which fixes an oval is isomorphic to S_5 and acts on the points of the oval as S_5 on five objects. Therefore, PG(2,4) has a collineation group G with an invariant oval which is isomorphic to the group of all linear permutations over GF(5). Clearly, G acts on such an oval as a sharply 2-transitive permutation group.

REFERENCES

[1] L.M. Abatangelo & G. Raguso, Una caratterizzazione degli archi chiusi giacenti su una conica irriducibile di un piano pascaliano di caratteristica due, *Rend. Mat. Appl.* (4) **1** (1981), pp. 39-45.

[2] V. Abatangelo, Doubly transitive (n+2)-arcs in projective plane of even order n, *J. Combin. Theory Ser.* A, **42** (1986) pp. 1-8.

[3] R. Artzy, Pascal's theorem on an oval, *Amer.Math. Monthly* **75** (1968) pp. 143-146.

[4] E.F. Assmus & H.E. Sacher, Ovals from the point of view of coding theory, in: *Higher Combinatorics* (Berlin, 1977) Reidel, 213-216.

[5] R. Baer, Polarities in finite projective planes, *Bull. Amer. Math. Soc.* **52** (1946), pp. 77-93.

[6] U. Bartocci, Una nuova classe di ovali proiettive finite, *Atti Accad. Naz. Lincei, Rend.* (8) **43** (1967), pp. 312-316.

[7] M. Biliotti & G. Korchmáros, Collineation groups strongly irreducible on an oval, *Ann. Discrete Math.* **30** (1986), pp. 85-98.

[8] M. Biliotti & G. Korchmáros, Collineation groups which are primitive on an oval of a projective plane of odd order, *J. London Math. Soc.* (2) **33** (1986), pp. 525-534.

[9] M. Biliotti & G. Korchmáros, Hyperovals with a transitive collineation group, *Geom. Dedicata* **24** (1987), pp. 269-281.

[10] M. Biliotti & G. Korchmáros, Collineation groups preserving an oval in a projective plane of odd order, *J. Austral. Math. Soc. A* **48** (1990) pp. 156-170.

[11] P. Biscarini & G. Korchmáros, Ovali di un piano di Galois di ordine pari dotate di un gruppo di collineazioni transitivo sui punti, *Rend. Mat. Brescia* **7** (1984), pp. 125-135.

[12] A.A. Bruen & J.C. Fisher, Arcs and ovals in derivable planes, *Math. Zeitschr.* **125** (1972), pp. 122-128.

[13] F. Buekenhout, Plans projectifs à ovoides pascaliens, *Arch. Math.* **17** (1966), pp. 89-93.

[14] F. Buekenhout, Etude intrinsèque des ovales, *Rend. Mat. Appl.* (5) **25** (1966), pp. 333-393

[15] W. Cherowitzo, Hyperovals in desarguesian planes of even order, *Ann. Discrete Math.* **30** (1986), pp. 87-94.

[16] W. Cherowitzo, Ovals in Figueroa Planes, *J. Geometry* **37** (1990) 84-86.

[17] W. Cherowitzo, Harmonic ovals of even order, in: Finite Geometries, *Lecture Notes in Pure and Applied Math.* **103** (1985), pp. 65-81

[18] W. Cherowitzo, On the projectivity of *B*-ovals, *J. Geometry* **27** ((1986), pp. 119-139.

[19] W. Cherowitzo & D.I. Kiel & R.B. Killgrove, Ovals and other configurations in the known planes of order nine, *Congr. Numerantium* **55** (1986), pp. 167-179.

[20] J. Cofman, Doubly transitivity in finite affine and projective planes, in: *Proc. Proj. Geometry Conf., Univ. Illinois, Chicago* (1974) pp. 16-19.

[21] H.S.M. Coxeter, Projective geometry, 2nd Edition, (University Press, Toronto, 1974).

[22] M. Crismale, (q^2+q+1)-insiemi di tipo (0,1,2,q+1) e ovali nel piano di Hall di ordine pari q^2, *Note Mat.* **1** (1981), pp. 127-136.

[23] P. Dembowski, Finite geometries (Springer Verlag, Berlin-Heidelberg-New York, 1968).

[24] R.H.F. Denniston, Some non-desarguesian translation ovals, *Ars Combin.* **7** (1979), pp. 221-222.

[25] M. de Resmini, On the semifield plane of order 16 with kern GF(2), *Ars Combin.* **24A** (1987), pp. 75-92.

[26] M. de Resmini & L. Puccio, Some combinatorial properties of the dual Lorimer planes, *Ars Combin.* **24A** (1987), pp. 131-148.

[27] M. de Resmini, Some combinatorial properties of a semi-translation plane, *Congr. Numerantium* **59** (1987), pp. 5-12.

[28] G. Faina, Sul doppio cappio associato ad un ovale, *Bollettino U.M.I.* (5) **15-A** (1978), pp. 440-443.

[29] G. Faina, Una estensione agli ovali astratti del teorema di Buekenhout sugli ovali pascaliani, *Algebra e Geometria, Bolletino U.M.I. Suppl.* **2** (1980), pp. 355-364.

[30] G. Faina, On the group of automorphisms of a finite Buekenhout oval which is not projective, *Simon Stevin* **57** (1983), pp. 99-102.

[31] G. Faina, Un esempio di ovale astratto non proiettivo a tangenti pascaliane il cui gruppo degli automorfismi e risolubile e due volte transitivo, *Rend. Mat. Brescia* **7** (1984), pp. 289-296.

[32] G. Faina, Pascalian configurations in projective planes, *Ann. Discrete Math.* **30** (1986), pp. 203-215.

[33] G. Faina, Graphs, codes and Buekenhout ovals, preprint.

[34] G. Faina & G. Korchmáros, Una caratterizzazione del gruppo lineare PGL(2,K), *Algebra e Geometria, Bolletino U.M.I. Suppl.* **2** (1980), pp. 297-309.

[35] G. Faina & G. Cecconi, A finite Buekenhout oval which is not projective, *Simon Stevin* **56** (1982), pp. 121-127.

[36] G. Faina & G. Korchmáros, Risultati intorno ad una congettura relativa agli archi chiusi, *Rend. Mat. Appl.* (7) **1** (1981), pp. 55-61.

[37] G. Faina & G. Korchmáros, Desarguesian configurations inscribed in an oval, *Ann. Discrete Math.* **14** (1982), pp. 207-210.

[38] G. Faina & G. Korchmáros, Il sottogruppo generato dalle involuzioni regolari di un B-ovale transitivo, *Rend. Sem. Mat. Univ. Padova* **74** (1985), pp. 139-145.

[39] O. Fernandes, On an oval with the four point pascalian property, *Canad. Math. Bull.* **27** (1984), pp. 295-300

[40] O. Fernandes, On Baer involution which maps an oval into itself, *Arch. Math.* **44** (1985), pp. 467-481.

[41] O. Fernandez, Harmonic points and the intersections of ovals, *Geom. Dedicata* **19** (1985), pp. 271-276.

[42] M.J. Ganley, Polarities in translation planes, *Geom. Dedicata* **1** (1972), pp. 103-106.

[43] C.W. Garner, Conics in finite proejctive planes, *J. Geometry* **12** (1979), 1-7.

[44] D.G. Glynn, Two new sequences of ovals in finite Desarguesian planes of even order, *Combinatorial Mathematics* X, *Lecture Notes in Math.* **1036**, Springer Verlag, 1983, pp. 217-229.

[45] D.G. Glynn, A condition for the existence of ovals in PG(2,q), q even, *Geom. Dedicata,* **32** (1989), pp. 247-252.

[46] D. Gorenstein, The classification of the finite simple groups (Plenum Press, New York 1983).

[47] M. Hall. jr., Ovals in the desarguesian plane of order 16, *Annali Mat. Pura Appl.* **102** (1975), pp. 159-176.

[48] A. Hartman & A. Rosa, Cyclic one-factorization of the complete graph, *Europ. J. Combin.* **6** (1985), pp. 45-49.

[49] C. Hering, On subgroups with trivial normalizer intersection, *J. Algebra* **20** (1972), pp. 622-29.

[50] C. Hering, On the structure of finite collineation groups of projective planes, *Abh. Math. Sem. Hamburg* **49** (1979), pp. 155-182.

[51] C. Hering, Finite collineation groups of projective planes containing nontrivial perspectivities, *Proc. Symposia Pure Math.* **37** (1980), pp. 473-477.

[52] C. Hering, On Beweglichkeit in affine planes, in: Finite geometries, *Lecture Notes Pure Appl. Math.* **82** (1983), pp. 197-209.

[53] J.W.P. Hirschfeld, Projective geometries over finite fields (Clarendon Press, Oxford, 1979).

[54] J.W.P. Hirschfeld, Ovals in a Desarguesian plane of even order, *Annali Mat. Pura Appl.* **102** (1975), pp. 79-89.

[55] C.E. Hofmann, Specializations of Pascal's theorem on an oval, *J. Geometry,* **1** (1971), pp. 143-153.

[56] W.M. Kantor, On unitary polarities of finite projective planes, *Canad. J. Math.* **23** (1971), pp. 1060-1077.

[57] W.M. Kantor, Symplectic groups, symmetric designs and line ovals, *J. Algebra* **33** (1975), pp. 43-58.

[58] H. Karzel & K. Sörensen, Projektive Ebenen mit einem pascalischen Oval, *Abh. Math. Sem. Hamb.* **35** (1970), pp. 89-93.

[59] P.B. Kirkpatrick, Collineation group which are sharply transitive on an oval, *Bull. Austral. Math. Soc.* **11** (1974), pp. 197-211.

[60] G. Korchmáros, Ovali nei piani di Hall d'ordine dispari, *Atti Accad. Naz. Lincei, Rend.* **56** (1974), pp. 315-317.

[61] G. Korchmáros, Su una classificazione delle ovali dotate di automorfismi, *Atti Accad. Naz. dei XL Rend.* **5** I-II (1975/76), pp. 77-86.

[62] G. Korchmáros, Ovali nei piani di Moulton di ordine dispari, *Atti dei Convegni Lincei* 7-II (1976), pp. 395-398.

[63] G. Korchmáros, Sulle ovali di traslazione in un piano di Galois di ordine pari, *Atti Accad. Naz. XL Rend.* 5-III (1977), pp. 55-65.

[64] G. Korchmáros, Una proprietà gruppale delle involuzioni planari che mutano in sé un'ovale di un piano proiettivo, *Annali Mat. Pura Appl.* (4) **116** (1978), pp. 189-205.

[65] G. Korchmáros, Gruppi di collineazioni transitivi sui punti di una ovale [(q+2)-arco] di $S_{2,q}$, q pari, *Atti Sem. Mat. Fis. Univ. Modena* **27** (1978), pp. 89-105.

[66] G. Korchmáros, Le ovali di linea del piano di Lüneburg d'ordine 2^{2r} che possono venir mutate in sé da un gruppo di collineazioni isomorfo al gruppo semplice $Sz(2^r)$ di Suzuki, *Atti Accad. Naz. Lincei, Memor. (8)* **15** (1979), pp. 295-315.

[67] G. Korchmáros, Una generalizzazione del teorema di F. Buekenhout sulle ovali pascaliane, *Bollettino U.M.I.* **18-B** (1981), pp. 673-687.

[68] G. Korchmáros, Inherited arcs in affine planes, *J. Combin. Theory Ser. A* **42** (1986) 140-143.

[69] G. Korchmáros, 2-transitive abstract ovals of odd order, *Aeq. Math.* **33** (1987), pp. 208-219.

[70] G. Korchmáros, Collineation groups doubly transitive on the points at infinity in an affine plane of order 2^r, *Arch. Math.* **37** (1981) pp. 572-576.

[71] G.Korchmáros, Cyclic one-factorization with an invariant one-factor of the complete graph, *Ars Combin.* **27** (1989) pp. 133-138.

[72] C.W.H. Lam & L. Thiel & S. Scwierz & J. McKay, The non-existence of ovals in a projective plane of order 10, *Discrete Math.* **45** (1983), pp. 319-321.

[73] N. Krier, A Buekenhout oval which in not projective, *Arch. Math.* **28** (1977), pp. 323-324.

[74] B. Larato, Sull'esistenza delle ovali di tipo $\mathcal{D}(x^k)$, *Atti Sem. Mat. Fis. Univ. Modena* **29** (1980), pp. 345-348.

[75] L. Lunelli & M. Sce, Considerazioni aritmetiche e risultati sperimentali sui {K;n}-archi, *Ist. Lomb. Acc. Sci. Rend. A* **98** (1964), pp.3-52.

[76] H. Lüneburg, Charakterisierungen der endlichen desarguesschen projektiven Ebenen, *Math. Zeitschr.* **85** (1964), pp. 419-450.

[77] H. Lüneburg, Translation planes (Springer Verlag, Berlin-Heidelberg- New York, 1980).

[78] G.E. Martin, Oval co-ordinates in a projective plane, in: *Atti Conv. Geometria Combinatoria, (Perugia, 1983)*, pp. 323-330.

[79] E. Mendelsohn & A. Rosa, One-factorizations of the complete graph - a survey, *J. Graph Theory* **9** (1985), pp. 43-65.

[80] C.M. O'Keefe, Ovals in desarguesian planes, *Australas. J. Combin.* **1** (1990), pp. 149-159.

[81] C.M. O'Keefe & A.A. Pascasio & T. Pentilla, Hyperovals in Hall planes, preprint.

[82] C.M. O'Keefe & T. Pentilla, Symmetries of arcs, *J. Geometry* (1990), to appear.

[83] C.M. O'Keefe & T. Pentilla, Hyperovals in PG(2,16), *Europ. J. Combin.* (1990), to appear.

[84] C.M. O'Keefe & T. Pentilla, Polynomials for hyperovals of Desarguesian planes, *J. Austral. Math. Soc. Ser. A*, to appear.

[85] C.M. O'Keefe & T. Pentilla, A new hyperoval in PG(2,32), preprint.

[86] C.M. O'Keefe & T. Pentilla & C.H. Praeger, Stabilizers of hyperovals in PG(2,32), preprint.

[87] T.G. Ostrom, Ovals, dualities, and Desargues's theorem, *Canad. J. Math.* 7 (1955), pp. 417-431.

[88] T.G. Ostrom, Concoids: Cone-like figures in non-pappian planes, in: *Geometry-von Staudt's point of view*, (Reidel, Dortrecht, Boston. 175-198.

[89] B. Polster, private cominication.

[90] S.E. Payne, A complete determination of translation ovoids in finite Desarguesian planes, *Atti Accad. Naz. Lincei, Rend.* (8) **51** (1971), pp. 328-331.

[91] S.E. Payne, A new infinite family of generalized quadrangles, *Congr. Numerantium* **49** (1985), pp. 115-128.

[92] J.F. Rigby, Pascal ovals in projective planes, *Canad. J. Math.* **21** (1969), pp. 1462-1476.

[93] L.A. Rosati, Insiemi di sostituzioni strettamente 3-transitivi e ovali, *Bollettino U.M.I.* **4** (1971), pp. 463-467.

[94] L.A. Rosati, Sulle ovali dei piani desarguesiani finiti d'ordine pari, *Ann. Discrete Math.* **18** (1983), pp. 713-720.

[95] T.G. Room, Polarities and ovals in the Hughes plane, *J. Austral. Math Soc.* **13** (1972), pp. 196-204.

[96] B. Segre, Sulle ovali nei piani lineari finiti, *Atti Accad. Naz. Lincei, Rend.* **17** (1954), pp. 1-2.

[97] B. Segre, Ovals in a finite projective plane, *Canad. J. Math* 7 (1955) pp. 414-416.

[98] B. Segre, Lectures on modern geometry (with an appendix by L. Lombardo Radice), (Cremonese, Roma, 1960).

[99] B. Segre, Ovali e curve σ nei piani di Galois di caratteristica due, *Atti Accad. Naz. Lincei, Rend.* **34** (1962) 785-790.

[100] B. Segre & U. Bartocci, Ovali ed altre curve nei piani di Galois di caratteristica due, *Acta Arith.* **8** (1971), pp. 423-449.

[101] B. Segre & G. Korchmáros, Una proprietà degli insiemi di punti di un piano di Galois caratterizzante quelli formati dai punti delle singole rette esterne ad una conica, *Atti Accad. Naz. Lincei, Rend.* (8) **62** (1977), pp. 363-369.

[102] G. Stroh, Über Gruppen mit 2-Sylow-Durchschnitten vom Rang ≤3. I, *J. Algebra* **43** (1976) 398-456.

[103] T. Szönyi, Complete arcs in non-desarguesian planes, *Ars Combin.* **25C** (1988), pp. 169-178.

[104] G. Tallini, Linear codes associated with geometric structures, *Results in Math.* **12** (1987), pp. 411-422.

[105] J.G.Thompson, Ovals in a projective plane of order 10, *London Math. Soc. Lecture Notes* **52** (1981), pp. 187-190.

[106] J.A. Thas & S.E. Payne & E. Gevaert, A family of ovals with few collineations, *Europ. J. Combin.* **9** (1988), pp. 353-362.

[107] O. Veblen & J.W. Young, Projective geometry, (Ginn and Co., Blaisdell, 1965).

[108] F. Wettl, Remarks on closed arcs, preprint.

Schubert Polynomials

I. G. MACDONALD

School of Mathematical Sciences
Queen Mary College
London E1 4NS

INTRODUCTION

The theory of Schubert polynomials has its origins in algebraic geometry, and in particular in the enumerative geometry of the flag manifolds. The reader of this article will however detect no trace of geometry or reference to these origins. In recent years A. Lascoux and M.-P. Schützenberger have developed an elegant and purely combinatorial theory of Schubert polynomials in a long series of articles [**L1**]- [**L3**], [**LS1**]- [**LS7**]. It seems likely that this theory will prove to be a useful addition to the existing weaponry for attacking combinatorial problems relating to permutations and symmetric groups.

Most of the results expounded here occur somewhere in the publications of Lascoux and Schützenberger, though not always accompanied by proof, and I have not attempted to give chapter and verse at each point. For lack of space, many proofs have been omitted, especially in the earlier sections, but I hope I have retained enough to convey the flavour of the subject. Complete proofs will be found for example in [**M2**].

1. PERMUTATIONS

In this first section we shall review briefly, without proofs, some facts and notions relating to permutations that will be used later. Proofs may be found e.g. in [**M2**].

For each integer $n \geq 1$, let S_n denote the symmetric group of degree n, namely the group of all permutations of the set $[1,n] = \{1,2,\dots n\}$. Each $w \in S_n$ is a mapping of $[1,n]$ onto itself; we shall write all mappings on the left of their arguments, so that the image of $i \in [1,n]$ under w is $w(i)$. We shall occasionally denote w by the sequence $(w(1),\dots,w(n))$.

For $1 \leq i < n$ let s_i denote the transposition that interchanges i and $i+1$, and fixes all other $j \in [1,n]$. The s_i generate the group S_n, subject to the relations

$$(1.1) \qquad \begin{cases} s_i^2 = 1, & \\ s_i s_j = s_j s_i & \text{if } |i-j| > 1, \\ s_i s_j s_i = s_j s_i s_j & \text{if } |i-j| = 1. \end{cases}$$

For each $w \in S_n$, the *length* $\ell(w)$ of w is the minimal length of a sequence $a = (a_1,\dots,a_p)$ such that $w = s_{a_1} \cdots s_{a_p}$. Such sequences of minimal length are called *reduced words* for w, and the set of reduced words for w will be denoted by $R(w)$.

The number $\ell(w)$ is also equal to the number of pairs (i,j) in $[1,n]$ such that $i < j$ and $w(i) > w(j)$. So for example the permutation $w_0 = (n, n-1, \dots, 2, 1) \in S_n$ has length $\ell(w_0) = \frac{1}{2}n(n-1)$; it is called the *longest element* of S_n.

If $m > n$, we shall identify S_n with the subgroup of permutations $w \in S_m$ that fix $n+1, n+2, \dots, m$. We may then form the group

$$S_\infty = \bigcup_{n=1}^{\infty} S_n$$

consisting of all permutations of the set of positive integers that fix all but a finite number of them. The length of $w \in S_n$ is the same whether w is regarded as an element of S_n or as an element of $S_m, m > n$, and hence $\ell(w)$ is well-defined for all $w \in S_\infty$. Likewise for the set $R(w)$ of reduced words for w.

Let $GR(w)$ denote the graph whose vertices are the reduced words for w, and in which a reduced word $a \in R(w)$ is joined by an edge to each of the words obtained from a by either interchanging two consecutive terms i, j such that $|i-j| > 1$, or by replacing three consecutive terms i, j, i such that $|i-j| = 1$ by j, i, j (compare the relations (1.1)). Then a basic fact is that

(1.2). *For each $w \in S_\infty$, the graph $GR(w)$ is connected.*

Bruhat order

Let $v, w \in S_\infty$. We shall write $v \to w$ to mean that $\ell(w) = \ell(v) + 1$ and that vw^{-1} (or equivalently $v^{-1}w$) is a transposition (i.e., fixes all but two elements of the set $[1, \infty)$ of positive integers).

The *Bruhat order*, denoted by $\leq$, is the partial order on S_∞ that is the transitive closure of the relation $\to$. In other words, $v \leq w$ means that there exists a sequence $(v_0, v_1, \ldots, v_r)$ of permutations such that

$$v = v_0 \to v_1 \to \cdots \to v_r = w$$

(so that $\ell(w) = \ell(v) + r$).

The following proposition gives an equivalent description of the Bruhat order.

(1.3). *Let $v, w \in S_\infty$ and let $a = (a_1, \ldots, a_p)$ be a reduced word for w. Then the following conditions are equivalent:*

(i) *$v \leq w$;*

(ii) *there is a subsequence $b = (b_1, \ldots, b_q)$ of a such that $v = s_{b_1} \cdots s_{b_q}$;*

(iii) *there is a subsequence b of a which is a reduced word for v.*

Thus the permutations v such that $v \leq w$ can all be obtained from any one reduced word for w.

Diagrams and codes

The *diagram* of a permutation $w \in S_n$ is the set $D(w)$ of points (i, j) in the lattice square $\Sigma_n = [1, n]^2$ such that $i < w^{-1}j$ and $j < wi$. In drawing such diagrams we adopt the convention (as for matrices) that the first coordinate increases from north to south, and the second coordinate from west to east. If we define the *graph* $G(w)$ of w to be the set of points $(i, w(i))$ $(1 \le i \le n)$, then the complement of $D(w)$ in the square Σ_n consists of all points $(i,j) \in \Sigma_n$ due south or due east of some point of $G(w)$, and hence is the union of the hooks with corners at the points of $G(w)$. For example, if $w = (365142) \in S_6$, $D(w)$ consists of the points circled in the picture below:

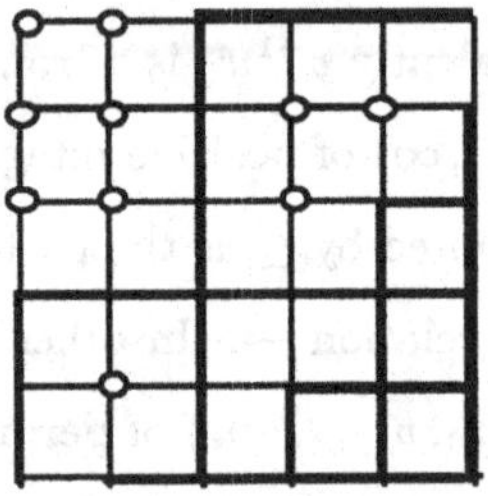

The diagram of w is the same whether we regard w as an element of S_n or as an element of $S_m, m > n$. Thus $D(w)$ is well-defined for all $w \in S_\infty$. From the definition it is clear that $D(w^{-1})$ is the transpose of $D(w)$ (i.e., we have $(i,j) \in D(w^{-1})$ if and only if $(j,i) \in D(w)$) and that Card $D(w) = \ell(w)$.

Next, let $w \in S_n$ and for each $i \ge 1$ let

$$c_i(w) = \text{Card}\{j : j > i \quad \text{and} \quad w(j) < w(i)\}.$$

Thus $c_i(w)$ is the number of points in the ith row of $D(w)$. The vector $c(w) = (c_1(w), \dots, c_n(w)) \in \mathbb{N}^n$ is called the *code* of w. As with partitions, we may disregard any string of zeros at the right-hand end of $c(w)$, and with this understanding the code $c(w)$ is well-defined for all $w \in S_\infty$.

The permutation w may be reconstructed from its code $c(w) = (c_1, c_2, \dots)$ as follows: for each integer $i \ge 1$, $w(i)$ is the $(c_i+1)^{th}$ element, in increasing order, of the sequence of positive integers from which $w(1)$, $w(2), \dots, w(i-1)$ have been deleted. The sum $|c| = c_1 + c_2 + \cdots$ is equal to $\ell(w)$. Each sequence $c = (c_1, c_2, \dots)$ of non-negative integers such that $|c| < \infty$ occurs as the code of a unique permutation $w \in S_\infty$.

Recall that a *partition* is any (finite or infinite) sequence $\lambda = (\lambda_1, \lambda_2, \dots)$ of non-negative integers such that $\lambda_1 \geq \lambda_2 \geq \dots$ and such that $|\lambda| = \lambda_1 + \lambda_2 + \cdots$ is finite (so that only finitely many of the λ_i are non zero). It is convenient not to distinguish between two partitions which differ only by a string of zeros at the end, so that for example $(2,1)$, $(2,1,0)$ and $(2,1,0,0,\dots)$ are to be regarded as the same partition. The number of non-zero terms λ_i is called the *length* $\ell(\lambda)$ of the partition λ. A partition λ may be described graphically by its *diagram* $D(\lambda)$, which is the set of lattice points $(i,j) \in \mathbb{Z}^2$ such that $1 \leq j \leq \lambda_i$. By reflecting $D(\lambda)$ in the main diagonal we obtain the diagram of a partition $\lambda' = (\lambda'_1, \lambda'_2, \dots)$ called the *conjugate* of λ. Thus λ'_i is the number of points in the ith *column* of $D(\lambda)$. Finally, if $\lambda = (\lambda_1, \lambda_2, \dots)$ and $\mu = (\mu_1, \mu_2, \dots)$ are two partitions, the relation $\lambda \geq \mu$ means that $|\lambda| = |\mu|$ and $\lambda_1 + \cdots + \lambda_i \geq \mu_1 + \cdots + \mu_i$ for all integers $i \geq 1$.

With this understood, we define the *shape* $\lambda(w)$ of a permutation w to be the partition obtained by arranging the integers $c_i(w)$ in weakly decreasing order. Thus $|\lambda(w)| = \text{Card } D(w) = \ell(w)$.

The shapes of w and w^{-1} are related by

$$(1.4) \qquad \lambda(w)' \geq \lambda(w^{-1}).$$

Vexillary permutations

Special interest attaches to those permutations $w \in S_\infty$ for which $\lambda(w)' = \lambda(w^{-1})$. They may be characterized in various ways:

(1.5). *The following conditions on a permutation $w \in S_\infty$ are equivalent:*

(i) *the set of rows of $D(w)$ is totally ordered by inclusion;*

(ii) *the set of columns of $D(w)$ is totally ordered by inclusion;*

(iii) *there do not exist a, b, c, d such that $1 \leq a < b < c < d$ and $w(b) < w(a) < w(d) < w(c)$;*

(iv) $\lambda(w)' = \lambda(w^{-1})$.

A permutation $w \in S_\infty$ is said to be *vexillary* if it satisfies these equivalent conditions. By (1.5) (iii), the first non-vexillary permutation is (2143) in S_4.

Amongst the vexillary permutations, there are two particular subclasses:

(i) *dominant* permutations, which are those $w \in S_\infty$ for which $c_1(w) \geq c_2(w) \geq \cdots$, so that $c(w)$ is a *partition*. Equivalent conditions are that $c(w^{-1})$ is a partition, or that $D(w)$ is the diagram of a partition.

(ii) *Grassmannian* permutations, which are those $w \in S_\infty$ having at most one descent (i.e., $w(r) > w(r+1)$ for at most one value of r). An equivalent condition is that $c_1(w) \leq \cdots \leq c_r(w)$ and $c_i(w) = 0$ for all $i > r$.

In terms of its code $c(w) = (c_1, c_2, \dots)$, a permutation $w \in S_\infty$ is vexillary if and only if the sequence $c(w)$ satisfies the following two conditions:

(V1) If $i < j$ and $c_i > c_j$, then

$$\operatorname{Card}\{k : i < k < j \quad \text{and} \quad c_k < c_j\} \leq c_i - c_j,$$

(V2) If $i < j$ and $c_i \leq c_j$, then $c_k \geq c_i$ whenever $i < k < j$.

Next, let w be a permutation with code $c(w) = (c_1, c_2, \dots)$. For each $i \geq 1$ such that $c_i \neq 0$ let $e_i = \max\{j : j \geq i \text{ and } c_j \geq c_i\}$. Arrange the numbers e_i in increasing order of magnitude, say $\varphi_1 \leq \varphi_2 \leq \cdots \leq \varphi_m$. The sequence

$$\varphi(w) = (\varphi_1, \varphi_2, \cdots, \varphi_m)$$

is called the *flag* of w. It is a sequence of length $\ell(\lambda)$, where λ is the shape of w.

(1.6). *A vexillary permutation $w \in S_\infty$ is uniquely determined by its shape $\lambda(w)$ and its flag $\varphi(w)$.*

Let us write $\lambda = \lambda(w)$ in the form

$$\lambda = (p_1^{m_1}, p_2^{m_2}, \ldots, p_k^{m_k})$$

where $p_1 > p_2 > \cdots > p_k > 0$ and each $m_i \geq 1$, and the notation $p_1^{m_1}$ means p_1 repeated m_1 times, and so on. For $1 \leq r \leq k$ let $f_r = \max\{j : c_j(w) \geq p_r\}$. Then $f_1 \leq f_2 \leq \cdots \leq f_k$ and the flag of w is

$$\varphi(w) = (f_1^{m_1}, f_2^{m_2}, \ldots, f_k^{m_k}) .$$

This holds whether or not w is vexillary, and the f_r satisfy

$$f_r \geq m_1 + m_2 + \cdots + m_r \qquad (1 \leq r \leq k) .$$

If now w is vexillary, it follows from the conditions (V1) and (V2) that the f_r must also satisfy the inequalities

$$0 \leq f_r - f_{r-1} \leq m_r + p_{r-1} - p_r \qquad (2 \leq r \leq k) .$$

The code $c(w)$ of w is now constructed as follows: first, the m_1 entries equal to p_1 are inserted at the right-hand end of the interval $[1, f_1]$; then the m_2 entries equal to p_2 are inserted in the rightmost available spaces in the interval $[1, f_2]$, and so on: for each $r \geq 1$, when all the terms greater than p_r in the sequence $c(w)$ have been inserted, the m_r entries equal to p_r are inserted in the rightmost available spaces in the interval $[1, f_r]$.

2. DIVIDED DIFFERENCES

Let $x_1, x_2, \ldots x_n, \ldots$ be independent variables, and let

$$P_n = \mathbb{Z}[x_1, x_2, \ldots, x_n]$$

for each $n \geq 1$, and

$$P_\infty = \mathbb{Z}[x_1, x_2, \ldots] = \bigcup_{n=1}^{\infty} P_n .$$

The group S_n (resp. S_∞) acts on P_n (resp. P_∞), by permuting the x_i.

If $f \in P_\infty$, the polynomial $f - s_i f$ vanishes when x_i is set equal to x_{i+1}, and hence is divisible in P_∞ by $x_i - x_{i+1}$. We may therefore define, for each integer $i \geq 1$, a linear operator $\partial_i : P_\infty \to P_\infty$ by

$$\partial_i = (x_i - x_{i+1})^{-1}(1 - s_i).$$

These are the "divided difference" operators. They have degree -1, in the sense that if f is a homogeneous polynomial of degree d, then $\partial_i f$ is homogeneous of degree $d-1$. Also they satisfy

$$(2.1) \qquad \begin{cases} \partial_i^2 = 0, & \\ \partial_i\partial_j = \partial_j\partial_i & \text{if } \ |i-j| > 1, \\ \partial_i\partial_j\partial_i = \partial_j\partial_i\partial_j & \text{if } \ |i-j| = 1. \end{cases}$$

If now $a = (a_1, \dots, a_p)$ is any sequence of positive integers, let

$$\partial_a = \partial_{a_1}\partial_{a_2}\cdots\partial_{a_p}.$$

From (2.1) and (1.2) it follows that $\partial_a = \partial_b$ if a, b are reduced words for the same permutation w, and we may therefore define

$$\partial_w = \partial_a$$

unambiguously, where a is any element of $R(w)$. The operator $\partial_w : P_\infty \to P_\infty$ has degree $-\ell(w)$.

If on the other hand $a = (a_1, \dots, a_p)$ is *not* reduced (i.e. if it is not a reduced word for some $w \in S_\infty$) then we have $\partial_a = 0$. Hence for $v, w \in S_\infty$ we have

$$(2.2) \qquad \partial_v\partial_w = \begin{cases} \partial_{vw} & \text{if } \ \ell(vw) = \ell(v) + \ell(w), \\ 0 & \textit{otherwise}. \end{cases}$$

Let $w_0 = (n, n-1, \dots, 2, 1)$ be the longest element of S_n, as in §1. Then we have

$$(2.3) \qquad \partial_{w_0} = a_\delta^{-1} \sum_{w \in S_n} \varepsilon(w) w,$$

where $a_\delta = \prod_{1\le i<j\le n}(x_i - x_j)$, *and* $\varepsilon(w) = (-1)^{\ell(w)}$ *is the sign of* $w \in S_n$.

If $f, g \in P_\infty$, the expression of $\partial_w(fg)$ as a sum of products $(\partial_u f)(\partial_v g)$ (i.e., the "Leibnitz formula" for ∂_w) is in general rather complicated. However, there is one case in which it is reasonably simple, namely when one of the factors f, g is linear:

(2.4). *If* $f = \sum \alpha_i x_i$ *then*

$$\partial_w(fg) = w(f)\partial_w g + \sum(\alpha_i - \alpha_j)\partial_{wt_{ij}} g$$

summed over all pairs $i < j$ *such that* $\ell(wt_{ij}) = \ell(w) - 1$, *where* t_{ij} *is the transposition that interchanges* i *and* j.

In the general case, the expression of $\partial_w(fg)$ in terms of f and g involves "relative" operators $\partial_{w/v}$, where v is another permutation. They are defined as follows. Let $a = (a_1, \ldots, a_p) \in R(w)$, and for each subsequence b of a define

$$\varphi(a,b) = \varphi_1(a,b)\cdots\varphi_p(a,b)$$

where

$$\varphi_i(a,b) = \begin{cases} s_{a_i} & \text{if } \ a_i \in b\,, \\ \partial_{a_i} & \text{if } \ a_i \notin b\,. \end{cases}$$

Then for each $v \in S_\infty$ we define

$$\partial_{w/v} = v^{-1}\sum_b \varphi(a,b)$$

summed over all subsequences b of a such that b is a reduced word for v. This sum is independent of the choice of reduced word $a \in R(w)$, and by (1.3) we have $\partial_{w/v} = 0$ unless $v \le w$ in the Bruhat order. If $v \le w$, $\partial_{w/v}$ is a linear operator on P_∞ of degree $-\ell(w) + \ell(v)$, and the Leibnitz formula reads

$$\partial_w(fg) = \sum_{v\le w} v(\partial_{w/v} f)\cdot \partial_v g\,. \tag{2.5}$$

If $v = 1$, then $\partial_{w/v} = \partial_w$. If $v \to w$, so that $w = vt_{ij}$ for some pair $i < j$, then

$$\partial_{w/v} = (x_i - x_j)^{-1}(1 - t_{ij})$$

is the divided difference operator relative to x_i and x_j.

3. MULTI-SCHUR FUNCTIONS

Let $X = (x_1, x_2, \dots)$ be an *alphabet*, that is to say a (finite or infinite) sequence of independent commuting variables. Recall [**M1**, Chap. I] that the *complete symmetric functions* $h_r(X)$ $(r \geq 0)$ are defined by the formal power series expansion

$$H(X,t) = \prod_{i \geq 1} (1 - x_i t)^{-1} = \sum_{r \geq 0} h_r(X) t^r$$

where t is another variable. More generally, if $Y = (y_1, y_2, \dots)$ is another alphabet, we define $h_r(X+Y)$ and $h_r(X-Y)$ by

$$\sum h_r(X+Y)t^r = H(X,t)H(Y,t),$$
$$\sum h_r(X-Y)t^r = H(X,t)H(Y,t)^{-1}.$$

Thus $X + Y$ is just the union of the alphabets X and Y; on the other hand, $X - Y$ is a formal difference of two alphabets, or (as we shall say) a *virtual* alphabet.

If $\lambda = (\lambda_1, \lambda_2, \dots)$ and $\mu = (\mu_1, \mu_2, \dots)$ are partitions of length $\leq n$, and $Z = X - Y$ is a virtual alphabet, the *skew Schur function* $s_{\lambda/\mu}(Z)$ is defined by the formula

$$s_{\lambda/\mu}(Z) = \det \left(h_{\lambda_i - \mu_j - i + j}(Z)\right)_{1 \leq i,j \leq n}.$$

We generalize this definition as follows: let $Z_i (1 \leq i \leq n)$ be virtual alphabets and let λ, μ be partitions of length $\leq n$; then the *multi-Schur function* $s_{\lambda/\mu}(Z_1, \dots, Z_n)$ is defined by

$$(3.1) \qquad s_{\lambda/\mu}(Z_1, \dots, Z_n) = \det \left(h_{\lambda_i - \mu_j - i + j}(Z_i)\right)_{1 \leq i,j \leq n}.$$

It vanishes identically unless $D(\lambda) \supset D(\mu)$ (i.e., $\lambda_i \geq \mu_i$ for all i).

When $\mu = 0$ we write s_λ in place of $s_{\lambda/0}$.

Let

$$X_i = (x_1, \dots, x_i), \quad Y_i = (y_1, \dots, y_i)$$

for each $i \geq 0$. Then we have

$$(3.2) \qquad s_\lambda(X_1 - Y_{\lambda_1}, \dots, X_n - Y_{\lambda_n}) = \prod_{(i,j)\in D(\lambda)} (x_i - y_j)$$

if $\lambda = (\lambda_1, \dots, \lambda_n)$ is a partition of length $\leq n$, and $D(\lambda)$ is its diagram, as in §1. In particular, when all the y_i are set equal to 0, we have

$$(3.3) \qquad s_\lambda(X_1, \dots, X_n) = \prod_{i=1}^{n} x_i^{\lambda_i} = x^\lambda$$

and when all the x_i are set equal to 0,

$$(3.4) \qquad s_\lambda(-Y_{\lambda_1}, \dots, -Y_{\lambda_n}) = (-1)^{|\lambda|} y^{\lambda'}$$

where λ' is the partition conjugate to λ. If we replace the y's by x's, and λ by λ', (3.4) takes the form

$$(3.4') \qquad s_{\lambda'}(-X_{\lambda'_1}, \dots, -X_{\lambda'_n}) = (-1)^{|\lambda|} x^\lambda .$$

Duality

The Schur functions $s_{\lambda/\mu}$ satisfy the relation [**M1**, Ch. I, §5]

$$s_{\lambda/\mu}(-Z) = (-1)^{|\lambda|-|\mu|} s_{\lambda'/\mu'}(Z) .$$

The multi-Schur functions (3.1) satisfy a similar relation. As in §1, let us write the partition λ in the form

$$\lambda = (p_1^{m_1}, \dots, p_k^{m_k})$$

where $p_1 > p_2 > \cdots > p_k > 0$, and each m_i is ≥ 1. Also let

$$\lambda' = (q_1^{n_1}, \dots, q_k^{n_k})$$

be the conjugate partition, and put $c_i = p_i - q_{k+1-i}$. Let Z be a (virtual) alphabet and let

$$Z_i = Z + X_{c_i}$$

for $1 \le i \le k$, where $X_{c_i} = (x_1, \ldots, x_{c_i})$ as above. Then we have

$$(3.5) \quad s_{\lambda/\mu}((-Z_1)^{m_1}, \ldots, (-Z_k)^{m_k}) = (-1)^{|\lambda|-|\mu|} s_{\lambda'/\mu'}(Z_k^{n_1}, \ldots Z_1^{n_k})$$

where the symbol $(-Z_1)^{\tilde{m}_1}$ means $(-Z_1, \ldots, -Z_1)$ to m_1 terms, and so on.

4. SCHUBERT POLYNOMIALS

Let $\delta = \delta_n$ be the partition $(n-1, n-2, \ldots, 1)$, so that $x^\delta = x_1^{n-1} x_2^{n-2} \cdots x_{n-1}$. For each $w \in S_n$ the *Schubert polynomial* $\mathfrak{S}_w$ is defined by

$$(4.1) \quad \mathfrak{S}_w = \partial_{w^{-1}w_0}(x^\delta)$$

where w_0 is the longest element of S_n. In particular, we have $\mathfrak{S}_{w_0} = x^\delta$ and $\mathfrak{S}_1 = 1$. In general, for each $w \in S_n$, $\mathfrak{S}_w$ is a nonzero homogeneous polynomial in $x_1, \ldots, x_{n-1}$ of degree $\ell(w)$, of the form

$$(4.2) \quad \mathfrak{S}_w = \sum_\alpha c_\alpha x^\alpha$$

summed over $\alpha = (\alpha_1, \ldots, \alpha_{n-1}) \in \mathbf{N}^{n-1}$ such that $\alpha \subset \delta$ (i.e., $\alpha_i \le n-i$ for each i) and $|\alpha| = \sum \alpha_i = \ell(w)$. (The coefficients c_α are in fact nonnegative integers: see (4.12) below.)

If $v, w \in S_n$ it follows from (2.2) that

$$(4.3) \quad \partial_v \mathfrak{S}_w = \begin{cases} \mathfrak{S}_{wv^{-1}} & \text{if } \ell(wv^{-1}) = \ell(w) - \ell(w), \\ 0 & \text{otherwise}. \end{cases}$$

By taking $v = s_i$, we see that $\mathfrak{S}_w$ is symmetrical in x_i and x_{i+1} if and only if $w(i) < w(i+1)$. In particular, $\mathfrak{S}_{s_i}$ is a homogeneous polynomial of degree 1, symmetric in $x_1, \ldots, x_i$, and such that $\partial_i \mathfrak{S}_{s_i} = 1$, so that

$$(4.4) \quad \mathfrak{S}_{s_i} = x_1 + \cdots + x_i \qquad (1 \le i \le n-1).$$

Let $m > n$. As explained in §1, we may regard S_n as a subgroup of S_m (fixing each of $n+1, n+2, \ldots, m$), i.e., we have an embedding j of S_n in S_m. A straightforward calculation shows that if $w \in S_n$ we have

$$\mathfrak{S}_w = \mathfrak{S}_{j(w)} \tag{4.5}$$

and hence that $\mathfrak{S}_w$ is a well-defined polynomial for all $w \in S_\infty$.

For certain classes of permutations, there are explicit formulas for $\mathfrak{S}_w$. First, if $u \in S_n$ and $v \in S_p$, we denote by $u \times v$ the permutation $(u(1), \ldots, u(n), v(1)+n, \ldots, v(p)+n)$ in S_{n+p}. We have then

$$\mathfrak{S}_{u\times v} = \mathfrak{S}_u \cdot \mathfrak{S}_{1_n \times v} \tag{4.6}$$

where 1_n is the identity element of S_n.

Next, if w is dominant (resp. Grassmannian), $\mathfrak{S}_w$ is a monomial (resp. a Schur function). More precisely:

(4.7). (i) *If w is dominant of shape λ, then*

$$\mathfrak{S}_w = x^\lambda .$$

(ii) If w is Grassmannian of shape λ, then

$$\mathfrak{S}_w = s_\lambda(x_1, \ldots, x_r)$$

where r is the unique descent of w (i.e., $w(r) > w(r+1)$).

More generally, suppose that w is vexillary, with shape $\lambda = (\lambda_1, \ldots, \lambda_m)$ and flag $\varphi(\varphi_1, \ldots, \varphi_m)$, as defined in §1. Then $\mathfrak{S}_w$ is a multi-Schur function, namely

$$\mathfrak{S}_w = s_\lambda(X_{\varphi_1}, \ldots, X_{\varphi_n}) \tag{4.8}$$

where as before $X_i = (x_1, \ldots, x_i)$ for each i.

Let H_n denote the additive subgroup of $P_n = \mathbb{Z}[x_1, \ldots, x_n]$ spanned by the monomials x^α such that $\alpha \subset \delta_n$. By (4.2) the Schubert polynomials

$\mathfrak{S}_w$, $w \in S_n$, lie in H_n, and it follows easily from (4.3) that they form a *$\mathbb{Z}$-basis* of H_n. By letting $n \to \infty$, we deduce that the $\mathfrak{S}_w$, $w \in S_\infty$, form a $\mathbb{Z}$-basis of the ring $P_\infty = \mathbb{Z}[x_1, x_2, \ldots]$.

Let $\eta : P_\infty \to \mathbb{Z}$ be the ring homomorphism defined by $\eta(x_i) = 0$ for all i, so that $\eta(f)$ is the constant term of the polynomial f. Then the expression of f in terms of Schubert polynomials is

$$f = \sum_w \eta(\partial_w f)\mathfrak{S}_w . \tag{4.9}$$

By linearity, it is only necessary to verify this formula when f is a Schubert polynomial $\mathfrak{S}_v$, in which case it follows from (4.3).

Next, let $f = \sum \alpha_i x_i$ be a homogeneous linear polynomial, and let $w \in S_\infty$. Then we have

$$f\mathfrak{S}_w = \sum (\alpha_i - \alpha_j)\mathfrak{S}_{wt_{ij}} \tag{4.10}$$

where t_{ij} is the transposition that interchanges i and j, and the sum is over pairs $i < j$ such that $\ell(wt_{ij}) = \ell(w) + 1$.

For by (4.9) the coefficient of $\mathfrak{S}_v$ in $f\mathfrak{S}_w$ is $\eta(\partial_v(f\mathfrak{S}_w))$, which can be evaluated by (2.4).

In view of (4.4), a particular case of (4.10) is *Monk's formula* [**Mo**]: for each $r \geq 1$ we have

$$\mathfrak{S}_{s_r}\mathfrak{S}_w = \sum \mathfrak{S}_{wt} \tag{4.11}$$

summed over transpositions $t = t_{ij}$ such that $1 \leq i \leq r < j$ and $\ell(wt) = \ell(w) + 1$.

This formula characterizes the algebra of Schubert polynomials, just as Pieri's formula expressing the product $s_\lambda h_r$ as a sum of Schur functions characterizes the algebra of Schur functions.

Another particular case of (4.10) is the following. Let w be a permutation with last descent r, so that $w(r) > w(r+1)$ and $w(r+1) < w(r+2) <$

.... Choose the largest $s > r$ such that $w(r) > w(s)$ and let $v = wt_{rs}$. Then (4.10) gives

$$x_r \mathfrak{S}_v = \mathfrak{S}_w - \sum_{w'} \mathfrak{S}_{w'} \tag{4.12}$$

summed over all permutations $w' = vt_{qr}$, where $q < r$ and $\ell(w') = \ell(v)+1 = \ell(w)$. A simple inductive argument, based on (4.12), then shows that

(4.13). *For each permutation w, $\mathfrak{S}_w$ is a polynomial in $x_1, x_2, \ldots$ with positive integral coefficients.*

In view of (4.13) it would be desirable to have a combinatorial rule for obtaining the monomials contained in a given Schubert polynomial. In this direction there is an attractive conjecture due to Axel Kohnert (which so far as I know has not yet been proved in general).

Let D be a "diagram", which for present purposes means any finite non-empty set of lattice points (i,j) in the positive (south-east) quadrant. Choose a point $p = (i,j) \in D$ which is rightmost in its row, and suppose that not all the points $(1,j), \ldots, (i-1,j)$ directly above p belong to D. If h is the largest integer less than i such that $(h,j) \notin D$, let D_1 denote the diagram obtained from D by replacing $p = (i,j)$ by (h,j). We can then repeat the process on D_1, by choosing the rightmost element in some row, and obtain a diagram D_2, and so on. Let $K(D)$ denote the set of all diagrams (including D itself) obtainable from D by a sequence of such moves.

Next, we associate with each diagram D a monomial

$$x^D = \prod_{i \geq 1} x_i^{a_i}$$

where a_i is the number of elements of D in the ith row, i.e., the number of j such that $(i,j) \in D$. With this notation established, Kohnert's conjecture [K] states that

(4.14?). *For each permutation w,*

$$\mathfrak{S}_w = \sum_{D \in K(D(w))} x^D$$

where $D(w)$ is the diagram (§1) of w.

5. ORTHOGONALITY

The symmetric group S_n acts on $P_n = \mathbb{Z}[x_1, \ldots, x_n]$ by permuting the x_i. The subring of invariants for this action is denoted by Λ_n: it is the ring of symmetric polynomials in $x_1, \ldots, x_n$. It is well-known (see e.g. E. Artin, Galois Theory, p. 41) that P_n is a free Λ_n-module with a basis consisting of the monomials x^α such that $\alpha \subset \delta$, where as usual $\delta = (n-1, n-2, \ldots, 1)$. In other words the multiplication map $P_n \otimes P_n \to P_n$ induces a linear isomorphism $H_n \otimes \Lambda_n \to P_n$, where (as in §4) H_n is the additive subgroup of P_n spanned by the monomials x^α, $\alpha \subset \delta$. Since the Schubert polynomials $\mathfrak{S}_w$, $w \in S_n$, form a $\mathbb{Z}$-basis of H_n, it follows that

(5.1). *The $\mathfrak{S}_w$, $w \in S_n$, form a Λ_n-basis of P_n.*

We define a scalar product on P_n, with values in Λ_n, by the rule

$$\langle f, g \rangle = \partial_{w_0}(fg) \qquad (f, g \in P_n) \tag{5.2}$$

where w_0 is the longest element of S_n. Each divided difference operator $\partial_i (1 \le i \le n-1)$ is Λ_n-linear, and hence so is ∂_{w_0}. It follows that the scalar product (5.2) is also Λ_n-linear.

We have

$$\langle \partial_w f, g \rangle = \langle f, \ \partial_{w^{-1}} g \rangle , \tag{5.3}$$

$$\langle w f, g \rangle = \varepsilon(w) \langle f, w^{-1} g \rangle \tag{5.4}$$

for all $w \in S_n$, where $\varepsilon(w) = (-1)^{\ell(w)}$. (It is enough to verify these formulas when $w = s_i$, and this is straightforward.)

Next, let $u, v \in S_n$. Then

$$(5.5) \qquad \langle w_0 \mathfrak{S}_u,\ \mathfrak{S}_{vw_0} \rangle = \varepsilon(v)\delta_{uv}$$

where δ_{uv} is the Kronecker delta.

For we have

$$\begin{aligned}\langle w_0 \mathfrak{S}_u,\ \mathfrak{S}_{vw_0} \rangle &= \langle w_0 \mathfrak{S}_u,\ \partial_{w_0 v^{-1} w_0} x^\delta \rangle \\ &= \langle \partial_{w_0 v w_0}(w_0 \mathfrak{S}_u), x^\delta \rangle \\ &= \varepsilon(v) \langle w_0 \partial_v \mathfrak{S}_u, x^\delta \rangle\end{aligned}$$

by (5.3). The scalar product is therefore zero unless $\ell(u) - \ell(v) = \ell(uv^{-1})$, and then it is equal to $\varepsilon(v)\langle w_0 \mathfrak{S}_{uv^{-1}}, x^\delta \rangle$. Now by (4.2) $\mathfrak{S}_{uv^{-1}}$ is a linear combination of monomials x^α such that $\alpha \subset \delta$ and $|\alpha| = \ell(u) - \ell(v)$. Hence $w_0(\mathfrak{S}_{uv^{-1}})x^\delta$ is a sum of monomials x^β where

$$\beta = w_0 \alpha + \delta \subset w_0 \delta + \delta = (n-1, \ldots, n-1).$$

Now $\partial_{w_0} x^\beta = 0$ unless all the components β_i of β are distinct, by (2.3); and since $0 \le \beta_i \le n-1$ for each i, it follows that $\partial_{w_0} x^\beta = 0$ unless $\beta = w\delta$ for some $w \in S_n$, and in that case $w_0 \alpha = \beta - \delta = w\delta - \delta$ must have all its components ≥ 0. So the only possibility that gives a non zero scalar product is $w = 1$, $\alpha = 0$, $u = v$, and in that case we have

$$\begin{aligned}\langle w_o \mathfrak{S}_u, \mathfrak{S}_{vw_0} \rangle &= \varepsilon(v) \langle 1, x^\delta \rangle \\ &= \varepsilon(v) \partial_{w_0}(x^\delta) = \varepsilon(v).\end{aligned}$$

Let $x = (x_1, \ldots, x_n)$ and $y = (y_1, \ldots, y_n)$ be two sequences of independent variables, and let

$$\Delta = \Delta(x, y) = \prod_{i+j \le n} (x_i - y_j).$$

Then we have

$$(5.6) \qquad \Delta(wx, x) = \begin{cases} 0 & \text{if } \ w \ne w_0, \\ \varepsilon(w_0) a_\delta(x) & \text{if } \ w = w_0, \end{cases}$$

where $a_\delta(x) = \prod_{1 \le i < j \le n} (x_i - x_j)$.

The polynomial $\Delta(x,y)$ is a linear combination of the monomials x^α, $\alpha \subset \delta$, with coefficients in $\mathbb{Z}[y_1, \ldots, y_n] = P_n(y)$, and hence can be written uniquely in the form

$$\Delta(x,y) = \sum_{w \in S_n} \mathfrak{S}_w(x) T_w(y)$$

with $T_w(y) \in P_n(y)$. By (5.4) we have

$$T_w(y) = \langle \Delta(x,y),\ w_0 \mathfrak{S}_{ww_0}(-x) \rangle_x$$

where the suffix x means that the scalar product is taken in the x variables. Hence

$$\begin{aligned} T_w(y) &= \partial_{w_0}(\Delta(x,y) w_0(\mathfrak{S}_{ww_0}(-x))) \\ &= a_\delta(x)^{-1} \sum_{v \in S_n} \varepsilon(v) \Delta(vx, y) v w_0(\mathfrak{S}_{ww_0}(-x)) \end{aligned}$$

by (2.3). Now this expression must be independent of $x_1, \ldots, x_n$. Hence we may set $x_i = y_i$ $(1 \le i \le n)$. But then (5.6) shows that the only non-zero term in the sum is that corresponding to $v = w_0$, and we obtain $T_w(y) = \mathfrak{S}_{ww_0}(-y)$. Hence we have proved

(5.7) $$\Delta(x,y) = \sum_{w \in S_n} \mathfrak{S}_w(x) \mathfrak{S}_{ww_0}(-y).$$

Let $(\mathfrak{S}^w)_{w \in S_n}$ be the Λ_n-basis of P_n dual to the basis $(\mathfrak{S}_w)$, relative to the scalar product (5.2). From (5.4) and (5.5) it follows that

(5.8) $$\mathfrak{S}^w(x) = w_0 \mathfrak{S}_{ww_0}(-x)$$

and hence from (5.7) that

(5.9) $$C(x,y) = \sum_{w \in S_n} \mathfrak{S}^w(x) \mathfrak{S}_w(y) = \prod_{i \le i < j \le n} (y_i - x_j).$$

This polynomial $C(x,y)$ is a "reproducing kernel" for the scalar product, i.e., we have

(5.10) $$\langle f(x),\ C(x,y) \rangle_x = f(y)$$

for $f \in H_n$, the $\mathbb{Z}$-span of the Schubert polynomials $\mathfrak{S}_w$. For (5.10) is clearly true for $f = \mathfrak{S}_w$, and the general case follows by linearity. (Notice that the $\mathbb{Z}$-span of the $\mathfrak{S}^w$ is not H_n, but $w_0 H_n$, by (5.8).)

Let $u \in S_n$ and let $(a_1, \ldots, a_p)$ be a reduced word for u, so that $\partial_u = \partial_{a_1} \cdots \partial_{a_p}$. Since $\partial_a = (x_a - x_{a+1})^{-1}(1 - s_a)$ for each $a \geq 1$, it follows that we may write

$$\partial_u = \varepsilon(w_0) a_\delta^{-1} \sum_{v \leq u} \alpha(u, v) v \tag{5.11}$$

where the sum is over $v \in S_n$ such that $v \leq u$ in the Bruhat order (1.3). The coefficients $\alpha(u, v)$ are in fact polynomials in $x_1, \ldots, x_n$, and $\alpha(u, u) \neq 0$ for each $u \in S_n$. Hence we can invert the equations (5.11), say

$$u = \sum_{v \leq u} \beta(u, v) \partial_v \tag{5.12}$$

and again the coefficients $\beta(u, v)$ are polynomials. It turns out that the α's and the β's satisfy the following relations:

(5.3). (i) $\beta(u, v) = \varepsilon(uv) \alpha(v w_0, u w_0)$,

(ii) $\alpha(u^{-1}, v^{-1}) = v^{-1}(\alpha(u, v))$,

(iii) $\alpha(w_0 u w_0, w_0 v w_0) = \varepsilon(u w_0) w_0(\alpha(u, v))$.

All this is a consequence of (5.10), but we have no space for the details, which will be found in **[M2]**.

6. DOUBLE SCHUBERT POLYNOMIALS

Let $x = (x_1, \ldots, x_n)$ and $y = (y_1, \ldots, y_n)$ be two sequences of independent variables, and as in §5 let

$$\Delta(x, y) = \prod_{i+j \leq n} (x_i - y_j).$$

For each $w \in S_n$ we define the *double Schubert polynomial* $\mathfrak{S}_w(x, y)$ to be

$$\mathfrak{S}_w(x, y) = \partial_{w^{-1} w_0} \Delta(x, y) \tag{6.1}$$

where $\partial_{w^{-1}w_0}$ acts on the x variables.

Since $\Delta(x,0) = x^\delta$ we have

$$\mathfrak{S}_w(x,0) = \mathfrak{S}_w(x)\,. \tag{6.2}$$

From (5.7) it follows that

$$\mathfrak{S}_w(x,y) = \sum_{v\in S_n} \partial_{w^{-1}w_0}\mathfrak{S}_{vw_0}(x)\mathfrak{S}_v(-y)$$

and hence by (4.2) we have

$$\mathfrak{S}_w(x,y) = \sum_{u,v} \mathfrak{S}_u(x)\mathfrak{S}_v(-y) \tag{6.3}$$

summed over all $u, v \in S_n$ such that $w = v^{-1}u$ and $\ell(w) = \ell(u) + \ell(v)$.

It follows from (6.3) that $\mathfrak{S}_w(x,y)$ is a homogeneous polynomial of degree $\ell(w)$ in $x_1, \dots, x_{n-1}, y_1, \dots, y_{n-1}$. In particular, $\mathfrak{S}_{w_0}(x,y) = \Delta(x,y)$ and $\mathfrak{S}_1(x,y) = 1$; moreover

$$\mathfrak{S}_{w^{-1}}(x,y) = \mathfrak{S}_w(-y,-x) = \varepsilon(w)\mathfrak{S}_w(y,x)\,, \tag{6.4}$$

(6.5). $\mathfrak{S}_w(x,x) = 0$ *for all* $w \in S_n$ *except* $w = 1$.

Suppose next that $m > n$, and let j be the embedding of S_n in S_m. Then it follows from (6.3) and (4.5) that

$$\mathfrak{S}_w(x,y) = \mathfrak{S}_{j(w)}(x,y)$$

for all $w \in S_n$, and hence that the double Schubert polynomials are well-defined for all permutations $w \in S_\infty$.

If K is any commutative ring, let $K(S_\infty)$ denote the K-module of all functions on S_∞ with values in K. We define a multiplication in $K(S_\infty)$ as follows: for $f, g \in K(S_\infty)$

$$(fg)(w) = \sum_{u,v} f(u)g(v)$$

summed over all $u, v \in S_\infty$ such that $uv = w$ and $\ell(u) + \ell(v) = \ell(w)$. For this multiplication, $K(S_\infty)$ is an associative (but not commutative) ring, with identity element **1** the characteristic function of the identity permutation 1. It carries an involution $f \mapsto f^*$, defined by

$$f^*(w) = f(w^{-1})$$

for all $w \in S_\infty$, and we have $(fg)^* = g^* f^*$ for all $f, g \in K(S_\infty)$.

Now let $\mathfrak{S}(x)$ (resp. $\mathfrak{S}(x,y)$) be the function on S_∞ whose value at a permutation w is $\mathfrak{S}_w(x)$ (resp. $\mathfrak{S}_w(x,y)$). (The coefficient ring K is the ring $\mathbb{Z}[x,y]$ of polynomials in the x's and y's.) These functions satisfy the following relations:

(6.6). (i) $\mathfrak{S}(x, 0) = s(x)$,

(ii) $\mathfrak{S}(x,x) = \mathbf{1}$,

(iii) $\mathfrak{S}(x,y)^* = \mathfrak{S}(-y,-x)$,

(iv) $\mathfrak{S}(x)^{-1} = \mathfrak{S}(0,x)$,

(v) $\mathfrak{S}(x)^* = \mathfrak{S}(-x)^{-1}$,

(vi) $\mathfrak{S}(x,y) = \mathfrak{S}(y)^{-1}\mathfrak{S}(x) = \mathfrak{S}(y,x)^{-1}$.

From the last of these formulas we have $\mathfrak{S}(x) = \mathfrak{S}(y)\mathfrak{S}(x,y)$, or explicitly

$$\mathfrak{S}_w(x) = \sum_{u,v} \mathfrak{S}_u(y)\mathfrak{S}_v(x,y)$$

summed over permutations u, v such that $uv = w$ and $\ell(u) + \ell(v) = \ell(w)$, so that $u = wv^{-1}$ and $\mathfrak{S}_u = \partial_v \mathfrak{S}_w$. Hence

$$\mathfrak{S}_w(x) = \sum_v \mathfrak{S}_v(x,y)\partial_v\mathfrak{S}_w(y)\,.$$

The sum here may be taken over all $v \in S_\infty$, since $\partial_v \mathfrak{S}_w = 0$ unless $\ell(wv^{-1}) = \ell(w) - \ell(v)$. By linearity it follows that

(6.7). *For all* $f \in P_n = \mathbb{Z}[x_1, \ldots, x_n]$ we have

$$f(x) = \sum_w \mathfrak{S}_w(x,y)\partial_w f(y)$$

summed over permutations w such that $w(n+1) < w(n+2) < \cdots$.

When $n = 1$, (6.7) is Newton's interpolation formula. Also when each $y_i = 0$, (6.7) reduces to (4.9).

If r is any integer, let $\mathfrak{S}_w(x, r)$ denote the polynomial obtained from $\mathfrak{S}_w(x, y)$ by setting $y_1 = y_2 = \cdots = r$. Since

$$\begin{aligned}\mathfrak{S}_{w_0}(x, r) = \Delta(x, r) &= \prod_{i=1}^{n-1} (x_i - r)^{n-i} \\ &= \mathfrak{S}_{w_0}(x - r)\end{aligned}$$

where $x - r$ means $(x_1 - r, x_2 - r, \dots)$, it follows from the definitions (6.1) and (4.1) that

$$\mathfrak{S}_w(x, r) = \mathfrak{S}_w(x - r)$$

for all permutations w. Hence, by (6.6) (vi),

$$\mathfrak{S}(x - r) = \mathfrak{S}(r)^{-1} \mathfrak{S}(x)$$

and in particular, for all integers q,

$$\mathfrak{S}(q - r) = \mathfrak{S}(r)^{-1} \mathfrak{S}(q)$$

from which it follows that

$$\mathfrak{S}(r) = \mathfrak{S}(1)^r \tag{6.8}$$

for all $r \in \mathbb{Z}$.

Since $\mathfrak{S}_w(x)$ is a sum of monomials with positive integral coefficients (4.13), $\mathfrak{S}_w(1)$ is the number of monomials in $\mathfrak{S}_w(x)$, each monomial being counted the number of times it occurs. By homogeneity, we have

$$\mathfrak{S}_w(r) = r^{\ell(w)} \mathfrak{S}_w(1). \tag{6.9}$$

From (6.6) (v) and (6.8) we obtain

$$\mathfrak{S}(1)^* = \mathfrak{S}(-1)^{-1} = \mathfrak{S}(1)$$

i.e., $\mathfrak{S}_w(1) = \mathfrak{S}_{w^{-1}}(1)$ for each permutation w, so that

(6.10). *For each $w \in S_\infty$, the number of monomials occurring in $\mathfrak{S}_w$ is equal to the number of monomials occurring in $\mathfrak{S}_{w^{-1}}$.*

Now consider the function $F = \mathfrak{S}(1) - \mathbf{1}$, whose value at $w \in S_\infty$ is

$$F(w) = \begin{cases} \text{number of monomials in } \mathfrak{S}_w, & \text{if } w \neq 1 \\ 0, & \text{if } w = 1. \end{cases}$$

For each positive integer p we have

$$\begin{aligned} F^p &= (\mathfrak{S}(1) - \mathbf{1})^p \\ &= \sum_{r=0}^{p} (-1)^r \binom{p}{r} \mathfrak{S}(1)^r \\ &= \sum_{r=0}^{p} (-1)^r \binom{p}{r} \mathfrak{S}(r) \qquad (1) \end{aligned}$$

by (6.8). The value of (1) at a permutation w of length p is by (6.9) equal to

$$\left(\sum_{r=0}^{p} (-1)^r \binom{p}{r} r^p \right) \mathfrak{S}_w(1)$$

which is equal to $p!\, \mathfrak{S}_w(1)$ (consider the coefficient of t^p in $(e^t - 1)^p$). On the other hand, $F^p(w)$ is by definition equal to

$$\sum_{w_1, \dots, w_p} F(w_1) \cdots F(w_p) \qquad (2)$$

summed over sequences $(w_1, \dots, w_p)$ of permutations such that $w_1 \cdots w_p = w$, $\ell(w_1) + \cdots + \ell(w_p) = \ell(w) = p$, and $w_i \neq 1 (1 \leq i \leq p)$. It follows that each w_i has length 1, hence $w_i = s_{a_i}$ say, and that $(a_1, \dots, a_p)$ is a reduced word for w. Since

$$\mathfrak{S}_{s_a} = x_1 + \cdots + x_a$$

we have $F(w_i) = \mathfrak{S}_{s_{a_i}}(1) = a_i$, and hence the sum (2) is equal to $\Sigma a_1 a_2 \cdots a_p$ summed over all $(a_1, \dots, a_p) \in R(w)$.

We have therefore proved that

(6.11). *The number of monomials in* $\mathfrak{S}_w$ *is*

$$\mathfrak{S}_w(1) = \frac{1}{p!}\sum a_1 a_2 \cdots a_p$$

summed over all $(a_1, \ldots, a_p) \in R(w)$, *where* $p = \ell(w)$.

Remarks. 1. The reduced words for the permutation $1_m \times w$, where $m \geq 1$, are $(m + a_1, \ldots, m + a_p)$ where $(a_1, \ldots, a_p) \in R(w)$. Hence from (6.11) and homogeneity we have

$$\mathfrak{S}_{1_m \times w}\left(\frac{1}{m}\right) = \frac{1}{p!}\sum \left(1 + \frac{a_1}{m}\right) \cdots \left(1 + \frac{a_p}{m}\right)$$

summed over $R(w)$ as before. Letting $m \to \infty$, we deduce that

$$\text{Card } R(w) = p! \lim_{m \to \infty} \mathfrak{S}_{1_m \times w}\left(\frac{1}{m}\right). \tag{6.12}$$

2. If w is dominant of length p, then $\mathfrak{S}_w$ is a monomial (4.7), and hence in this case

$$\sum_{R(w)} a_1 a_2 \cdots a_p = p!$$

3. Suppose that w is vexillary of length p. Then by (4.8) we have

$$\mathfrak{S}_w = s_\lambda(X_{\varphi_1}, \ldots, X_{\varphi_r})$$

where λ is the shape of w and $\varphi = (\varphi_1, \ldots, \varphi_r)$ the flag of w. Hence

$$\mathfrak{S}_{1_m \times w} = s_\lambda(X_{\varphi_1 + m}, \ldots, X_{\varphi_r + m})$$

for each $m \geq 1$. If we now set each x_i equal to $1/m$ and then let $m \to \infty$, we shall obtain in the lmit the Schur function s_λ for the series e^t (**[M1]**, Ch. I, §3, Ex. 5), which is equal to $h(\lambda)^{-1}$, where $h(\lambda)$ is the product of the hook-lengths of the partition λ. Hence it follows from (6.12) that if w is vexillary of length p, then

$$\text{Card } R(w) = p!/h(\lambda) \tag{6.13}$$

where λ is the shape of w. In other words, the number of reduced words for a vexillary permutation of length p and shape λ is equal to the degree of the irreducible representation of S_p indexed by λ.

4. It seems likely that there is a q-analogue of (6.11). Some experimental evidence suggests the following

Conjecture. $\mathfrak{S}_w(1,q,q^2,\dots) = \Sigma q^{\varphi(a)} \frac{(1-q^{a_1})\cdots(1-q^{a_p})}{(1-q)\cdots(1-q^p)}$ summed as in (6.11) over all reduced words $a = (a_1,\dots,a_p)$ for w, where

$$\varphi(a) = \sum \{i : a_i < a_{i+1}\}.$$

When w is vexillary the double Schubert polynomial $\mathfrak{S}_w(x,y)$ can be expressed as a multi-Schur function, just as in the case of (single) Schubert polynomials (§4). First we have

(6.14). (i) *If w is dominant of shape λ, then*

$$\begin{aligned}\mathfrak{S}_w(x,y) &= \prod_{(i,j)\in D(\lambda)} (x_i - y_j) \\ &= s_\lambda(X_1 - Y_{\lambda_1},\dots,X_m - Y_{\lambda_m})\end{aligned}$$

where $m = \ell(\lambda)$ and $X_i = (x_1,\dots,x_i)$, $Y_i = (y_1,\dots,y_i)$ for $i \geq 1$.

(ii) *If w is Grassmannian of shape λ, then*

$$\mathfrak{S}_w(x,y) = s_\lambda(X_m - Y_{\lambda_1+m-1},\dots,X_m - Y_{\lambda_m}).$$

If now w is vexillary with shape $\lambda(w) = (p_1^{m_1},\dots,p_k^{m_k})$ and flag $\varphi(w) = (f_1^{m_1},\dots,f_k^{m_k})$ as in §4, then w^{-1} is also vexillary, with shape $\lambda(w^{-1}) = \lambda(w)' = (q_1^{n_1},\dots,q_k^{n_k})$ the conjugate of $\lambda(w)$, and flag $\varphi(w^{-1}) = (g_1^{n_1},\dots,g_k^{n_k})$, where

$$g_i + q_i = f_{k+1-i} + p_{k+1-i} \qquad (1 \leq i \leq k).$$

With this notation we have

$$\mathfrak{S}_w(x,y) = s_\lambda((X_{f_1} - Y_{g_k})^{m_1},\dots,(X_{f_k} - Y_{g_1})^{m_k}). \tag{6.15}$$

The relation $\mathfrak{S}_{w^{-1}}(x,y) = \varepsilon(w)\mathfrak{S}_w(y,x)$ (6.4) when w is vexillary gives precisely the duality theorem (3.5) (with $\mu = 0$).

References

[K] A. Kohnert, 1990, Thesis, Bayreuth.

[L1] A. Lascoux, 1974, "Puissances extérieures, déterminants et cycles de Schubert,", *Bull. Soc. Math. France* **102**, 161-179.

[L2] A. Lascoux, 1982, "Classes de Chern des variétés de drapeaux", *C.R. Acad. Sci. Paris* **295**, 393-398.

[L3] A. Lascoux, 1988, "Anneau de Grothendieck de la variété des drapeaux", *preprint.*

[LS1] A. Lascoux and M.-P. Schützenberger, 1982, "Polynômes de Schubert,", *C.R. Acad. Sci. Paris* **294**, 447-450.

[LS2] A. Lascoux and M.-P. Schützenberger, 1982, "Structure de Hopf de l'anneau de cohomologie et de l'anneau de Grothendieck d'une variété de drapeaux", *C.R. Acad. Sci. Paris* **295**, 629-633.

[LS3] A. Lascoux and M.-P. Schützenberger, 1983, "Symmetry and flag manifolds", *Springer Lecture Notes* **996**, 118-144.

[LS4] A. Lascoux and M.-P. Schützenberger, 1985, "Schubert polynomials and the Littlewood-Richardson rule", *Letters in Math. Physics* **10**, 111-124.

[LS5] A. Lascoux and M.-P. Schützenberger, 1985, "Interpolation de Newton à plusieurs variables", Sém. d'Algèbre M.P. Malliavin 1983-84, *Springer Lecture Notes* **1146**, 161-175.

[LS6] A. Lascoux and M.-P. Schützenberger, 1987, "Symmetrization operators on polynomial rings", *Funkt. Anal.* **21**, 77-78.

[LS7] A. Lascoux and M.-P. Schützenberger, 1988, "Schubert and Grothendieck polynomials", *preprint.*

[LS8] A. Lascoux and M.-P. Schützenberger, 1989, "Tableaux and non-commutative Schubert polynomials", *to appear in Funkt. Anal.*

[M1] I. G. Macdonald, 1979, "Symmetric functions and Hall polynomials," *O.U.P.*

[M2] I. G. Macdonald, 1991, "Notes on Schubert polynomials", *preprint.*

[Mo] D. Monk, 1959, "The geometry of flag manifolds", *Proc. London Math. Soc.* (3) **9**, 253-286.

Computational Methods in Design Theory

RUDOLF MATHON

Department of Computer Science
University of Toronto
Toronto, Ontario, Canada M5S1A4

1 Introduction

Enumeration theory, which aims to count the number of distinct (non-equivalent) elements in a given class of combinatorial objects, constitutes a significant area in combinatorial analysis. The object of constructive enumeration consists of creating a complete list of configurations with given properties [5,8]. There are several reasons which stimulate research in constructive enumeration. Classical methods are not applicable to many interesting classes of objects such as strongly regular graphs, combinatorial designs, error correcting codes, etc. At present, the only available way to count them is by using algorithmic techniques for fixed values of parameters. Lists of objects are important for generating and testing various hypotheses about invariants, characterization, etc. Moreover, examples of designs with given properties are needed in many areas of applied combinatorics such as coding and experiment planning theories, network reliability and cryptography. Algorithms for constructive enumeration frequently require searching in high dimensional spaces and employ sophisticated techniques to identify partial (final) solutions. Such methods may be of independent interest in artificial intelligence, computer vision, neural networks and combinatorial optimization.

There are several common algorithmic approaches which are used to search for combinatorial configurations with particular properties. These can be divided into two broad classes depending on whether or not they search for all possible solutions in a systematic manner. Among the exhaustive techniques backtracking plays a prominent role. A backtrack algorithm attempts to find a solution vector by recursively building up partial solutions one element at a time. The vectors are examined in lexicographical order until all possible candidates for a component have been exhausted after which one backtracks. The search tree is pruned by rejecting partial solutions which are not admissible or have been generated before (so called isomorph rejection). One of the first systematic searches was carried out in 1919 by Cole, Cummings and White [4] to classify Steiner triple systems of order 15. Since then many researchers have used backtrack search to enumerate designs with small parameters [3,6,8]. However, simple exhaustive techniques quickly run out of steam for larger size problems due to exponential growth in search complexity. One way of delaying the combinatorial explosion is to look only for some restricted classes of "nice" configurations such as cyclic or transitive designs. By assuming certain automorphisms it has been possible to either construct or prove nonexistence of classes of graphs, designs and projective planes. The use of group actions has also found its way into algorithmic techniques for generating designs of larger size. Most current construction methods employ a tk-matrix or an incidence matrix as a design representation. A group action induces a fusion of a tk-matrix or a tactical decomposition of an incidence matrix which results in a significant reduction of computational effort. Methods based on tk-matrices and tactical decompositions will be discussed in Sections 2 and 4. In Section 3 we will use tk-matrices to enumerate completely 2-(6,3,2q) designs for arbitrary values of q.

The other class of methods is nonexhaustive in nature and frequently employs some probabilistic techniques. The search is usually represented in the form of an optimization problem in which the "goodness" of an approximate solution is measured in terms of an objective or cost function. Several randomized search procedures have been applied in the construction of combinatorial objects. These include hill-climbing, local search and simulated annealing [2,9,20]. In hill-climbing we keep randomly extending a partial solution until either a final solution is found or a set of partial solutions is reached which can not be extended, in which case we repeat the search from another initial guess. In local search one defines a set of configurations, a simple transformation and a cost function on these configurations. The goal is to find a configuration of minimum cost by applying randomized transformations. A new configuration is accepted if its cost does not exceed the cost of its pre-image. Simulated annealing is a local search in which poor local minima are avoided by allowing an occasional uphill move. This is accomplished in a randomized fashion by using "temperature" as a control parameter [9]. Randomized search techniques will be discussed in Section 4.

2 Tk-matrices

A *t-(v,k,λ) design*, $0<t<k<v$, is a pair (X,B) where B is a collection of k-subsets (called blocks) from a v-set V such that each t-subset of X is contained in exactly λ blocks of B. A design is called *simple* if it has no repeated blocks and *trivial* if it

is a multiple of all k-subsets of X. For $0<s<v$ denote by X_s the set of all s-subsets of X, $|X_s| = \binom{v}{s}$. Let $A_{tk} = (a_{ij})$ be a $|X_t|$ by $|X_k|$ matrix with rows and columns indexed by t- and k-subsets of X, respectively, $a_{ij} = 1$ if the i-th t-set is contained in the j-th k-set and $a_{ij} = 0$ otherwise. Hence, A_{tk} is a t- versus k-subset incidence matrix, where incidence is set inclusion. It is easy to see that the existence of a t-(v,k,λ) design is equivalent to the existence of a solution $\mathbf{x}$ to the matrix equation $A_{tk}\mathbf{x} = \lambda\mathbf{j}$, where $\mathbf{x}$ is a $|X_k|$-vector with non-negative integer entries and $\mathbf{j}$ is the all one $|X_t|$-vector. This formulation converts the problem of finding a design to an integer programming problem. Unfortunately, for most designs of interest the tk-matrix A_{tk} is prohibitively large. So, for example, the smallest parameter set 2-(22,8,4) for which a design is not known to exist gives rise to a 231 by 319770 tk-matrix. The size of A_{tk} can be reduced by assuming a group action on the set X. A group G acting on X induces an action on both X_t and X_k. Let ρ_t and ρ_k be the number of orbits of G on X_t and X_k, respectively. Denote by $A_{tk}(G)$ a ρ_t by ρ_k matrix with integer entries a_{ij} counting the number of k-sets in the j-th orbit of X_k containing a representative t-set in the i-th orbit of X_t. Hence $A_{tk}(G)$ is obtained from A_{tk} by adding the columns in each k-orbit and keeping one representative row from each t-orbit. The new integer system $A_{tk}(G)\mathbf{x} = \lambda\mathbf{j}$ requires finding a non-negative integer ρ_k-vector $\mathbf{x}$, where $\mathbf{j}$ is a ρ_t-vector of all one's. If a solution is found then the corresponding design has G as a subgroup of its automorphism group. Tk-matrices under group action were introduced by Kramer and Mesner [10] and led to a series of significant discoveries in the theory of t-designs. Magliveras and Leavitt [15] used projective linear groups to find the first examples of simple 6-designs. Later, Kramer, Leavitt and Magliveras [12] generated many simple 6-designs on 19 and 20 points based on the groups PSL_2 and PGL_2. To solve the linear system $A_{tk}(G)\mathbf{x} = \lambda\mathbf{j}$ they employed an ingenious exhaustive algorithm due to Leavitt in which new equations are added to the system to restrict the search space. Leavitt's algorithm can handle general tk-matrices with typically 30-50 rows and up to a few hundred columns.

An integer programing approach for computing (0,1)-vectors $\mathbf{x}$ satisfying $A_{tk}(G)\mathbf{x} = \lambda\mathbf{j}$ was proposed by Kreher and Radziszowski [14]. They observed that if $\mathbf{x}$ is an integer solution then $[\mathbf{x}, \mathbf{0}]^T$ is a vector in the integer lattice L spanned by the columns (integer linear combinations) of the matrix

$$B = \begin{bmatrix} I & \mathbf{0} \\ A_{tk}(G) & -\mathbf{j} \end{bmatrix}$$

They also noticed that integer solutions $\mathbf{x}$ satisfying $A_{tk}(G)\mathbf{x} = \lambda\mathbf{j}$ are often short vectors in L and used a modified basis reduction algorithm for finding short vectors in some reduced lattice. These heuristics proved successful in the construction of many nice t-designs with specified automorphism groups, most notably a simple 6-(14,7,4) design [14].

3 Enumeration of 2-(6,3,2q) designs

In order to illustrate the tk-matrix approach we will completely enumerate the one parameter family of 2-(6,3,2q) designs . A 2-(6,3,2q) design (X,B) has 10q blocks of size 3, each element is contained in half of them and each unordered pair of elements appears in 2q blocks. The tk-matrix A_{23} has 15 rows and 20 columns indexed by the unordered pairs and triples from the element set X={1,2,3,4,5,6} (see Table 1).

Table 1. A_{23} matrix for the set X={1,2,3,4,5,6}

A_{23}	123	124	125	126	134	135	136	145	146	156	234	235	236	245	246	256	345	346	356	456
12	1	1	1	1	0	0	0	0	0	0	0	0	0	0	0	0	0	0	0	0
13	1	0	0	0	1	1	1	0	0	0	0	0	0	0	0	0	0	0	0	0
14	0	1	0	0	1	0	0	1	1	0	0	0	0	0	0	0	0	0	0	0
15	0	0	1	0	0	1	0	1	0	1	0	0	0	0	0	0	0	0	0	0
16	0	0	0	1	0	0	1	0	1	1	0	0	0	0	0	0	0	0	0	0
23	1	0	0	0	0	0	0	0	0	0	1	1	1	0	0	0	0	0	0	0
24	0	1	0	0	0	0	0	0	0	0	1	0	0	1	1	0	0	0	0	0
25	0	0	1	0	0	0	0	0	0	0	0	1	0	1	0	1	0	0	0	0
26	0	0	0	1	0	0	0	0	0	0	0	0	1	0	1	1	0	0	0	0
34	0	0	0	0	1	0	0	0	0	0	1	0	0	0	0	0	1	1	0	0
35	0	0	0	0	0	1	0	0	0	0	0	1	0	0	0	0	1	0	1	0
36	0	0	0	0	0	0	1	0	0	0	0	0	1	0	0	0	0	1	1	0
45	0	0	0	0	0	0	0	1	0	0	0	0	0	1	0	0	1	0	0	1
46	0	0	0	0	0	0	0	0	1	0	0	0	0	0	1	0	0	1	0	1
56	0	0	0	0	0	0	0	0	0	1	0	0	0	0	0	1	0	0	1	1

Then a 2-(6,3,2q) design exists if and only if $A_{23}\mathbf{x} = 2q\mathbf{j}$ for an integer 20-vector of block multiplicities $\mathbf{x}\geq 0$. Since $A_{23}A_{23}{}^{T}= 4I+B$, where B is a symmetric adjacency matrix of a strongly regular graph (15,8,4,4) with eigenvalues 8,2,–2 of multiplicity 1,5,9, resp., we see that $\text{rank}(A_{23}A_{23}{}^{T}) = 15$. Hence, we can write $\mathbf{x} = A_{23}{}^{T}\mathbf{y}+Z^{T}\mathbf{z}$, where $A_{23}Z^{T}= 0$ and $\text{rank}(ZZ^{T}) = 5$, $\mathbf{y}$ and $\mathbf{z}$ are real vectors of length 15 and 5 resp. Since $A_{23}\mathbf{x} = AA_{23}A_{23}{}^{T}\mathbf{y} = 2q\mathbf{j}$ is constant, we see that $\mathbf{y} = q/6\mathbf{j}$ and so $\mathbf{x} = Z^{T}\mathbf{z}+q/2\mathbf{j}$ for any real 5-vector $\mathbf{z}$. The rows 10+i and 11–i of Z add up to 0 which implies that $x_{10+i}=q-x_{11-i}$, $1\leq i\leq 10$, and so $\mathbf{0}\leq\mathbf{x}\leq q\mathbf{j}$.

It is possible to express **z** in terms of a special subset of the components of **x** and derive a special relation between them

$$\begin{bmatrix} z_1 \\ z_3 \\ z_4 \\ z_2 \\ z_5 \end{bmatrix} = \begin{bmatrix} 0&0&1&1&0 \\ 0&0&0&1&1 \\ 1&0&0&0&1 \\ 1&1&0&0&0 \\ 0&1&1&0&0 \end{bmatrix} \begin{bmatrix} x_1 \\ x_2 \\ x_9 \\ x_{10} \\ x_6 \end{bmatrix} - \frac{q}{2} \begin{bmatrix} 1 \\ 1 \\ 1 \\ 1 \\ 1 \end{bmatrix}, \quad \begin{bmatrix} x_8 \\ x_7 \\ x_3 \\ x_5 \\ x_4 \end{bmatrix} = \begin{bmatrix} 1&0&-1&-1&0 \\ 0&1&0&-1&-1 \\ -1&0&1&0&-1 \\ -1&-1&0&1&0 \\ 0&-1&-1&0&1 \end{bmatrix} \begin{bmatrix} x_1 \\ x_2 \\ x_9 \\ x_{10} \\ x_6 \end{bmatrix} + \begin{bmatrix} q \\ q \\ q \\ q \\ q \end{bmatrix},$$

where $x_{10+i}=q-x_{11-i}$, $1\le i\le 10$. This can be used to find all distinct designs for a particular value of q. We count the number of distinct selections of x_1, x_2, x_6, x_9, x_{10}, between 0 and q such that x_3,x_4,x_5,x_7,x_8 have values also between 0 and q. If $x_1=q-k$, $x_6=q-i$ and $x_9=q-j$ then the admissible values of x_2 and x_{10} are given in the following two diagrams:

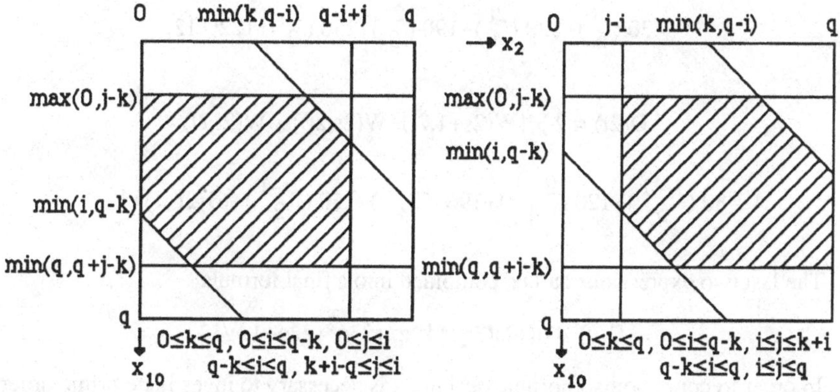

Here the sides which are not parallel to the coordinate axes have slope −1. The number of distinct designs is the sum of integer points in the hatched areas over admissible values of i, j and k.

Let $\delta = q-i+j$ and denote by $V(k,j,\delta,q)$ the hatched area in the first diagram for $0\le k,j\le\delta$, $0\le\delta\le q$. If $W(\delta,q) = \sum_{k=0}^{\delta}\sum_{j=0}^{\delta} V(k,j,\delta,q)$, then

$$W(2s+1,q) = 4\sum_{k=0}^{s}\sum_{j=0}^{k}\left[(2s+2)(q-k+j+1) - 2\binom{2s+2-k}{2}\right] + 4\sum_{k=0}^{s-1}\sum_{j=k+1}^{s}\left[(2s+2)(q+k-j+1)\right.$$

$$\left.- 2\binom{2s+2-j}{2}\right] = 24\,(2q+7)\binom{s+2}{3} - 240\binom{s+3}{4} - 16\binom{s+1}{2} + 8q(s+1),$$

$$W(2s,q) = 4\sum_{k=0}^{s-1}\sum_{j=0}^{k}[(2s+1)(q-k+j+1) - 2\binom{2s+1-k}{2}] + 4\sum_{k=0}^{s-2}\sum_{j=k+1}^{s-1}[(2s+1)(q+k-j+1) -$$

$$2\binom{2s+1-j}{2}] + 4\sum_{k=0}^{s-1}[(2s+1)(q+k-s+1) - 2\binom{s+1}{2}] + (2s+1)(q+1) - 2\binom{s+1}{2} =$$

$$48(q+1)\binom{s+2}{3} - 240\binom{s+2}{4} - (24q+70)\binom{s+1}{2} + (q+1)(2s+1).$$

Noting that the second diagram can be obtained from the first by adding j–i to x_2, we obtain a formula for the number D(q) of distinct 2-(6,3,2q) designs:

$$D(2t+1) = 2\sum_{s=0}^{t}[W(2s+1,2t+1)+W(2s,2t+1)] - W(2t+1,2t+1) =$$

$$30\binom{2t}{5}+120\binom{2t}{4}+190\binom{2t}{3}+150\binom{2t}{2}+122t+12,$$

$$D(2t) = 2\sum_{s=0}^{t}[W(2s+1,2t)+W(2s,2t)] - W(2t,2t) =$$

$$30\binom{2t-1}{5}+120\binom{2t-1}{4}+190\binom{2t-1}{3}+150\binom{2t-1}{2}+61(2t-1)+12.$$

The last two expressions can be combined into a final formula

$$D(q) = (q+1)(3q^4+12q^3+23q^2+22q+12)/12.$$

In order to count nonisomorphic designs it is necessary to investigate prime order automorphisms. The primes dividing $|S_6|$ are 5, 3 and 2.

(α) Automorphisms of order 5

The permutation σ = (2 3 4 5 6) generates 4 block-orbits and 3 orbits of pairs. Solving $A_{23}(\sigma)x = 2q\mathbf{j}$ yields $x_2=x_3=q-x_1$, $x_4=x_1$ in terms of x_1. If $x_2=x_3$ then $|G| = 60$. If q=2t+1 then the number of of distinct designs $D_{60}(2t+1) = 2t+2$. If q=2t then the case $x_1=x_2=x_3=x_4=t$ must be excluded (a t-multiple of a complete design) and so $D_{60}(2t) = 2t$. Dividing by the index $r_{60}=2$ of the stabilizer in the block-orbit group counts the nonisomorphic designs $N_{60}(2t+1) = N_{60}(2t+2) = t+1$. Finally, the t-multiple of the complete design is unique, $N_{720}(2t+1) = 0$, $N_{720}(2t+2) = 1$.

(β) Automorphisms of order 3

Both permutations $\sigma_1 = (1\ 2\ 3)(4\ 5\ 6)$ and $\sigma_2 = (1\ 2\ 3)$ fix 2 blocks and partition the remaining blocks into 6 orbits of size 3. For σ_1 we obtain the solution $x_1=2q-x_2-x_3-x_4$, and $x_{i+4}=q-x_i$, $1\le i\le 4$, where all x_is' are between 0 and q inclusive. The total number of solutions can easily be counted with the aid of a diagram in which the admissible values for x_3 and x_4 are plotted for a given value of $x_2=i$. Summation over all i, $0<i<q$ yields $T(q) = (q+1)(2q^2+4q+3)/3$. Designs with $|G| = 3s$, $s>1$ correspond to either $x_k=x_l$, or to $x_k=q-x_l$, for some $k<l$. Summation over all values i, $0\le i\le q$ and $k,l\in\{2,3,4\}$, $k<l$, and using inclusion-exclusion, yields $P(2t+1) = 18t^2+24t+6-2Q(2t+1)$, $P(2t+2) = 18t^2+36t+21-2Q(2t+2)$, where $Q(s) = |\{x|\ s\le 3x \le 2s\}|$ is the number of solutions with $x_2=x_3=x_4$. The number of distinct designs with $|G| = 3$ is $D_3(q) = T(q)-P(q)$. Dividing by the index $r_3=12$ yields the number of nonisomorphic designs $N_3(2t+1) = (t(t+1)(8t+11)+3Q(2t+1))/18$, $N_3(2t+2) = (8t^3+9t^2+t-3+3Q(2t+2))/18$. If $x_2=x_3=x_4\ne q/2$ then $|G| = 36$ and $r_{36}=2$. Since $D_{36}(2t+1) = Q(2t+1)$, $D_{36}(2t+2) = Q(2t+2)-1$, where $Q(3t+i) = t+i-1$, we see that $N_{36} = D_{36}/2$. If $x_k=x_l\ne q/2$ for some $k<l$ and $x_m\ne x_k$, $x_m\ne q-x_k$ then $|G| = 6$ and $r_3=6$. The number of distinct designs is easily counted. We have $D_6(2t+1) = 6t(t+1)-3Q(2t+1)$, $D_6(2t+2) = 6t^2+9t+6-3Q(2t+2)$ and so is $N_6(2t+i) = t(t+1)-N_{36}(2t+i)$, $i=1,2$. If $x_k=q-x_l$, $x_k\ne x_l$ and $x_m\ne x_k$, $x_m\ne x_l$, $x_m\ne q/2$ for some distinct $k,l,m\in\{2,3,4\}$ then $|G| = 12$ and $r_{12}=12$. Counting distinct designs yields $D_{12}(2t+i) = 12t(t+1)$ and so $N_{12}(2t+i) = t(t+1)$, $i=1, 2$. If $q=2t$ is even and either $x_k=x_l=t$ and $x_m\ne t$, or $x_k=q-x_l$, $x_k\ne x_l$ and $x_m=t$ for some distinct $k,l,m\in\{2,3,4\}$ then $|G| = 24$ and $r_{24}=6$. Then $D_{24}(2t+2) = 12(t+1)$ and so $N_{24}(2t+1) = 0$, $N_{24}(2t+2) = t+1$. Finally, if $x_k=x_l=q-x_m\ne q/2$ then $|G| = 60$ and if $x_2=x_3=x_4=t$ and q is even then $|G| = 720$. These cases have been considered before. For σ_2 we obtain the solutions $x_1=3x_8-q$, $x_5=q-x_1$, $x_2=x_3=x_4=q-x_8$, $x_6=x_7=x_8$ which are contained in those for σ_1.

(γ) Automorphisms of order 2

There are 3 types of permutations $\sigma_1 = (1\ 2)(3\ 4)(5\ 6)$, $\sigma_2 = (1\ 2)(3\ 4)$ and $\sigma_3 = (1\ 2)$ fixing 0, 2 and 4 elements, respectively. The solutions of the tk-matrix equations for σ_1 are contained in those for σ_2. We will treat each of the remaining cases separately.

Case 1 (2 fixed points) There are 4 fixed blocks and 8 orbits of size 2. Solving the tk-matrix equations yields $x_1=x_9+x_{11}+x_{12}-q$, $x_2=q+x_9-x_{11}-x_{12}$, $x_3=x_{10}=q-x_9$, $x_4=x_9$, $x_5=q-x_{11}$, $x_6=q-x_{12}$, $x_7=q-x_1$ and $x_8=q-x_2$, where $0\le x_i\le q$ for all i. Using a diagram to sum the numbers of admissible values of x_{11} and x_{12} for a fixed $x_9=i$ and then summing over all i, $0\le i\le q$ yields $T(2t+1) = 2(t+1)(5t^2+10t+6)/3$, $T(2t+2) = (10t^3+45t^2+71t-39)/3$. Designs with $|G| = 2s$, $s>1$ satisfy at least one of the following conditions: $x_9=x_{11}$, $x_9=x_{12}$, $x_9=q-x_{11}$, $x_9=q-x_{12}$, $x_{11}=x_{12}$, $x_{11}=q-x_{12}$, $x_9=q/2$. Using inclusion-exclusion we can sum over this subset of values to obtain $P(4t+1) = 52t^2+36t+4-8Q(1)$, $P(4t+2) = 64t^2+54t+13-8Q(2)$, $P(4t+3) = 52t^2+88t+36-8Q(3)$, $P(4t+4) = 64t^2+118t+55-8Q(4)$, where $Q(i) = N_{36}(4t+i)$. The number of distinct designs with $|G| = 2$ is $D_2{}^1(q) = T(Q)-P(Q)$. Since the index is 8, we have $N_2{}^1(q) = D_2{}^1(q)/8$. If $x_{11}=x_{12}$ and $x_9 \neq q/2$, or if $x_{11} \neq x_{12}$, $x_9=q/2$ and x_{11}, $x_{12} \neq q/2$, $x_{11}+x_{12} \neq q$, $x_{11}+2x_{12} \neq 3q/2$, $2x_{11}+x_{12} \neq 3q/2$ then $|G| = 4$ and the index is 4 in the first case and 12 in the second case. Using inclusion-exclusion we can sum over this subset of values to obtain the number of distinct designs $D_4(4t+1) = 4t(t+1)-4Q(1)$, $D_4(4t+3) = 4(t+1)^2-4Q(3)$, $D_4(4t+2) = 4t^2-4Q(2)+[12t^2]$, $D_4(4t+4) = 4t(t+1)-4Q(3)-[12t(t+1)]$, where $Q(i)= N_{36}(4t+i)$. The number of nonisomorphic designs can be obtained from D_4 by dividing the terms in the square brackets by 12 and all other terms by 4. Finally, if $q=2t$, $x_9=t$, $x_{11} \neq t$ and either $x_{11}=x_{12}$, or $x_{11}+2x_{12}=3t$, or $2x_{11}+x_{12}=3t$, then $|G| = 8$ and the index is 8. An easy count yields $D_8(4t) = D_8(4t+2) = 6t$ and so $N_8(4t+2) = t$, $N_8(4t+4) = t+1$. All other values of x_9, x_{11} and x_{12} have been treated before.

Case 2 (4 fixed points) There are 8 fixed blocks and 6 orbits of size 2. Solving the tk-matrix equations yields $x_1=x_{12}+x_{13}+x_{14}-q$, $x_2=q+x_{12}-x_{13}-x_{14}$, $x_3=q-x_{12}+x_{13}-x_{14}$, $x_4=q-x_{12}-x_{13}+x_{14}$, $x_{4+i}=q-x_{11+i}$, $1 \leq i \leq 3$, $x_{7+j}=q-x_j$, $1 \leq j \leq 4$, where $0 \leq x_i \leq q$ for all i. Using a diagram to sum the numbers of admissible values of x_{13} and x_{14} for a fixed $x_{12}=i$ and then summing over all i, $0 \leq i \leq q$ yields $T(2t+1) = 4t(t+1)(t+2)/3$, $T(2t+2) = (4t^3+18t^2+32t+21)/3$. Designs with $|G| = 2s$, $s>1$ satisfy at least one of the following conditions: $x_k=x_l$, $x_k=q-x_l$, for some $k<l$, $k,l \in \{12,13,14\}$. Summation over all i, $0 \leq i \leq q$ yields $P(4t+1)= 24t(t+1)-16Q(1)$, $P(4t+3) = 24(t+1)^2-16Q(3)$, $P(4t+2) = 24t^2+24t+7-16Q(2)$, $P(4t+4) = 24t^2+48t+25-16Q(4)$, where again $Q(i) = N_{36}(4t+i)$. The number of distinct designs is $D_2{}^2(q) = T(Q)-P(Q)$. Since the index is 24, we have $N_2{}^2(q) = D_2{}^2(q)/24$. All other cases have been considered before.

The number $N_1(q)$ of nonisomorphic designs with a trivial automorphism group is then given by

$$N_1(q) = [\, D(q) - 720 \sum_g N_g(q)/g \,]/ 720,$$

where g=2,3,4,6,8,12,24,36,60 and 720. The total number N(q) of nonisomorphic 2-(6,3,2q) designs is the sum of all the $N_g(q)$'s and can be expressed as a polynomial of degree 5 (see Tables 2 and 3)

$$N(12t+i) = \sum_{s=0}^{5} a_k(i)\, t^k,$$

with coefficients $a_k(i)$ depending on the residue class i modulo 12, $1 \leq i \leq 12$.

Table 2. Coefficients of N(12t+i) enumerating nonisomorphic 2-(6,3,2q) designs

i	$10a_5$	$10a_4$	$10a_3$	$10a_2$	$10a_1$	$10a_0$
1	864	720	1400	620	136	10
2	864	1080	1700	1005	321	40
3	864	1440	2120	1480	476	60
4	864	1800	2660	2075	821	130
5	864	2160	3320	2820	1176	190
6	864	2520	4100	3745	1771	340
7	864	2880	5000	4880	2436	480
8	864	3240	6020	6255	3411	760
9	864	3600	7160	7900	4536	1050
10	864	3960	8420	9845	6061	1530
11	864	4320	9800	12120	7836	2060
12	864	4680	11300	14755	10121	2860

Table 3. Coefficients of $N_g(12t+i)$ enumerating nonisomorphic 2-(6,3,2q) designs with automorphism group of order g

g	720	60		36		24		12			8		6			4		
i	a_0	a_1	a_0	a_1	a_0	a_1	a_0	a_2	a_1	a_0	a_1	a_0	a_2	a_1	a_0	a_2	a_1	a_0
1	0	6	1	2	0	0	0	36	6	0	0	0	36	4	0	9	1	0
2	1	6	1	2	0	12	2	36	6	0	3	0	36	4	0	18	–2	0
3	0	6	2	2	1	0	0	36	18	2	0	0	36	16	1	9	4	0
4	1	6	2	2	0	12	4	36	18	2	3	1	36	16	2	18	4	0
5	0	6	3	2	1	0	0	36	30	6	0	0	36	28	5	9	7	1
6	1	6	3	2	1	12	6	36	30	6	3	1	36	28	5	18	10	1
7	0	6	4	2	1	0	0	36	42	12	0	0	36	40	11	9	10	3
8	1	6	4	2	1	12	8	36	42	12	3	2	36	40	11	18	16	3
9	0	6	5	2	2	0	0	36	54	20	0	0	36	52	18	9	13	4
10	1	6	5	2	1	12	10	36	54	20	3	2	36	52	19	18	22	7
11	0	6	6	2	2	0	0	36	66	30	0	0	36	64	28	9	16	7
12	1	6	6	2	2	12	12	36	66	30	3	3	36	64	28	18	28	10

g	3				2				1					
i	a_3	a_2	a_1	a_0	$2a_3$	$2a_2$	$2a_1$	$2a_0$	$10a_5$	$10a_4$	$10a_3$	$10a_2$	$10a_1$	$10a_0$
1	96	–6	–3	0	204	–33	–9	0	864	720	–580	35	21	0
2	96	18	1	0	204	–9	1	0	864	1080	–280	–30	–4	0
3	96	42	3	0	204	69	–3	0	864	1440	140	–95	1	0
4	96	66	15	1	204	93	15	0	864	1800	680	50	–14	0
5	96	90	25	2	204	171	37	2	864	2160	1340	255	11	0
6	96	114	45	6	204	195	63	8	864	2520	2120	730	96	0
7	96	138	63	9	204	273	111	12	864	2880	3020	1325	251	20
8	96	162	91	17	204	297	145	24	864	3240	4040	2250	566	50
9	96	186	117	24	204	375	219	42	864	3600	5180	3355	1001	110
10	96	210	153	37	204	399	261	56	864	3960	6440	4850	1716	230
11	96	234	187	49	204	477	361	88	864	4320	7820	6585	2621	400
12	96	258	231	69	204	501	411	114	864	4680	9320	8770	3946	680

In Table 4 we list the numbers of nonisomorphic 2-(6,3,2q) designs for all admissible orders of the automorphism group and values of q ranging from 1 to 20. It is worth noting that, while N_{720} is constant, N_{60}, N_{36}, N_{24}, N_8 are linear, N_{12}, N_6, N_4 are quadratic, N_3, N_2 are cubic polynomials. N_1, N and D/ 720 are fifth degree polynomials with the same leading coefficients. Consequently, N(q) can be approximated from below by D(q)/ 720.

Table 4. Number N_g of nonisomorphic 2-(6,3,2q) designs with group of order g

q	N_{720}	N_{60}	N_{36}	N_{24}	N_{12}	N_8	N_6	N_4	N_3	N_2	N_1	N
1	0	1	0	0	0	0	0	0	0	0	0	1
2	1	1	0	2	0	0	0	0	0	0	0	4
3	0	2	1	0	2	0	1	0	0	0	0	6
4	1	2	0	4	2	1	2	0	1	0	0	13
5	0	3	1	0	6	0	5	1	2	1	0	19
6	1	3	1	6	6	1	5	1	6	4	0	34
7	0	4	1	0	12	0	11	3	9	6	2	48
8	1	4	1	8	12	2	11	3	17	12	5	76
9	0	5	2	0	20	0	18	4	24	21	11	105
10	1	5	1	10	20	2	19	7	37	28	23	153
11	0	6	2	0	30	0	28	7	49	44	40	206
12	1	6	2	12	30	3	28	10	69	57	68	286

13	0	7	2	0	42	0	40	10	87	81	106	375
14	1	7	2	14	42	3	40	16	115	98	163	501
15	0	8	3	0	56	0	53	13	141	135	235	644
16	1	8	2	16	56	4	54	22	178	156	338	835
17	0	9	3	0	72	0	69	17	213	207	463	1053
18	1	9	3	18	72	4	69	29	261	235	633	1334
19	0	10	3	0	90	0	87	22	306	300	836	1654
20	1	10	3	20	90	5	87	37	366	335	1101	2055

The counting formulae for 2-(6,3,2q) designs have been known to the author since 1984 and have been used to obtain the values listed in [9]. They are published here for the first time. Since then Engel and Gronau [7] independently derived the formula for the number of distinct designs D(q) by interpolating the known numbers for small values of q.

4 Tactical Decompositions

A majority of algorithms for the construction of 2-(v,k,λ) designs are based on incidence matrices. An element-block incidence matrix A of a 2-design satisfies the matrix equation $AA^T = (r-\lambda)I - \lambda J$, where $r = \lambda(v-1)/(k-1)$ is the repetition number and I, J are the unit and all one matrices of order v, respectively. A backtrack algorithm can be employed on a block by block or element by element basis. In either case its efficiency depends critically on the ability to recognize suitable partial solutions.The following inexpensive but discriminating inequality has been utilized in most of our recent searches.

Lemma. Let D(X,B) be a 2-(v,k,λ) design with b blocks and repetition number r. For $Y \in X$ and $C \in B$ let E be the corresponding induced configuration with w elements and c blocks. Denote by s the sum of all blocksizes in E and by p the sum of pairs they cover. Then

$$(wr-s)d + p \leq \binom{w}{2}\lambda + \binom{d+1}{2}, \quad d = \left\lfloor \frac{wr-s}{b-c} \right\rfloor.$$

Proof. The sum of blocksizes of the remaining b–c blocks B\C of D induced by Y is wr–s. To derive a lower bound on the total number of pairs they might cover we note, that from all distributions of n non-negative integers k_i with a fixed sum σ, the one having the smallest sum of squares has the k_i's as close to each other as possible or, equivalently, has the smallest maximum difference $|k_i-k_j|$ over all i and j. In our case, this corresponds to $l = wr-s-d(b-c)$ blocks of size $d = \lfloor (wr-s)/(b-c) \rfloor$ and $b-c-l$ blocks of size d+1 covering $ld(d+1)/2+(b-c-l)d(d-1)/2$ pairs.

The inequality then follows from the fact that this lower bound cannot exceed the actual number of pairs $\lambda w(w+1)/2-p$ of Y which must be covered in B\C.

Another ingredient of a successful backtrack is an efficient isomorph rejection. Ideally, we would like to prune the search tree by eliminating any partial configuration which has been encountered before. One variation of backtracking called an *orderly* search generates all configurations in canonical form, thus eliminating the need of checking for duplicates [2,8]. Unfortunately, for many design problems it appears difficult to use canonical forms in a recursive way. A popular technique employs only a partial isomorph rejection by utilizing selected automorphisms of the configuration under construction. For example, in the enumeration of all 2-(9,3,3) designs [18] we have used neighborhood graphs to reduce significantly the production of isomorphic solutions. For the 22521 nonisomorphic designs we have generated a total of about 50000 solutions.

If G is a group of automorphisms of a 2-(v,k,λ) design D(X,B), let $X_1, X_2, \ldots, X_m$ be the point orbits and $B_1, B_2, \ldots, B_n$ be the block orbits of G. A *tactical decomposition* of D with respect to G is an m by n matrix $T(G) = (t_{ij})$ where t_{ij} is the number of elements from X_i contained in a block from B_j. In most cases G is a single automorphism of D of prime order generating orbits of size p and 1. From the matrix equation governing A we obtain relations for the entries of T(G)

$$\sum_{q=1}^{n} t_{iq}\, s_q = r_i, \quad \sum_{q=1}^{n} t_{iq}\, t_{jq}\, s_q = \lambda r_i\, r_j\,,$$

where $r_i = |X_i|$, $s_j = |B_j|$, and $1 \le i,j \le m$. The search for 2-(v,k,λ) designs proceeds in two stages, each of which requires solving a smaller problem than the original. In the first stage we backtrack for all tactical decompositions T(G) of the matrix A. For large designs we might want to impose a group action on T itself, thus inserting another stage. In any case, we end up with a set of feasible decompositions from which we keep only those which are nonisomorphic, and determine their automorphism groups. In the second stage we search for the actual circulants with valencies t_{ij} in the T-decomposition of A using the group of T for partial isomorph rejection. As in the case of tk-matrices, each of the resulting 2-(v,k,λ) designs has G as a subgroup of its automorphism group. We have used an efficient implementation of both of these stages to enumerate all 2-(25,4,1) and 2-(31,10,3) designs with nontrivial automorphism groups [11,16] as well as many other families of designs possessing a given automorphism [19].

As an illustration of a tactical decomposition search we will enumerate all symmetric 2-(45,12,3) designs with an automorphism of order 11. A simple analysis shows that both X and B decompose into 4 orbits of length 11 and a fixed point. After completing the first stage we end up with 2 nonisomorphic T matrices

$$\begin{bmatrix} 1 & 4 & 4 & 3 \\ 4 & 1 & 4 & 3 \\ 4 & 4 & 1 & 3 \\ 3 & 3 & 3 & 2 \end{bmatrix} \quad \begin{bmatrix} 5 & 2 & 2 & 3 \\ 2 & 5 & 2 & 3 \\ 2 & 2 & 5 & 3 \\ 3 & 3 & 3 & 2 \end{bmatrix}$$

(omitting the row and column corresponding to the fixed point and block). The first matrix yields a single self-dual design with group of order 11 and base blocks

$$\begin{array}{ll} (0_0\,0_1\,1_1\,4_1\,5_1\,0_2\,1_2\,3_2\,9_2\,0_3\,2_3\,7_3) & (0_0\,1_0\,3_0\,4_0\,10_1\,0_2\,4_2\,6_2\,10_2\,4_3\,7_3\,9_3) \\ (0_0\,1_0\,5_0\,7_0\,3_1\,5_1\,6_1\,8_1\,9_2\,6_3\,9_3\,10_3) & (0_0\,2_0\,5_0\,2_1\,4_1\,8_1\,6_2\,7_2\,10_2\,1_3\,2_3\,\infty) \\ (0_3\,1_3\,2_3\,3_3\,4_3\,5_3\,6_3\,7_3\,8_3\,9_3\,10_3\,\infty) & \qquad \bmod (11) \end{array}$$

The second matrix does not generate any solutions. The total CPU time for both stages and the analysis has not exceeded 5 minutes on an Apple MacII.

5 Randomized Search

The methods discussed in this section are all based on the concept of *local optimization.* A combinatorial optimization problem can be specified as a set Σ of solutions (or states) together with a cost function c that assigns a value to each solution. An optimal solution corresponds to a global minimum of the cost function over the set Σ (this might not be unique). Given an arbitrary solution $S \in \Sigma$ we define a set T_S of transformations (moves), each of which can be used to change S to another solution S'. The set of solutions that can be reached from S by applying T_S is called the *neighborhood* of S and denoted by N(S). In local optimization we repeatedly search through the neighborhood of S for a solution S' of lower cost $c(S')<c(S)$. Starting from some initial guess this procedure ends up in a locally optimal solution in the sense that none of its neighbors has lower cost. The hope is that some local minima will be good enough. There are many ways of selecting the next candidate in the neighborhood. If N(S) is small in size we can choose S' of minimum cost in N(S) which results in a steepest descent type algorithm. If N(S) is large we can keep selecting new candidates at random until one is found which has lower cost. It is sometimes desirable to accept a new S' if $c(S')\leq c(S)$, to increase the chance of reaching a global minimum, i.e. we accept not only "downhill moves" but also "sideway" moves. A basic property which a set Σ with the associated neighborhood and cost functions should have is that from any $S \in \Sigma$ it should be possible to reach a global minimum by a finite sequence of moves (some of which might be "uphill"). A computational requirement on $\Sigma(T,c)$ is that it should be easy to perturb solutions and to evaluate

their cost. We are now in a position to explain a particular instance of local search for combinatorial designs which has enjoyed success in some cases.

A *hill-climbing* search for a t-(v,k,λ) design (X,B) is a local optimization which uses partial designs as the solution set Σ. A partial t-(v,k,λ) design (X,B') is a set B' of blocks of size k from X such that any t-subset of X is contained in at most λ blocks. The cost associated with a partial design (X,B') is the difference |B|-|B'|, where the final number of blocks |B| can be expressed in terms of the parameters [2]. From a partial design (X,B'), we move to another one by adding a new block β to B' and possibly deleting some other blocks which violate the definition of a partial design. To make this efficient we should avoid creating too many conflicts by choosing only those new blocks which contain a large number of unused t-subsets. A successful hill-climbing algorithm was used by Stinson [20] for Steiner triple systems (which are 2-(v,3,1) designs). A new triple (x,y,z) is formed by selecting at random two unused pairs (x,y) and (x,z). If the third pair (y,z) already appears in the partial design we delete the unique block containing it. After a move the cost goes down by one or remains the same. Implementations of this algorithm exist in which the computational complexity of a move is a constant independent of v. Stinson's algorithm works well for all triple systems and related configurations such as Latin squares, etc. It has been used to generate large samples of random triple systems with various specified properties [17]. Hill-climbing does not work well if $k>3$ or $t>2$. One possible explanation of this is that almost all partial designs created during the search contain a large number of subconfigurations which are "forbidden" in the final design. This causes the algorithm to get stuck by failing to make any progress. In order to find a successful and generally applicable hill-climbing search we must learn how to efficiently recognize and avoid bad partial solutions.

In *simulated annealing* we employ local search augmented with a mechanism to avoid entrapment in poor local minima. This is accomplished by allowing an occasional uphill move with the help of a random number generator and a control parameter T called the "temperature". An uphill move of size Δ is accepted with probability $\exp(-\Delta/T)$ which diminishes as the temperature declines. Note that, for a fixed T, small uphill moves have higher probabilities than large ones. A typical implementation of simulated annealing uses a pair of nested loops governed by two parameters, a cooling ratio q, $0<q<1$ and an integer temperature length L (see [9]). In each instance of the inner loop, which is executed L times at constant temperature T, we pick a random neighbor S' of the current solution S and calculate the $\Delta=c(S')-c(S)$. If $\Delta\leq 0$ we set S:=S'. If $\Delta>0$ we accept S' with probability $\exp(-\Delta/T)$. In each pass of the outer loop the temperature is reduced by a factor $q<1$, T:=qT. The outer loop starts from some initial temperature T_0

and terminates when no further progress is being made, which corresponds to a "frozen" state.

We have implemented two algorithms which use simulated annealing to search for 2-(v,k,λ) designs. Both algorithms attempt to build a v by b incidence matrix with row sums r, column sums k and the inner products of two distint rows λ. In the first algorithm the solution space Σ consists of all v by b 0-1 matrices. A move in Σ consists in changing the parity of a randomly chosen entry $a_{ij}:=1-a_{ij}$. In the block design a move corresponds to adding element i to block j if it is not there and deleting it if it is there. The cost funtion is a weighted sum of squares of the deviations of row sums from r, column sums from k and products of rows from λ. The weights corresponding to rows, columns and row products should be chosen in the ratio v:b:1, respectively. In the second algorithm we have worked with 0-1 matrices with constant row sums r and column sums k. A move consists in changing the parity of all entries in a 2 by 2 submatrix with row and column sums 1. In addition to preserving row and column sums it is known that any two matrices in this space can be obtained from each other by a finite sequence of such transformations. In the block design a move corresponds to swapping two elements belonging to two different blocks. The cost function is the sum of squares of deviations from λ of the v(v-1)/2 row products. To set the various control parameters in our algorithms we followed some generally accepted rules [9]. The recommended cooling ratio is q=0.95. The temperature length L should be a fixed multiple of the size of a typical neighborhood. The neighborhoods of the two proposed spaces are vb and $v(v-1)(r-\lambda)^2/2$, respectively. The size factors ranged from 10 to 50 in the first algorithm and from 1 to 5 in the second algorithm depending on the overall size of the incidence matrix and the available computer resources. The initial temperature 1 worked well for unit sums of weights in the first case and with no restrictions in the second case. The initial success rate (the ratio of accepted solutions in the inner loop to L) should be about 30% to insure adequate mixing of the solutions. Finally, the search was terminated if for 5 consecutive instances of the outer loop the cost of the best solution remained the same.

We now describe the results of our experiments. To our surprise, the first algorithm performed considerably worse than the second. Its only advantage is its much wider applicability to designs which are either pairwise balanced (k and r are variable) or (r,λ) designs (k variable). To be successful it required large size factors and more overall trials. The second algorithm performed well on small designs (with vb<400). It sucessfully generated symmetric 2-(4t–1,2t–1,t–1) for t<6 and $2\text{-}(k^2+k+1,k+1,1)$ designs for k<9. For designs with small block size k a local search in which sideways moves were allowed worked best. This corresponded to setting the initial temperature to 0 (frozen state). For k=3 our algorithm worked every time, usually terminating during the first pass of the outer loop with about the same efficiency as hill-climbing. The results for k=4 are summarized in Table 5.

Table 5. Average number of conflicts in local search for 2-(v,4,λ) designs

v\λ	1	2	3	4	v\λ	1	2	3	4
13	0	0	0	0	28	8.8	3.4	0.3	0
16	0	0	0	0	29			0.3	
17			0		31		5.1		0
19		0.3		0	32			0.6	
20			0		33			1.2	
21			0		34		4.8		0
22		2.3		0	36			0.3	
24			0		37	15	6	0.6	0
25	6.5	3	0	0	40	19.4	7.1	1.6	0

Each entry in this table is the average number of conflicts based on ten runs for each value of the parameters. For a fixed λ the conflicts grow linearly as a function of v, the slope declining as λ increases. For λ≥4 our local search succeeds every time in a similar fashion to that for k=3. There seems to be a general pattern. For every k there seems to be a $\Lambda = \Lambda(k)$ such that for every $\lambda \geq \Lambda$ local search is perfect ($\Lambda(3) = 1$, $\Lambda(4) = 3$, $\Lambda(5) = 6$, etc.)

Acknowledgement

Research of the author is supported by the Natural Sciences and Engineering Research Council of Canada under grant number A8651.

References

[1] T. Beth, D. Jungnickel, H. Lenz, *Design Theory*, Bibliographishes Institut, Manheim-Wien-Zurich, 1985.

[2] M.J. Colbourn, Algorithmic aspects of combinatorial designs: A survey, *Ann. Discrete Math.* **26** (1985), 67-136.

[3] M.J. Colbourn, R.A. Mathon, On cyclic Steiner 2-designs, *Ann. Discrete Math.* **7** (1980), 215-253.

[4] F.N.Cole, L.D. Cummings, H.S. White, Complete classification of the triad systems on fifteen elements, *Memoires of the National Academy of Sciences U.S.A.* **14**, Second memoir (1919), 1-89.

[5] I.A. Faradzev, Constructive enumeration of combinatorial objects, Problèmes Combinatoires et Théorie des Graphes, CNRS, Paris, 1977, 131-135.

[6] P.B. Gibbons, Computing techniques for the construction and analysis of block designs, PhD Thesis, University of Toronto (1976).

[7] K. Engel, H.-D.O.F. Gronau, On 2-(6,3,λ) designs, *Rostock. Math. Kolloq.* **34** (1988), 37-48.

[8] A.V. Ivanov, Constructive enumeration of incidence systems, *Ann. Discrete Math.* **26** (1985), 227-246.

[9] D.S. Johnson, C.A. Aragon, L.A. McGeoch, C. Schevon, Optimization by simulated annealing: An experimental evaluation. Part I (Graph partitioning), Part II (Graph coloring and number partitioning), to appear in *Operations Research.*

[10] E.S. Kramer, D.M. Mesner, t-designs on hypergraphs, *Discrete Math.* **15** (1976), 263-296.

[11] E.S. Kramer, S.S. Magliveras, R. Mathon, The Steiner systems S(2,4,25) with nontrivial automorphism group, *Discrete Math.* **77** (1989), 137-157.

[12] E.S. Kramer, D.W. Leavitt, S.S. Magliveras, Construction procedures for t-designs and the existence of new simple 6-designs, *Ann. Discrete Math.* **26** (1985), 247-274.

[13] D.L. Kreher, S.P. Radziszowski, Finding simple t-designs by using basis reduction, *Congressus Numerantium* **55** (1986), 235-244.

[14] D.L. Kreher, S.P. Radziszowski, The existence of simple 6-(14,7,4) designs, *J. Combinat. Theory A* (to appear).

[15] S.S. Magliveras, D.W. Leavitt, Simple 6-(33,8,36) Designs from $P\Gamma L_2(32)$, Proceedings of the Durham Computational Group Theory Symposium, Academic Press, (1983), 337-352.

[16] R. Mathon, Symmetric (31,10,3) designs with nontrivial automorphism group, *Ars Combinatoria* **25** (1988), 171-183.

[17] R. Mathon, Nonisomorphic designs by simulated annealing (to appear).

[18] R. Mathon, D. Lomas, A census of 2-(9,3,3) designs (to appear).

[19] R. Mathon, A. Rosa, Tables of parameters of BIBD's with $r \leq 41$ including existence, enumeration, and resolvability results, *Ann. Discrete Math.* **26** (1985), 275-308.

[20] D.R. Stinson, Hill-climbing algorithms for the construction of combinatorial designs, *Ann. Discrete Math.* **26** (1985), 321-334.

Fast and Slow Growing
(A combinatorial study of unprovability)

Martin Loebl
and
Jaroslav Nešetřil

Department of Applied Mathematics
Charles University
Malostranské nám. 25
11800 Praha 1
Czechoslovakia

Institut für Diskrete Mathemetik
Universität Bonn
Nassestraße 2
5300 Bonn
Germany

ABSTRACT

We survey the recent research related to the hierarchies of fast growing functions. We concentrate on two main areas: hierarchies of primitive recursive functions and hierarchies of (in Peano Arithmetic) provably total functions. In each of these areas we give a theoretical background and give all the main examples and results. These include examples from combinatorics, geometry, number theory and computer science. In the last part we outline some recently found techniques which enable a unified treatment for results in both areas. We give several sample proofs.

Contents

I. Prologue
II. Primitive recursive hierarchies
III. Examples of Ackermann growth
IV. Provably total Peano hierarchies
V. Examples of unprovable growth
VI. Techniques - bounds

I. Prologue

"Who-will-say-the-larger-number" is a universal children's game. Up to a certain age, the number 7 (of course) is the winning number. Later 13 is the winning number, and then millions and billions and billions of billions... . A very detailed analysis of this development is available (in the work of Piaget and his school).

Adult tricks like "one plus the largest number which you ever thought of" are not legal and thus the game soon loses its interest. However some interest is maintained if we deal with functions defined for all natural numbers. One is then led to examples of functions which have a fast growth rate. Examples include:

linear functions $f(x) = cx$
exponential functions $f(x) = 2^{cx}$
tower functions $f(x) = 2^{2^{\cdot^{\cdot^{\cdot^{2}}}}}\Big\}x$
wow functions (defined e.g. in [GRS])

and other exotic functions with other exotic names. It is perhaps interesting that under certain natural restrictions one can find fastest growing functions and this is the main subject of this paper.

II. Primitive recursive hierarchies

Denote by $\mathbb{N}$ the set of all natural numbers. We consider the following model.
All our functions will map $\mathbb{N}^k$ into $\mathbb{N}$, i.e. we consider *total* functions of k variables, k arbitrary. The *basis functions* include all constant functions, the successor function $n \mapsto n+1$ and k-ary projections $\pi_i(n_1, \ldots, n_k) = n_i$. From basis functions we shall generate functions by means of *substitution* (composition), which produces the function

$$f(g_1, \ldots, g_k)$$

provided that f is k-ary and $g_1, \ldots, g_k$ have the same arity ; and by means of *primitive recursion* (from functions f and g) :

$$h(0, x_2, \ldots, x_k) = f(x_2, \ldots, x_k)$$

$$h(n+1, x_2, \ldots, x_k) = g(n, h(n, x_2, \ldots, x_k), x_2, \ldots, x_k).$$

All functions which we obtain from basis functions by repeatedly applying substitution and primitive recursion are called ***primitive recursive functions.*** This definition captures our idea of building functions from simple ones by composition and induction.

All primitive recursive functions are total and all our functions mentioned in the Prologue are primitive recursive.

In particular, if $f(n)$ is primitive recursive (of one variable) the following function h is again primitive recursive:

$$h(0) = f(0)$$
$$h(n) = \underbrace{f \circ f \circ \cdots \circ f}_{n+1}(0) = f^{(n+1)}(0).$$

(This follows easily by primitive recursion applied to the functions f and $g(n,m) = f(m)$.)

In the following sense this construction - iterated composition - of the function h captures all growth rates of primitive recursive functions:

For all natural numbers define functions $F_0, F_1, \ldots, F_n, \ldots$ as follows:

$$F_0(x) = x + 1$$
$$F_{n+1}(x) = F_n^{(x+1)}(x) = \underbrace{F_n \circ \cdots \circ F_n}_{x+1}(x).$$

Thus

$$F_1(x) = 2x + 1$$
$$F_2(x) \geq 2^x$$
$$\vdots$$

This set of functions is called the *Grzegorczyk hierarchy* and has the following important property:

Theorem 1 [Grz] . *Let $f : \mathbb{N} \to \mathbb{N}$ be a primitive recursive function. Then there exist n and N such that*

$$f(x) \leq F_n(x)$$

for all $x \geq N$.

In this situation we also say that the functions F_n *eventually majorize* primitive recursive function f. Thus the functions from the Grzegorczyk hierarchy eventually majorize all primitive recursive functions.

Observe that all the functions F_n are strictly increasing and that, for every x and every $n < m$,

$$F_n(x) < F_m(x).$$

(For it suffices to prove $F_n(x) < F_{n+1}(x)$ which follows from definition.) Thus for $n < m$, F_m fails to be eventually majorized by F_n.

Now define the (diagonal) function $F(x)$ as follows:

$$F(n) = F_n(n).$$

Clearly the function $F(x)$ eventually dominates every function F_n ($F(x) \geq F_n(x)$ for every $x \geq n$) and thus F fails to be primitive recursive.

An example of a function with this property was constructed first by Ackermann half a century ago [A] . By now it is customary to call a function constructed in a similar way an *Ackermann-type function.*

Of course there are other hierarchies and Ackermann-type functions derived from them. For example [PH] uses functions

$$f_0(x) = x + 2$$
$$f_{n+1}(x) = (f_n)^{(x)}(2)$$

where again $g^{(x)}$ denotes the function g composed with itself x times. [HS] uses the following definition of functions $A_1, A_2, \ldots$:

$$A_1(n) = 2n$$
$$A_k(n) = A_{k-1}^{(n)}(1)$$
$$A(n) = A_n(n).$$

Thus $A_k(1) = 2$ for every k, $A_2(n) = 2^n$ and $A_3(n) = 2^{2^{\cdot^{\cdot^{\cdot^{2}}}}}\Big\}^n$. It follows that $A(1) = 2$, $A(2) = 4$, $A(3) = 16$ and $A(4)$ is the tower of 2's of height 65536. [Wi] defines functions $C_1, C_2, \ldots$ as follows:

$$C_1(n) = 1 \quad \text{for } n \geq 0$$
$$C_k(0) = 2 \quad \text{for } k \geq 2$$
$$C_k(m) = C_k(m-1) \cdot C_{k-1}(C_k(m-1)) \quad \text{for } k \geq 2, \text{ and } m \geq 1.$$

It follows that

$$C_2(m) = 2$$
$$C_k(1) = 2 \cdot C_{k-1}(2)$$
$$C_3(m) = 2^{m+1}.$$

A compact definition which is close to the original motivation is given by [T1],[T2] . Tarjan defines a 2-variable function $A(i,j)$ for $i,j \geq 0$:

$$A(i,0) = 0$$
$$A(0,j) = 2j \quad \text{for } j \geq 1$$
$$A(i,1) = A(i-1,2) \quad \text{for } i \geq 1$$
$$A(i,j) = A(i-1, A(i,j-1)) \quad \text{for } i \geq 1, j \geq 2.$$

$A(i,j)$ (and already $A(i,4)$) is a variant of the Ackermann function. The recent paper [Lo2] contains another such function:

$$L(1,n) = n$$
$$L(k,n) = 2L(k,n-1) + L(k-1, L(k,n-1)).$$

The function $L(k) = L(k,k)$ is again an Ackermann type.

All these various definitions share the basic property of the Grzegorczyk Hierarchy expressed by Theorem 1. The definitions are tailored to suit recursive procedures in applications some of which will be listed now. (The notation for the functions introduced in this section will be preserved.)

We use the standard notations for comparison of functions. Thus for functions $f, g : \mathbb{N} \to \mathbb{N}$ the symbol $f = O(g)$ means that there exists a constant c such that $f(n) \leq cg(n)$ for all sufficiently large n. If both $f = O(g)$ and $g = O(f)$ hold then we also say that f and g are of the same growth rate. This we denote by $f \sim g$. If $f(x) \sim x$ then f is said to be a function of *linear* growth or shortly a linear function.

III. Examples of Ackermann growth

The last two definitions of Ackermann-type functions given above involve double induction. This is a typical situation and in many instances the double induction leads to bounds on the Ackermann function.

First, let us mention two examples where the Ackermann upper bounds were recently avoided:

(a) The Van der Waerden Function

Denote by $VW(n)$ the minimal N such that for every partition $\{1, \ldots, N\}$ into two parts there exists an arithmetic progression with k terms in one of the parts.

Theorem 2 (Van der Waerden). *For every positive integer n the number $VW(n)$ exists.*

The original proof of Van der Waerden [vW] as well as its combinatorial axiomatization due to Hales-Jewett [HJ] proceed by double induction and do not produce primitive recursive upper bounds. These were the only constructive proofs. All the deep ergodic-theoretic generalizations of Theorem 2 by Furstenberg and others [F] are nonconstructive (and for some of their results no combinatorial proofs are known). These facts suggested the conjecture that the function $VW(n)$ has Ackermann growth. There was therefore great surprise when Shelah [Sh] found a new proof of the Van der Waerden theorem which yielded a primitive recursive upper bound. The Shelah upper bound is of order F_4 (i.e. an iterated tower function). The best known lower bound is of order F_2 (i.e. exponential). Graham offers \$1000 for an upper bound of order F_3.

(b) Complex graphs with large girth

P. Erdős in his seminal paper [E] on probabilistic methods in combinatorics proved the following:

Theorem 3 (Erdős). *For every k, l there exists a graph $G_{k,l}$ with the following properties:*
(i) the chromatic number $\chi(G_{k,l})$ of $G_{k,l}$ is $\geq k$
(ii) the girth $g(G_{k,l})$ of $G_{k,l}$ is $\geq l$.

This proof provides no construction for $G_{k,l}$ but guarantees the existence of $G_{k,l}$ with $O(k^l)$ vertices. No construction was known for several years (despite serious efforts) until Lovász [L] found a complicated construction using double induction which constructed $G_{k,l}$. However this gave no primitive recursive upper bound. The same was true for further constructive improvements [NR], [Kř]. Only recently an algebraic-type construction of graphs $G_{k,l}$ has been found by Margulis and, independently, by [LPS] . However, some problems still remain (such as the search for a primitive recursive construction for a generalization of Theorem 3 for hypergraphs).

(c) Epidemiography

We review some results of [FN] ,[FLN] and we state the simplest interesting case only.

Epidemiography $\mathcal{E}(n)$ is a game in which two players move alternately and each move consists in changing a list of positive integers. The game starts with a list containing just one entry, the positive integer n. In the i-th move, the player making the move selects a positive integer x from the list, deletes it from the list and in the case $x > 1$ replaces it by $i + 1$ numbers each equal to $x - 1$. The player who first faces the empty list (i.e. who first cannot move) is the loser, and his opponent is the winner.

Epidemiography $\mathcal{E}(n)$ is both easy and hard to analyse:

Theorem 4. *For every n the first player has a winning strategy for $\mathcal{E}(n)$.*

Theorem 5. *Let $e_{\max}(n)$ and $e_{\min}(n)$ denote respectively the maximal and minimal number of moves in $\mathcal{E}(n)$. Then*
(i) $e_{\max}(n) \geq F_m(n)$ *for all* $n \geq m + 2$;
(ii) $2^{2^{n-2}} \leq e_{\min}(n) \leq 2^{2^{n-1}}$ *for all* $n \geq 2$;
(iii) the first player can always force a win of $\mathcal{E}(n)$ in at most $F_3(2n)$ moves.

These results (proved in [FN] , [FLN]) imply that Epidemiography is a pathological game. It is easy to define and analyse, but even for modest values of n the winner will never live long enough to collect his / her prize. This complements a result of Rabin and Jones (see [J]) about the opposite pathology: there are games which end in a small number of steps (such as 6 steps) and for which it is very hard to find a winning strategy.

(d) The Set-Union Problem

This is an aspect of our subject related to Computer Science and it has rightly received much attention.

Let X be a set (universe) with n elements. We consider two types of instructions for on line manipulation of a family of disjoint subsets of X: $FIND(x)$ computes the name of the unique set in the family which contains an element x, and $UNION\ (A, B)$ combines sets A, B in the family into a new set $A \cup B$.

The Set-Union Problem consists in designing a data structure for an efficient execution of a sequence of instructions of this type. A variant of this problem has been studied by Galler, Fisher [GF] , Hopcroft, Ullman [HU] and Tarjan [T1] (see [T4]for details of the development). They introduced a data structure consisting of the following operations performed on a forest each of whose components is a rooted tree:

(i) If A, B are two components of the forest, the operation $UNION(A, B)$ adds an edge joining the roots of A and B, thus replacing the two components A and B by a single component whose root is taken by the root of A.

(ii) If X is the set of vertices of the forest and $x \in X$, the operation $FIND(x)$ consists in locating x and then following the path P_x from x to the root of T_x. In addition we collapse T_x by making all vertices of P_x the sons of the root: see Fig.1 .

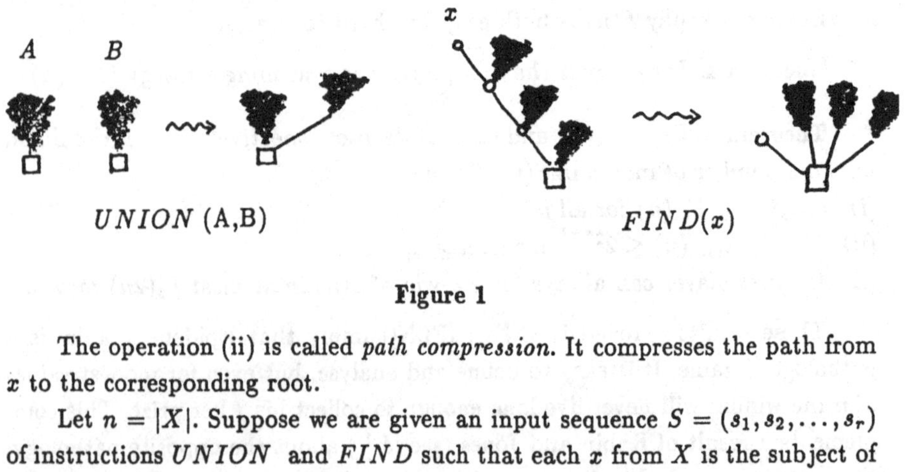

Figure 1

The operation (ii) is called *path compression*. It compresses the path from x to the corresponding root.

Let $n = |X|$. Suppose we are given an input sequence $S = (s_1, s_2, \ldots, s_r)$ of instructions $UNION$ and $FIND$ such that each x from X is the subject of exactly one operation $FIND(x)$ and S contains exactly $n-1$ $UNIONs$ (thus $r = 2n-1$). (These assumptions are made here for the sake of simplicity.) Denote by $T(S)$ the total time (= number of steps) required to execute this sequence. Let $T(n) = \max T(S)$, where the maximum is taken over all sequences S of $2n-1$ instructions the foregoing kind. It has been proved that $T(n) = O(n \cdot \log n)$.

The process of executing a sequence of instructions S of the above kind is known as the *set union algorithm.*

However, there is an important refinement. This involves assuming that the components of the forest have been assigned distinct "sizes" and replacing the operation $UNION$ by an operation called "*UNION by SIZE*", in which the root of the tree formed from A and B is taken to be the root of the larger of A and B, i.e. the root of the smaller of these trees becomes a son of the root of the larger. The resulting version of the set union algorithm (involving the two operations of Path Compression and *UNION by SIZE*) is extremely difficult to analyse. In his seminal paper [T1] Tarjan proved the following:

Let $T'(n) = \max T'(S)$ where $T'(S)$ is the total time required to execute the sequence S (with Path Compressions and *UNION by SIZE*). Then $T'(n) \sim O(n \cdot \alpha(n))$, where

$$\alpha(n) = \min\{i \geq 1 : A(i,4) > \log_2 n\}.$$

We say that $\alpha(n)$ is a *functional inverse* to the function $A(i,4)$.

As $A(i,4)$ is an Ackermann type function, $\alpha(n)$ is an extremely slowly growing function (e.g. $\alpha(n) = 1$ for $n < 2^{16}$). However α is an unbounded function and thus our SET - $UNION$ algorithm is nonlinear, i.e requires a time not bounded above by any linear function of n. (The existence of a linear algorithm is still an open problem.)

The main result of [T1] has been improved and modified in several other papers.

Also, manipulation of systems of disjoint sets is a key ingredient of many combinatorial algorithms and subsequently Tarjan's analysis was successfully used for other purposes. Let us just list the papers [AHU], [G1], [G2], [B], [W].

(e) Davenport-Schinzel sequences

Let n, s be positive integers. A sequence

$$S = (u_1, \ldots, u_m)$$

of integers is an (n,s) *Davenport-Schinzel sequence* (a $DS(n,s)$ sequence for short) if it satisfies the following conditions:

(i) $1 \leq u_i \leq n$ for each i;

(ii) For each $i < m$ we have $u_i \neq u_{i+1}$;

(iii) There do not exist $s+2$ indices $1 \leq i_1 < i_2 < \ldots < i_{s+2} \leq m$ such that $u_{i_1} = u_{i_3} = \ldots = a \neq b = u_{i_2} = u_{i_4} = \ldots$

The condition (iii) may be rephrased by saying that S does not contain a sequence of the form *ababab*... with $s+2$ terms as a (dispersed) subsequence (i.e. as a subword), see Fig.2 .

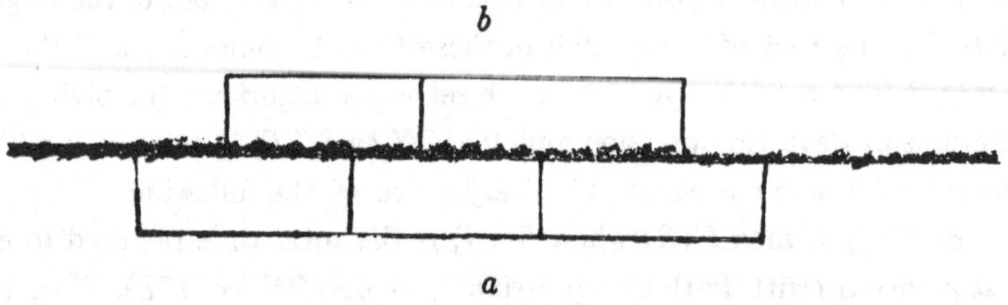

Figure 2

For example, a $DS(n,3)$ sequence is characterized by conditions (i), (ii) and by the absence of a subsequence of the form *ababa*, $a \neq b$.

Define $\lambda_s(n)$ as the maximal number of terms of a $DS(n,s)$ sequence. It is indeed easy to see that $\lambda_s(n)$ is defined. However, finding a good approximation to $\lambda_s(n)$ was a hard problem which was attacked by several people including Davenport and Schinzel who first propounded this problem [DS].

It is of course easy to prove $\lambda_1(n) = n$ and $\lambda_2(n) = 2n-1$. Further $\lambda_s(n) \geq s \cdot n - s + 1$ holds which is about all that is straightforward. In his early and deep paper [S] Szemeredi established for every s the upper bound

$$\lambda_s(n) = O(n \cdot \log^* n).$$

A nontrivial lower bound, and in particular an answer to the question whether $\lambda_s(n)$ has a linear upper bound, were not discovered until much later by Hart and Sharir [HS]who showed the following:

Theorem 6. *Let $A(n)$ be the Ackermann function defined in Section 1. Let $\alpha'(n)$ denote its functional inverse given by*

$$\alpha'(n) = \min\{k : \ A(k) \geq n\}.$$

Then there are positive constants c_1 and c_2 such that

$$c_1 \cdot n \cdot \alpha'(n) \leq \lambda_3(n) \leq c_2 \cdot n \cdot \alpha'(n).$$

General Remark: The definitions of fast growing hierarchies of functions are tailored to particular (double) induction procedures. Thus the definitions vary. However, owing to the spectacular growth of Ackermann-type functions their functional inverses are extremely slow growing and all these functions are of the same growth rate. In particular, the functions $\alpha(n)$ and $\alpha'(n)$ satisfy

$$\frac{1}{4}\alpha'(n) \leq \alpha(n) \leq 2\alpha'(n)$$

(see [HS]). A similar comment applies to other fast growing hierarchies of functions used in the literature.

Hart and Sharir's original proof was based on an analogy with Tarjan's result on path compressions (one has to consider generalized path compressions in a fixed order - postorder). Direct and simpler proofs of the non-linear growth of $\lambda_s(n)$ were subsequently found (see e.g. [Ko], [Wi]).

(f) Davenport-Schinzel Theory

A surprising number of variants of the original results concerning Davenport-Schinzel sequences have been found and the whole area seems to be forming a theory. Possibly this is due to the fact that the concept of Davenport-Schinzel sequence was discovered in a geometrical context (independently) by Davenport and Schinzel [DS] and by Atallah [At]. The setting is natural and easy to describe:

Let $f_1, \ldots, f_n$ be real continuous functions defined on the interval (say) $[0,1]$. Suppose that whenever $i \neq j$ the equation $f_i(x) = f_j(x)$ has at most 3 solutions (see Figure 3.). Let g be the pointwise minimum of the functions f_i, i.e. $g(x) = \min\{f_i(x) : i = 1, ..., n\}$. Now let m be the minimal number of intervals $I_1, ..., I_m$ which partition $[0,1]$ so that for each I_j there exists an index $i(j)$ such that $g(x) = f_{i(j)}(x)$ for all $x \in I_j$. It is easy to check that the sequence $i(1), i(2), ..., i(m)$ is a $DS(n,3)$ sequence.

In other words, the indices $i(1), ..., i(m)$ identify portions of the graphs of $f_1, \ldots, f_n$ which constitute the graph of g.

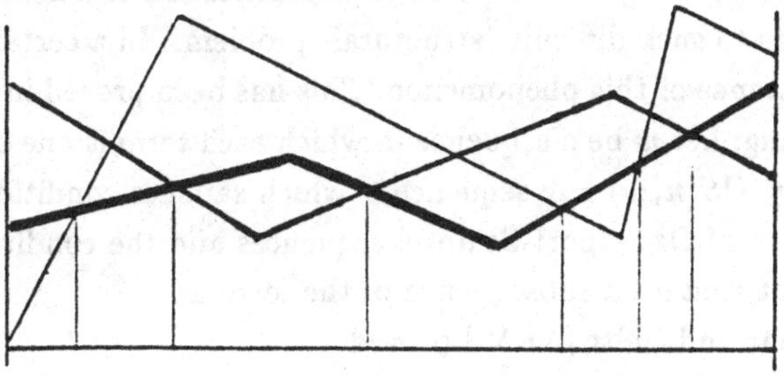

Figure 3

This (easily visualized) setting is relevant to several questions in computational geometry. We refer to [SCKLP] for a survey of these applications.

The original setting of Davenport-Schinzel sequences has been generalized in various directions. We list three of them:

A.

The idea of the lower envelope of functions has been refined in [Wi] and [WS] by Wiernik who constructed a family of n line segments such that the corresponding lower envelope has at least $n \cdot \alpha(n)$ linear pieces. This has been still further generalized in [EGHPPSSS] and [KLPS] to curves which intersect in at most 2 points (with linear upper bound) and to n Jordan arcs in the upper half plane which start and end on the x-axis such that any pair of them intersect in at most 3 points. In the latter case the total number of subarcs which appear on the boundary is of order $n \cdot \alpha(n)$.

B.

Davenport-Schinzel Theory has been generalized to matrices in [FH] . Denote by $f(n, C)$ the maximum number of 1's in an $n \times n$ matrix M which does not contain the configuration C (a configuration C is a partial matrix with 1's and blanks as its entries). It is proved in [FH] that for $C = \begin{pmatrix} 1 & & 1 & \\ & 1 & & 1 \end{pmatrix}$

and $C = \begin{pmatrix} 1 & & \\ & & 1 \\ & 1 & 1 \end{pmatrix}$ the quantity $f(n, C)$ is of order $n \cdot \alpha(n)$. However there are open problems even for configurations with four 1's and the characterization of configurations C with a linear upper bound on $f(n, C)$ is at present unknown.

C.

It is very surprising that the elementary definition of Davenport- Schinzel sequences leads to such difficult (structural) problems. In a certain sense, there is no other instance of this phenomenon. This has been proved in [AKV] in the following setting: Let w be a sequence in which each term is one of two symbols a, b. Denote by $DS(n, w)$ any sequence S which satisfies conditions (i) and (ii) in the definition of Davenport-Schinzel sequences and the condition
(iii′) S does not contain a subsequence of the form w.
Adamec, Klazar and Valtr [AKV] proved

Theorem 7. *The maximum number of terms of a $DS(n, w)$ sequence is bounded above by some linear function of n iff w does not contain ababa as a subsequence.*

The question of what replaces Theorem 7 when w contains three distinct letters is not yet completely solved.

(g) Superconcentrators and other special networks

An *n-network* is an acyclic directed graph with n distinguished vertices called *inputs* and n other distinguished vertices called *outputs*. The depth of an n-network is the maximal length of a directed path which it contains. An n-network is called an *n-superconcentrator* if, for any subset A of the inputs and any subset B of the outputs such that $|A| = |B|$, there exist $|A|$ vertex-disjoint paths joining the chosen inputs to the chosen outputs. Denote by $d(n)$ the minimal number of arcs of an n-superconcentrator and by $d_k(n)$ the minimal number of arcs of an n-superconcentrator of depth k. The quantity $d(n)$ is a subject of intensive study and it is known to be linear in n. In contrast, the quantity $d_k(n)$ is non-linear by the following result in [DDPW].

Theorem 8. *For $k \geq 2$ holds $d_{2k}(n) \sim n \cdot \lambda(k, n)$ where the function $\lambda(i, x)$ is the functional inverse to the function $A(i, n)$ defined in Section II. Explicitly: $\lambda(i, x) = \min\{j;\ A(i, j) \geq x\}$.*

It follows that the minimal possible depth of an n-superconcentrator with linear number of edges is asymptotically of order $\beta(n)$, where $\beta(n)$ is the functional inverse to the function $A(i, i)$ given by the formula $\beta(x) = \min\{i;\ A(i, i) \geq x\}$.

Recently, similar results were proved concerning boolean circuits of bounded depth which compute shifts: see [PS].

IV. Provably total Peano Hierarchies

Is it possible to construct functions which grow ("essentially") more rapidly than Ackermann's function ?

What role does Ackermann's function play in combinatorics ?

An answer to the first question is easier and will be given in this section. The answer to the second question is more complicated and some reasons for this may be inferred from sections III and V of the present article.

Of course the function A' defined by

$$A'(x) = \underbrace{A \circ \cdots \circ A}_{x}(x)$$

grows faster than any function A_i and also faster than the Ackermann function A. By repeating this process, we may define further functions A'', A''' and so forth. But soon we shall run into difficulties with notation and "run out of indices".

A natural generalization of Grzegorczyk 's and other Ackermann-type hierarchies was found by Schwichtenberg [Sch] and Wainer [Wa2]. They used a set-theoretical construction based on ordinal numbers. The aim of their study was to classify provability in Peano Arithmetic and we turn to this aspect later. First we develop the relevant part of the theory of ordinal numbers.

Denote by ω the set of all natural numbers; ω itself is the first infinite ordinal number. We shall need only ordinals from an initial segment of the class of all ordinal numbers. These numbers can be uniquely expressed in the form of exponential polynomials in ω. This form is called *Cantor Normal Form.* A typical example is

$$\omega^{\omega^{\omega}+1} + \omega^2 + \omega + \omega.$$

More precisely, these ordinal numbers and a natural linear ordering " $\leq$ " on them may be recursively constructed as follows: we start with all natural numbers and the symbol ω with the natural relation " $\leq$ " on natural numbers together with the relations $i < \omega$ for every natural number i. Further, if we have nonzero ordinal numbers $\gamma_1, ...\gamma_k$, $k \geq 1$, with the property $\gamma_k > \gamma_{k-1} > ... > \gamma_1$ then we may construct the ordinal numbers of the form

$$\omega^{\gamma_k} m_k + \omega^{\gamma_{k-1}} m_{k-1} + \ldots + \omega^{\gamma_1} m_1 + m_0,$$

where $m_0, m_1, \ldots, m_k$ are integers and $m_1, \ldots, m_k$ are positive and $m_0 \geq 0$. Moreover for two ordinal numbers α, β of the form

$$\alpha = \omega^{\gamma_k} m_k + \omega^{\gamma_{k-1}} m_{k-1} + \ldots + \omega^{\gamma_1} m_1 + m_0$$

$$\beta = \omega^{\delta_l} n_l + \omega^{\delta_{l-1}} n_{l-1} + \ldots + \omega^{\delta_1} n_1 + n_0$$

we put $\alpha < \beta$ if one of the following two cases holds:

(i) There exists an index i satisfying $(\gamma_{k-i}, m_{k-i}) \neq (\delta_{l-i}, n_{l-i})$ and the least such index i_0 satisfies either $\gamma_{k-i_0} < \delta_{l-i_0}$ or $\gamma_{k-i_0} = \delta_{l-i_0}$ and $m_{k-i_0} < n_{l-i_0}$.

(ii) $l \geq k$ and for each non-negative integer $i \leq k$, the pairs (γ_{k-i}, m_{k-i}) and (δ_{l-i}, n_{l-i}) are equal, and in addition either $l > k$ or $n_0 > m_0$.

Thus we consider these ordinal numbers as well defined symbols generalizing natural numbers and these symbols are ordered by a lexicographic-type ordering.

A more usual, set-theoretic, development of these ordinals is as follows: Let A be a linearly ordered set. A is called *well-ordered* if every nonempty subset of A has a first element, or equivalently every descending chain is finite. Two well-ordered sets which are isomorphic are said to have the same *order type*. An *ordinal number* is then defined to be the order type of any well-ordered set. This definition e.g. permits a convenient set-theoretical definition of arithmetical operations on the ordinal numbers. For instance, if α and β are the order types of the disjoint well-ordered sets A and B respectively, then $\alpha + \beta$ is defined to be the order type of the well-ordered set which consists of A followed by B.

In this way one can define ordinals $\alpha+1, \alpha+n, \alpha n, \alpha\omega, \omega^\alpha$ as order types of conveniently chosen sets. In this setting the set of all exponential polynomials in ω (in Cantor Normal Form) coincides with the set of all ordinal numbers less than

$$\epsilon_0 = \omega^{\omega^{\omega^{\cdot^{\cdot^{\cdot}}}}} .$$

For our purposes we need just the basics of ordinal arithmetic: if α is an ordinal number and m_1, m_2 positive integers, then

$$\alpha \cdot 0 = 0,\ \alpha^0 = 1 \ \text{ and } \alpha \cdot m_1 + \alpha \cdot m_2 = \alpha(m_1 + m_2).$$

An ordinal number is called a *successor* if it has the form $\beta + 1$ where β is another ordinal number. An ordinal number which is not a successor is called a *limit*.

The Hierarchy of fast growing Functions F_α, $\alpha < \epsilon_0$, is defined recursively. When α is a successor, F_α is defined using iteration, on the lines of the definition of the function F_{n+1} of the Grzegorczyk hierarchy. When α is a limit, we first define the so-called *fundamental sequence* of α, which is a sequence $(\alpha)_0 < (\alpha)_1 < (\alpha)_2 < \ldots$ of ordinals approaching α, and then define F_α by a process that might be described as "diagonalization over the fundamental sequence of α" and is reminiscent of the definition $F(n) = F_n(n)$ of the Ackermann-type function F in Section II. Fundamental sequences are defined as follows:

Given a limit $\alpha < \epsilon_0$ of the form

$$\alpha = \omega^{\gamma_k} m_k + \omega^{\gamma_{k-1}} m_{k-1} + \ldots + \omega^{\gamma_1} m_1$$

where $\gamma_k > \gamma_{k-1} > \ldots > \gamma_1 > 0$ and $m_1, \ldots, m_k$ are positive integers, the *fundamental sequence* $(\alpha)_0 < (\alpha)_1 < (\alpha)_2 < \ldots$ of α is recursively defined by

(i) If $\gamma_1 = \beta + 1$ then

$$(\alpha)_i = \omega^{\gamma_k} m_k + \omega^{\gamma_{k-1}} m_{k-1} + \ldots + \omega^{\gamma_2} m_2 + \omega^{\gamma_1}(m_1 - 1) + \omega^{\beta}(i+1).$$

(ii) If γ_1 is a limit and $(\gamma_1)_1 < (\gamma_1)_2 < \ldots$ is the fundamental sequence of γ_1 then

$$(\alpha)_i = \omega^{\gamma_k} m_k + \omega^{\gamma_{k-1}} m_{k-1} + \ldots + \omega^{\gamma_2} m_2 + \omega^{\gamma_1}(m_1 - 1) + \omega^{(\gamma_1)_i}.$$

Examples:

$(\omega)_i = i + 1$
$(\omega^\omega)_i = \omega^{i+1}$
$(\omega^\omega + \omega^{10})_i = \omega^\omega + \omega^9(i+1).$

With all this indexing machinery the Hierarchy of Fast Growing Functions is defined easily as follows:

$$F_0(n) = n + 1$$
$$F_{\alpha+1}(n) = F_\alpha^{(n+1)}(n)$$
$$F_\alpha(n) = F_{(\alpha)_n}(n) \text{ if } \alpha \text{ is a limit.}$$

This extends the Grzegorczyk hierarchy: we observe that $F(n+1) = F_\omega(n)$.

This hierarchy, like the Grzegorczyk hierarchy, has several variants. A closely related and sometimes more suitable one is the *Hardy Hierarchy* (see [Wal]) defined as follows.

$$H_0(n) = n$$
$$H_{\alpha+1}(n) = H_\alpha(n+1)$$
$$H_\alpha(n) = H_{(\alpha)_n}(n) \text{ if } \alpha \text{ is a limit.}$$

Thus $H_1(n) = n + 1$, $H_k(n) = n + k$ and $H_\omega(n) = 2n + 1$. Consequently $H_\omega(n) = F_1(n)$. This is one instance of a more general result:

Proposition 9. $H_{\omega^\alpha} = F_\alpha$ for every $\alpha < \epsilon_0$.

Proof. Following the proof in [BW] we show that $H_{\omega^\alpha} = F_\alpha$ in two steps. In both steps we use a simple transfinite induction argument.

Step 1. Let $\alpha = \omega^{\beta_k} m_k + \omega^{\beta_{k-1}} m_{k-1} + \ldots + \omega^{\beta_0} m_0$ and let $\gamma \leq \omega^{\beta_0}$. Then $H_{\alpha+\gamma} = H_\alpha \circ H_\gamma$.

This is shown by transfinite induction on γ as follows.

A. $H_{\alpha+1}(x) = H_\alpha(x+1) = H_\alpha \circ H_1(x)$.

B. Let $\gamma = \beta + 1$. Then

$$H_{\alpha+\beta+1}(x) = H_{\alpha+\beta}(x+1) = H_\alpha \circ H_\beta(x+1) = H_\alpha \circ H_{\beta+1}(x).$$

C. Let γ be a limit. Then

$$H_{\alpha+\gamma}(x) = H_{\alpha+(\gamma)_x}(x) = H_\alpha \circ H_{(\gamma)_x}(x) = H_\alpha \circ H_\gamma(x).$$

This finishes the proof of Step 1.

Step 2. $H_{\omega^\alpha} = F_\alpha$.

This is shown by transfinite induction on α as follows.

A. $(\alpha = 1) : H_\omega(x) = H_{x+1}(x) = 2x + 1 = \underbrace{F_0 \circ \ldots \circ F_0}_{x+1}(x) = F_1(x)$.

B. $(\alpha = \beta + 1)$:

$$H_{\omega^{\beta+1}}(x) = H_{\omega^\beta (x+1)}(x) = \underbrace{H_{\omega^\beta} \circ \ldots \circ H_{\omega^\beta}}_{x+1}(x) = \underbrace{F_\beta \circ \ldots \circ F_\beta}_{x+1}(x)_{x+1} = F_{\beta+1}(x).$$

C. (α is a limit):

$$H_{\omega^\alpha}(x) = H_{\omega^{(\alpha)_x}}(x) = F_{(\alpha)_x}(x) = F_\alpha(x).$$

This finishes the proof of Step 2 and also of the whole proposition.

□

Let us now return to our motivating question of this section: are there functions which grow essentially more rapidly than Ackermann's function? The answer to this question lies in the rôle played by the Hierarchy of Fast Growing Functions in mathematical logic (see [P1], [PH]):

Denote by $\mathbb{N}$ the set of all natural numbers together with addition + and multiplication · and constants 0 and 1. An *arithmetic term* is an expression built up from these 4 symbols and variables $x, y, z, \ldots$. An *arithmetic formula* is an expression built up from the atomic formulae $p = q$, $p < q$ where p, q are arithmetic terms by means of parentheses and the logical connectives $\wedge$ (and) , $\vee$ (or) , $\neg$ (not), $\rightarrow$ (implies), $\leftrightarrow$ (if and only if) and the quantifiers $\forall$, $\exists$. So for example

$$\forall n\ \exists z\ (n < z \wedge \forall x\ \forall y\ (x \cdot y = z \rightarrow (x = z \vee y = z)))$$

is a true arithmetic formula. An *arithmetic statement* is an arithmetic formula where all the variables are quantified (our example is an arithmetic statement). The following set of arithmetic statements is known as the Peano Axioms:

$\forall x\ (x + 0 = x)$

$\forall x\ \forall y\ (x + (y + 1) = (x + y) + 1)$

$\forall x\ (x \cdot 0 = 0)$

$\forall x\ \forall y\ (x(y + 1) = (xy) + x)$

$\forall x\ \forall y\ (x < y \leftrightarrow \exists z((x + z) + 1 = y))$

together with the "induction schema":

$\forall x\ (\varphi(x, 0) \wedge \forall y\ (\varphi(x, y) \rightarrow \varphi(x, y + 1)) \rightarrow \forall y(\varphi(x, y)))$, where $\varphi(x, y)$ is an arithmetic formula.

The theory determined by the Peano Axioms is called Peano Arithmetic (PA for short). For combinatorialists this is a very natural theory as it is known to be equivalent to the Theory of Finite Sets, i.e. to the environment in which Finite Combinatorics is usually treated.

When this theory was created, it was believed that the Peano axioms imply all arithmetic statements which are true for all natural numbers. This belief was destroyed in 1931 by K. Gödel who found a true arithmetic formula which cannot be derived from the set of Peano Axioms. The original examples of such formulae have been simplified, but not until 1977 were the first "mathematically interesting" ones found by Jeff Paris [P1]. This discovery (which we shall discuss more fully in the next section) inspired much related research. As we shall see a particularly useful (and illuminating) tool proved to be a result due to Schwichtenberg [Sch] and Wainer [Wa2] which extends Theorem 1 to the class of recursive functions (i.e. functions which are computable by an algorithm). More precisely we have:

Theorem 10 (Fast Growing Functions). *Let $f : \mathbb{N} \to \mathbb{N}$ be a recursive function given by an arithmetic statement*

$$\forall n\ \exists k\ (f(n) = k).$$

Then this statement is provable from the Peano Axioms if and only if there exists an ordinal $\alpha < \epsilon_0$ such that f is eventually majorized by F_α (i.e. $F_\alpha(i) > f(i)$ provided that i is large enough).

It follows from Proposition 9 that H_α can replace F_α in Theorem 10.

Below we shall use our theorem in the negative form: If a recursive function $f(x)$ has a spectactular growth rate(i.e. if it fails to be eventually majorized by a function F_α) then it cannot be given by an arithmetic statement provable from the Peano Axioms. (The latter we abreviate by saying that f cannot be *provably total.*) In other words if $\forall x \exists y (f(x) = y)$ is a true arithmetic statement then we thus have an example of a true statement which is unprovable from the Peano Axioms.

In this setting the above Theorem provides a combinatorial understanding of unprovability.

Examples will follow now.

V. Combinatorial Examples of Unprovable Statements

We list basically all the main examples of combinatorial statements which are true yet unprovable from the axioms of Peano Arithmetic or, equivalently, from the axioms of Theory of Finite Sets. We follow the chronology of the individual discoveries.

(a) The Iterated Ramsey Theorem

This is historically the first example. It involves both the finite and infinite Ramsey theorems which we recall in the following form:

Finite Ramsey Theorem 11. *For every choice of positive integers k, p, n there exists an integer $N = r(p, k, n)$ with the following property:*
For every partition $a_1 \cup \ldots \cup a_k$ of the set $\binom{N}{p}$ of all p-element subsets of the set $N = \{0, 1, ..., N-1\}$ there exist a class a_i and subset $Y \subseteq N$ such that $|Y| = n$ and $\binom{Y}{p} \subseteq a_i$.

The validity of this statement for particular values k, p, n, N is indicated by writing $N \to (n)^p_k$. Y is called a *homogeneous* set.

Infinite Ramsey Theorem 11′. *For every choice of positive integers p, k the following holds:*
For every partition $a_1 \cup \ldots \cup a_k$ of the set $\binom{\omega}{p}$ of all p−element subsets of $\omega = \{0, 1, \ldots\}$ there exists an i and an infinite subset $Y \subseteq \omega$ such that $\binom{Y}{p} \subseteq a_i$.

The validity of this statement is indicated by writing $\omega \to (\omega)^p_k$.

Paris [P2] introduced the following definitions: Let $X \subseteq \mathbb{N}$ be finite. Define any such set to be *0-dense*. X is said to be *1-dense* if $X \neq 0$ and $3+\min X < |X|$. Define X to be $(x+1)$*-dense* if for every partition $\binom{X}{3} = a_1 \cup a_2$ there exists $Y \subseteq X$ such that Y is x-dense and homogeneous (i.e. $\binom{Y}{3} \subseteq a_1$ or $\binom{Y}{3} \subseteq a_2$).

The following is then proved in [P2]:

Theorem 12. *The statement "For every x there exists y such that $\{0, 1, \ldots, y\}$ is x-dense" is true and unprovable from the Peano Axioms.*

The unprovability part of Theorem 12 was established by means of model theory methods (and no combinatorial proof of unprovability has hitherto been published). The fact that the statement is true is a consequence of the infinite Ramsey theorem. (We know by unprovability that it cannot be derived from the finite Ramsey theorem.) This we prove in the following slightly stronger form:

Proposition 13. *For every x and for every infinite sequence $y_1, y_2, \ldots, y_n, \ldots$ there exists n such that $\{y_1, \ldots, y_n\}$ is x-dense.*

Proof. To obtain a contradiction, suppose that x is such that the statement does not hold. Choose x minimal with this property and a corresponding sequence $y_1, y_2, \ldots, y_n, \ldots$. Clearly $x > 1$. This means that for every n there exists a *bad partition* of the set $\binom{Y_n}{3}$, $Y_n = \{y_0, y_1, \ldots, y_n\}$, which violates the definition of an x-dense set. As there are only finitely many partitions of a finite set and as a restriction of a bad partition is bad again, we can use these bad partitions to find a partition $\binom{\mathbb{N}}{3} = a_1 \cup a_2$ such that for every n the restriction of (a_1, a_2) to $\binom{Y_n}{3}$ is bad again. Applying the infinite Ramsey theorem we find an infinite subset $Z \subseteq Y$, $Z = \{z_1, z_2, \ldots\} \subseteq \{y_1, y_2, \ldots\}$ such that $\binom{Z}{3} \subseteq a_i$ for either $i = 1$ or $i = 2$. By minimality of x there exists n such that $Z_n = \{z_1, z_2, \ldots, z_n\}$ is $(x-1)$-dense. However this contradicts the fact that the restriction of the partition (a_1, a_2) to the set $\binom{\{y_1,\ldots,y_N=z_n\}}{3}$ was bad.

□

A similar statement of course holds for partitions $\binom{X}{p} = a_1 \cup a_2$ for every p (not only for $p = 3$). However for $p = 1$ the statement is easy to prove in the theory of finite sets by giving an easy upper bound. For $p = 2$ it is not known whether the statement is unprovable from the Peano Axioms.

(b) The Paris-Harrington Theorem

The following perhaps is the most popular example of a combinatorial unprovable result:

Theorem 14 (Paris and Harrington). *For every choice of positive integers k, p, n there exists an integer $N = r^*(p, k, n)$ with the following property: For every partition $a_1 \cup \ldots \cup a_k$ of the set $\binom{N}{p}$ of all p-element subsets of N there exists a class a_i and a subset $Y \subseteq N$ such that $|Y| \geq \min Y$, $|Y| \geq n$ and $\binom{Y}{p} \subseteq a_i$.*

By an argument like the proof of Proposition 13, we may show that this is a true statement. The proof again involves the infinite Ramsey theorem. Two proofs of the unprovability of the Theorem 14 from the Peano Axioms are given in [PH]. One of them is model theoretic (along the lines of the corresponding proof related to our first example (a)), and the other uses methods of mathematical logic. Actually the following combinatorial statement holds (see [P1], [PH]):

Theorem 15. *The existence for all k of numbers $r^*(k, k, k+1)$ and even $r^*(k, 3, k+1)$ is unprovable from the Peano Axioms.*

One can also see easily ([LN1]) that the existence for all k of numbers $r^*(k, 2, n)$ is unprovable from the Peano Axioms, but the status of numbers $r^*(k, 2, k+1)$ is at present unknown (compare this with the problem mentioned above in (a)).

It follows from methods of Paris [P2] that the function $r^*(k, k, k+1)$ fails to be bounded by a provably total recursive function. This gives rise to a paradoxical situation which defies a common combinatorial methodology. In a typical combinatorial situation one singles out a combinatorial property P and uses it to define a function $f : \mathbb{N} \to \mathbb{N}$ (such as the minimal number of edges of a graph with n vertices having P). We then want to estimate function f. What is tacitly assumed (and in most cases clear) is that f is bounded above by a function which can be produced by some elementary operations or even primitive recursion. However all functions thus produced are recursive and provably total. Thus, paradoxically, the problem "find an upper bound for $r^*(k,k,k+1)$" has no reasonable solution (at least in the sense in which the question is asked).

This does not indicate that one cannot study these bounds. For example Erdős and Mills [EM] give the following tight bounds for the case relevant to graphs:

$$2^{2^{c'n}} \le r^*(2,2,n) \le 2^{2^{cn}}$$

(for suitable constants c and c'). Also, as indicated in [PH] , for every fixed p the function $r^*(p,2,n)$ is bounded by a recursive provably total function.

This line of research was elaborated in an important paper [SoK] where Solovay and Ketonen gave a new proof of unprovability from the Peano Axioms of the existence of $r^*(k,k,k+1)$ by means of Wainer's Theorem. This gave a combinatorial proof of unprovability and in a sense indicated why a simple combinatorial statement may fail to be provable.

The paper [SoK] contains an interesting way of measuring finite sets by means of (possibly infinite) ordinals. This proceeds as follows:

For a finite set $X = \{x_1, \dots, x_n\}_<$ of natural numbers with $x_1 < \dots < x_n$ and an ordinal $\alpha < \epsilon_0$ put

$\{\alpha\}(\emptyset) = \alpha$

$\{0\}(X) = 0$

$\{\alpha+1\}(X) = \{\alpha\}(x_2, \dots, x_n)$

$\{\alpha\}(X) = \{(\alpha)_{x_1}\}(x_2, \dots, x_n)$ for limit α

(here (α_{x_1}) is the x_1-term of the fundamental sequence of α.

We say that the set X is *α-large* if $\{\alpha\}(X) = 0$. Clearly every set X with n elements is n-large. X is ω-large iff $\min X < |X|$.

The set $\{0,1,3,5,7,9,11,12,13,14,15,16,17\}$ is $\omega 2$-large.

Solovay and Ketonen related the minimal N such that the set $\{k, k+1, k+2, \dots, N\}$ is α-large to the function $H_\alpha(k)$ of the Hardy hierarchy and they proceed by showing how this is related to the function $r^*(p,k,n)$. The proof is rather technical and complicated.

A simple combinatorial proof of unprovability from the Peano axioms of the existence of $r^*(p,k,n)$ was recently given by the present authors [LN1] (along the lines in Section VI of this paper).

(c) Hercules versus Hydra

The battle between Hercules and Hydra yields perhaps the most beautiful unprovable result. It was introduced by Kirby and Paris in [KP]. The battle may be described as follows: At the beginning, Hydra is a rooted finite tree and it is modified by moves of Hercules and selfprotecting moves of Hydra to transform it into other rooted finite trees. A *head* of a rooted tree is a vertex which has no sons. The *predecessor* of a head v is the vertex of the unique path from v to the root, which has distance 2 from v. The *throat* of a head v is the subtree formed by the predecessor of v, father of v, and all sucessors of the father of v different from v.

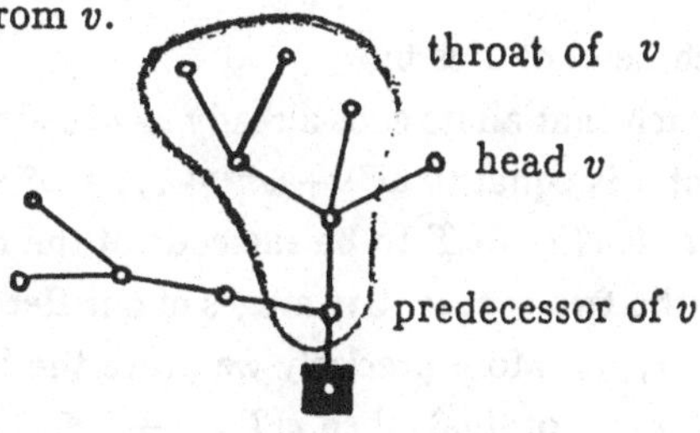

Fig. 4

Now the moves may be described as follows: In his n-th move (n is an integer) Hercules deletes a head of Hydra. In return, Hydra grows n new replicas of the throat of the head which was deleted. (If the deleted head was a son of the root, Hydra responds with an "empty" move which leaves the tree unchanged.) Hercules and Hydra alternate moves. Hercules wins if after a finite number of moves the only remaining vertex of Hydra is the root. Otherwise Hydra wins. An example of a battle is depicted in Fig. 5.

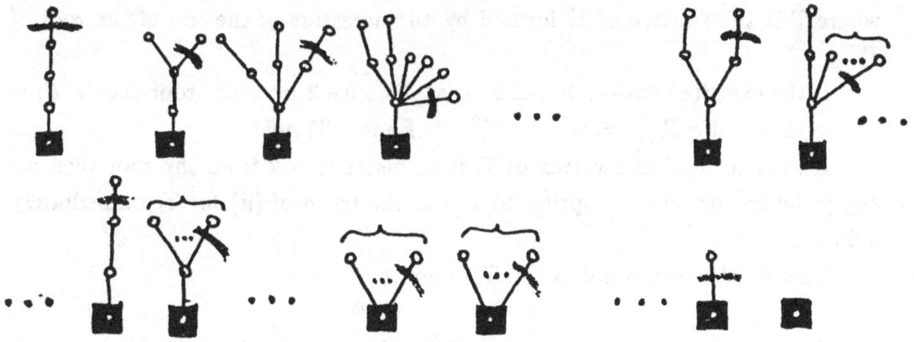

Fig. 5

Theorem 16 (Paris, Kirby). *Hercules always wins, no matter how he plays.*

Proof. Consider a particular Hercules-Hydra game which starts with a tree T and which leads to trees $T = T_1, T_2, T_3, \ldots$. We would like to prove that this sequence (or *trajectory of the game*) is finite. We use the *well ordering method*: we assign to each tree T_i an ordinal $c(T_i)$ (the *code* of T_i) in such a way that the codes form a strictly decreasing sequence $c(T_0) > c(T_1) > c(T_2) > \ldots$ It follows then that the sequence $T_0, T_1, \ldots$ has to be finite.

For a finite rooted tree T we define recursively the code of each of its vertices:

1. the code of each head of T is 0;
2. if v is a vertex such that all its sons already have codes $\alpha_1 \geq \alpha_2 \geq \ldots \geq \alpha_n$, then the code of v is equal to $\omega^{\alpha_1} + \omega^{\alpha_2} + \ldots + \omega^{\alpha_n}$.

We now define the *code* $c(T)$ of T to be the code of the root of T.

Let T_i and T_{i+1} be two consecutive stages of our Hercules-Hydra game. We show that $c(T_i) > c(T_{i+1})$. More precisely we prove the following:

(i) If $c(T_i)$ is a successor ordinal, then $c(T_{i+1}) + 1 \leq c(T_i)$.

(ii) If $c(T_i)$ is a limit ordinal, then $c(T_{i+1}) \leq (c(T_i))_i$, where the notation $(\ )_i$ relates to the fundamental sequence (see Section IV).

We prove (i) and (ii) by induction on $|T_i|$.

If $|T_i| = 2$, i.e. T_i is an edge, then $c(T_i) = 1$ and $c(T_{i+1}) = 0$, and hence (i) holds. In the induction step we distinguish 3 cases.

If the root of T_i has more than one son then we apply the induction assumption to the branch of T_i which contains the end-vertex executed by Hercules. (i) and (ii) then follow simply from the definition of the code of a tree.

Now suppose that the root of T_i has exactly one son. Then $c(T_i) = \omega^{c(\tilde{T})}$, where $\tilde{T}$ is the subtree of T_i formed by all successors of the son of the root of T_i.

If the executed end-vertex of T_i is at distance 2 from the root then $c(\tilde{T})$ is a successor and $c(T_{i+1}) = (i+1)\omega^{c(\tilde{T})-1}$. Hence (ii) holds.

If the executed end-vertex of T_i is at distance > 2 from the root then we apply the induction assumption to $\tilde{T}$ and the truth of (ii) for T_i immediately follows.

This finishes the proof of the Theorem.

□

The strong connection between the trajectory of a Hercules-Hydra game and the decreasing sequence of its ordinal codes is very useful in deriving both upper and lower bounds on the length of the game. This will be illustrated in section VI.

The Hercules-Hydra game is a prototype of a situation leading to an unprovable combinatorial result. It has recently become apparent that other such results may be obtained similarly.

Hercules-Hydra is a finite game where Hercules always wins. However it is a very long game. This has been shown first by Paris and Kirby [KP] who related this game to the concept of α-large sets and used [SoK]. They proved that the statement "every recursive strategy of Hercules is winning" is unprovable in the Theory of Finite Sets. This has been refined by Loebl [Lo1], by first specifying one head (called the *maxhead*) in each tree as follows:

1. If the tree has just one head, take it as maxhead.
2. Otherwise, let w be the nearest vertex to the root with more than one son. Let v be a son of w with minimal code. Then the maxhead of the tree is equal to the maxhead of its subtree formed by all successors of v.

We say that Hercules proceeds by the *strategy* MAX if he always deletes the maxhead. We remark that MAX can be visualized as follows: suppose that the trees appearing during a battle are put on the paper so that the new copies of a throat are the rightmost subtrees. Then MAX says: "always delete the rightmost head". Clearly MAX is a recursive strategy.

It has been proved by Loebl [Lo1] that the MAX strategy provides the longest battle. This also follows from the following refinement of the above proof:

Corollary 17. *Let $T = T_0, T_1, \ldots, T_n$ be the trajectory of a Hercules-Hydra game with MAX strategy. Then for every $i < n$ the following holds:*
(i′) If $c(T_i)$ is a successor ordinal then $c(T_{i+1}) + 1 = c(T_i)$.
(ii′) If $c(T_i)$ is a limit ordinal then $c(T_{i+1}) = c(T_i)_i$.

Proof. Simply check the above proof of Theorem 16.

□

From this it follows (and this will be proved in Section VI) that it is unprovable in Theory of Finite Sets that (the very simple) strategy MAX leads to a finite game.

(d) Goodstein Sequences

Let a be an integer and let n be a number such that $a < (n+1)^{\cdot^{\cdot^{\cdot (n+1)}}}\Big\}{n+1}$. Then the *complete base* $n+1$ *form of* a is the "standard" expression of a by means of the arithmetic operations $+$, $\cdot$, exponent, and numbers $\leq n+1$. It will be denoted by $CBF_{n+1}(a)$. *The Goodstein Sequence starting with* a is the sequence of numbers $a_0, a_1, \ldots,$ defined as follows: let n be the least number such that $a < (n+1)^{(n+1)^{\cdot^{\cdot^{\cdot (n+1)}}}}\Big\}{n+1}$. Put $a_0 = a-1$, and for $k \geq 1$, let a_{k+1} be obtained from $CBF_{n+k+1}(a_k)$ by replacing each $(n+k+1)$ by $(n+k+2)$ and then subtracting 1. The sequence terminates when it reaches 0. For example for $a = 266 < 3^{3^3}$ we have

$a = 3^5 + 3^2 + 3^2 + 3 + 2 = 3^{3+2} + 2 \cdot 3^2 + 3 + 2$
$a_0 = 3^{3+2} + 2 \cdot 3^2 + 3 + 1$
$a_1 = 4^{4+2} + 2 \cdot 4^2 + 4 + 1 - 1 = 4^{4+2} + 2 \cdot 4^2 + 4$
$a_2 = 5^{5+2} + 2 \cdot 5^2 + 5 - 1 = 5^{5+2} + 2 \cdot 5^2 + 4$
etc.

Such examples may (as in the case of the Hercules-Hydra game) give the impression that the sequence $a_0, a_1, a_2, \ldots$ never terminates. Nevertheless we have

Theorem 18 (Goodstein). *For every a there exists a non-negative integer r such that $a_r = 0$.*

Proof. Let n be the least number such that $a < (n+1)^{\cdot^{\cdot^{\cdot (n+1)}}}\Big\}{n+1}$. Consider $CBF_{n+1}(a)$ and replace each entry of $n+1$ by ω. We thus obtain an ordinal number α in the Cantor Normal Form which will be called the code $c(a)$ of a. One sees easily that the codes $c(a_0), c(a_1), \ldots$ form a strictly decreasing sequence of ordinals. This in turn yields the finiteness of the sequence a_i.

□

A relationship between Goodstein sequences and the Hercules-Hydra game (stressed already in the original paper [KP]) is manifested by the following construction, which defines by induction a tree T_α for each $\alpha < \epsilon_0$:

1. $T_k =$ 

2. $T_{\alpha+\beta} =$ 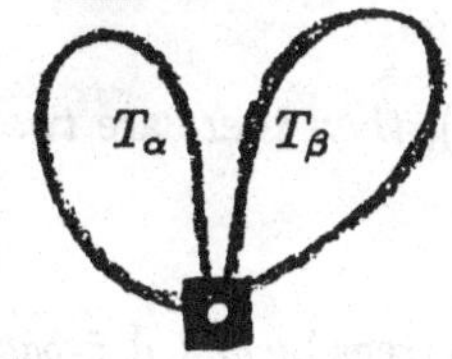

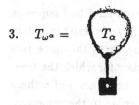

3. $T_{\omega^\alpha} =$ 

We remark that for a given finite rooted tree T there exists a unique ordinal $\alpha < \epsilon_0$ such that $T = T_\alpha$. This may be shown by the *coding procedure* which we introduced in the proof of Theorem 16. One can also check that the MAX strategy performed on a tree $T_{c(a_i)}$ gives $T_{c(a_{i+1})}$, which (since the finiteness of the game resulting from this strategy is unprovable in Finite Set Theory) implies the following result of Kirby and Paris [KP]:

Theorem 19 (Kirby, Paris). *The statement "Each Goodstein Sequence is finite" cannot be proved in Finite Set Theory.*

Cichon [C] gives another short proof of this result.

(e) Finite Miniaturizations of Kruskal's Theorem

Let T_1 and T_2 be trees. T_1 is called a *minor* of T_2 if T_1 may be obtained from a subtree of T_2 by contraction of some edges. The following is one of the fundamental combinatorial results [Kru]:

Theorem 20 (Kruskal). *Let $T_1, T_2, \ldots$ be an infinite sequence of finite trees. Then there exist two indices $i < j$ such that T_i is a minor of T_j.*

H. Friedman suggested in 1982 that it would be of interest to study finite miniaturizations of Kruskal's theorem based on the following observation:

Lemma 21. *Let $f : \mathbb{N} \to \mathbb{N}$ be a function. Then there exists a natural number $n(f)$ with the following property:*

If $T_1, \ldots, T_{n(f)}$ are finite trees, $|T_i| \le f(i)$, then there are two indices $i < j$ such that T_i is a minor of T_j.

Proof. Let a finite sequence $T_1, \ldots, T_n$ of finite trees be called *f-bad* if $|T_i| \le f(i)$ for $i = 1, \ldots, n$ and there is no pair i, j such that $1 \le i < j \le n$ and T_i is a minor of T_j. To obtain a contradiction, assume that $f : \mathbb{N} \to \mathbb{N}$ is a counterexample to Lemma 21. Then for every natural number n there exists an f-bad sequence $T_1^n, T_2^n, \ldots, T_n^n$ of finite trees. Since $|T_i^n| \le f(1)$ for all n, i, there exist infinitely many values of n for which the trees T_1^n are all isomorphic to the same tree T_1, and amongst these values of n there are infinitely many for which the trees T_2^n are all isomorphic to the same tree T_2, and amongst these values of n there are infinitely many for which the trees T_3^n are all isomorphic to the same tree T_3, and so forth. This yields an infinite sequence of trees $T_1, T_2, \ldots$ every initial segment of which is an f-bad sequence. It follows that there is no pair i, j such that $i < j$ and T_i is a minor of T_j; and this contradicts Theorem 20.

□

Let $S(f)$ denote the statement that there exists a natural number $n(f)$ with the property described in Lemma 21. Thus $S(f)$ is true for every f. Friedman proved the following:

Theorem 22. *Put $f_k(n) = k + n$. Then the statement $\forall k\ S(f_k)$ cannot be proved in Theory of Finite Sets.*

Loebl and Matoušek [LM2] strengthened this as follows:

Theorem 23. *Put $g_k(n) = k + 4\log n$.*
Then the statement $\forall k\ S(g_k)$ cannot be proved in Theory of Finite Sets.

Theorem 23 is nearly best possible because it is known that if $h_k(n) = k + \frac{1}{2}\log n$, then the statement $\forall k\ S(h_k)$ is provable in Theory of Finite Sets.

While Friedman's proof uses methods of mathematical logic, Loebl and Matoušek proceed combinatorially by constructing long bad sequences. We shall return to this in Section VI where we shall prove that for functions $\varphi_k(n) = k \cdot n!$ the statement $\forall\ k\ S(\varphi_k)$ is unprovable in the Theory of Finite Sets.

Finally let us remark that Friedman's theorem is independent of the axioms of a larger fragment of second order arithmetic than Peano Arithmetic (and this was the original motivation for his research). We do not discuss this line of research here.

(f) Approximations of functions

Now we state what may be the simplest known example of a combinatorial fact which constitutes a statement unprovable from the Peano Axioms. This discussion is based on unpublished work of Pudlák (see [HP]). For integers $x < y$ we put $[x,y] = \{x, x+1, \ldots, y\}, [x,y) = \{x, x+1, \ldots, y-1\}$ and $(x, \infty) = \{z : z > x\}$. We say that a function f is a *partial function on* X if the domain of f is a subset of X. For a set Y put $f(Y) = \{z : f(y) = z$ for some $y \in Y\}$. Now let $A = \{a_1, \ldots, a_n\}$ be a set of positive integers with $a_1 < a_2 < \ldots < a_n$ and let f be a partial function on $[0, a_n]$. We say that A is an *approximation of* f if $f([0, a_i)) \subseteq [0, a_{i+1}) \bigcup (a_n, \infty)$ for $i = 1, 2, \ldots, n$. We now define recursively the notion of a k-strong set X (Pudlák called these sets "k-dense"; but we use a different terminology to avoid confusion with the concept discussed above in (c)). Every finite set is 0-*strong.*A set X is 1-*strong* iff $|X| \geq 3$. For $k \geq 2$ a set X is said to be k-*strong* if for every partial function f on $[0, maxX]$ there is an approximation A of f such that $A \subseteq X$ and A is $(k-1)$-strong. Let $S(k,m)$ denote the statement "there exists $n \geq m$ such that $[m,n]$ is k-strong." Pudlák proved (see [HP]) the following:

Theorem 24. *The statement* $\forall k \ \forall m \ S(k,m)$ *is unprovable from the Peano Axioms.*

The proof uses the original model-theoretic method of Paris. No combinatorial proof has hitherto been published.

(g) Complexity analysis (in computer science)

In sections III (d) and (e) Ackermann-type functions proved relevant to two topics: Davenport-Schinzel sequences and the SET-UNION problem. These two (seemingly very different) topics are conceptually related in that they study behaviour of particular systems of path compressions on finite trees. In this section we relate questions of this type to unprovability.

Let us recall that in III (d) we introduced two operations to handle the data structure (forests) for the SET-UNION problem: PATH COMPRESSIONS and UNION by SIZE. Using these two operations, the maximal time $T'(n)$ required to execute a sequence of $n-1$ UNIONs and n FINDs is of order $n\alpha(n)$. We might expect that a further restriction on the input sequence S of UNIONs and FINDs will produce even better bounds.

An interesting special case in which this occurs is when FINDs are performed in a particular order known as a POSTORDER. A POSTORDER is the linear ordering of leaves induced by a drawing of a tree in the plane. (Equivalently, it is the linear order of leaves induced by orderings of sons of every vertex). The following is proved in [LN4]:

Theorem 25 (Linearity of POSTORDER). *Let S be an input sequence on a forest with n leaves where the FIND operations are listed in a POSTORDER. Then $t(S) \leq 5n$.*

In [LN2] ,[LN3] the authors introduced more complicated restrictions which are called LOCAL POSTORDER. The details are too complicated to be included here. Let us just state two main results which are relevant to our main theme:

Let S_n be an input sequence (of UNIONs and FINDs) subject to the restrictions forming LOCAL POSTORDER. Denote by $t(S_n)$ the total number of steps of the LOCAL POSTORDER procedure which realize S_n. Then we have [LN3]:

Theorem 26 (non-linearity). *For every k there is an n and there exists a LOCAL POSTORDER input sequence S_N on an universum of size n leaves such that*

$$t(S_N) \geq kn.$$

Theorem 27 (unprovability). *Let S_n be a LOCAL POSTORDER input sequence on an universum of size n. Then*

$$t(S_N) \leq kn\epsilon(n)$$

for a positive constant $k > 0$ (independent of n), where ϵ is the functional inverse to the function f_{ϵ_0}:

$$\epsilon(n) = min\{l : f_{\epsilon_0}(l) \geq n\}.$$

Theorem 27 could be rephrased by saying that the complexity analysis of an (almost linear) algorithm for LOCAL POSTORDER cannot be carried out in full in Theory of Finite Sets. This is not the only occurrence of this phenomenon in computer science. Other examples appear in [JY], [O], [KOR] and are mostly concerned with formal-language problems.

VI. Techniques - bounds

Estimating a given combinatorial quantity is perhaps the most typical kind of problem in combinatorics. Various techniques are available: direct counting, sieve formulas (and other tricks), probabilistic methods, algebraic methods etc.. See also [SV].

In dealing with sets of very large size we cannot hope to estimate the sizes by some explicit ("nice") formulas and a less direct approach is necessary.

We might say that this section concerns "how to show that a given set contains a finite number of elements". An implicit method, which leads to upper and lower bounds involving fast growing functions, may be described as follows: given a sequence $A = a_0, a_1, \ldots$, we would like to find out whether this sequence is finite. The answer is positive if we manage to find a mapping f from A to a well-ordered set $(B, \leq)$ with the property that for $i < j$, $f(a_i) > f(a_j)$ (for the definition of well-ordering see Section IV).

In Section V we have already seen two illustrations of this method (when proving finiteness of the Hercules-Hydra game and of Goodstein sequences). Let us first reinforce these two examples by a more careful analysis.

What bounds could be expected when we use coding by descending ordinal numbers $< \epsilon_0$? Let us consider the following special case of this problem: given an ordinal number $\alpha < \epsilon_0$ and a positive integer p, we define the operation $(\ )^+$ by

$(\alpha, p)^+ = (\alpha - 1, p + 1)$ if α is a successor ordinal,

$(\alpha, p)^+ = ((\alpha)_p, p + 1)$ if α is a limit ordinal,

(where $(\alpha)_p$ is the p-term of the fundamental sequence for α).

Now fix a pair (α, p) and consider the sequence

$$(\alpha, p), (\alpha, p)^+, ((\alpha, p)^+)^+, \ldots, ((\ldots(\alpha, p)^+ \ldots)^+)^+,$$

where we continue iterating the operation $(\ )^+$ until we reach an ordered pair whose first component is zero. Denote this sequence by $L_\alpha(p)$ and let $l_\alpha(p)$ be the number of its terms.

It is easy to see for every $\alpha < \epsilon_0$ that $p < q$ implies $l_\alpha(p) < l_\alpha(q)$. It is important that we can derive both upper and lower bounds for $l_\alpha(p)$. We start with a lower bound which will be repeatedly used:

Basic lemma 28 (lower bound). *If $\alpha < \epsilon_0$ and $p \in \mathbb{N}$ then*

$$l_\alpha(p) > H_\alpha(p) - p.$$

Proof. We proceed by transfinite induction on α.

Obviously, for every natural number n, $l_n(p)$ is the length of the sequence $(n,p),(n-1,p+1),\ldots,(0,p+n)$. Thus $l_n(p) = n+1 > H_n(p) - p = n$. In the induction step we have either

$$\begin{aligned} l_{\alpha+1}(p) &= 1 + l_\alpha(p+1) > 1 + H_\alpha(p+1) - p - 1 \\ &= H_{\alpha+1}(p) - p; \text{ or (in a limit case)} \end{aligned}$$

$$\begin{aligned} l_\alpha(p) &= 1 + l_{(\alpha)_p}(p+1) > 1 + H_{(\alpha)_p}(p+1) - p - 1 \\ &> H_{(\alpha)_p}(p) - p = H_\alpha(p) - p \end{aligned}$$

□

What this means is that the quantity $l_\alpha(p)$ fails to be a provably total function.

Corollary 29. *Let $T = T_0, T_1, \ldots, T_n$ be a trajectory of the Hercules-Hydra game under the MAX strategy. Put $c(T_p) = \alpha$. Then $n \geq H_\alpha(p)$.*

Proof. Since statements (i′) and (ii′) of Corollary 17 correspond exactly to the definition of $(\ )^+$, it follows that $n = l_\alpha(p) + p - 1$ by Corollary 17, and hence $n \geq H_\alpha(p)$ by Lemma 28.

□

A similar type of argument establishes the unprovability of the finiteness of Goodstein sequences (see [C]).

As another application let us sketch a proof of unprovability of a Friedman-type result. Define a function $\varphi_k : \mathbb{N} \to \mathbb{N}$ by $\varphi_k(n) = k \cdot n!$ If T_1 and T_2 are rooted trees, T_1 is called a ***rooted minor*** of T_2 if T_1 may be obtained from a rooted subtree of T_2 (i.e. a subtree with the same root as T_2) by contraction of some edges.

Corollary 30. *The statement $\forall k\ S(\varphi_k)$ cannot be proved in Finite Set Theory.*

Proof. If $\alpha < \beta < \epsilon_0$ are ordinals, and T_α, T_β are rooted trees with codes α and β (constructed as in Section V (c)), then T_β fails to be a rooted minor of T_α. This (a crucial observation of [LM2]) follows easily as both the operations of taking rooted subtrees and contraction decrease the code of a tree.

This means that the sequences $L_\alpha(p)$ define very long sequences of trees in which no tree is a minor of a later one. To see this, let $\alpha = \alpha_1, \alpha_2, \ldots$ be the ordinals appearing in the pairs of $L_\alpha(p)$. Let $\tilde{T}_{\alpha_i}$ be obtained from T_{α_i} by adding a path of length $|T_\alpha|$ below its root. Now let $i < j$ be two indices. As $\alpha_i > \alpha_j$, it follows that T_{α_i} is not a rooted minor of T_{α_j} and from this it follows easily that $\tilde{T}_{\alpha_i}$ is not a minor of $\tilde{T}_{\alpha_j}$. On the other hand it is not very hard to prove that $|\tilde{T}_{\alpha_i}| \leq |\tilde{T}_{\alpha_1}| \cdot i!$. Put $k = |\tilde{T}_\alpha|$. Applying Lemma 28 we get that for every p holds $n(\varphi_k) > l_\alpha(p) > H_\alpha(p) - p$. Thus the function $n(\varphi_k)$ fails to be bounded by a function from Hardy hierarchy. The statement now follows from Theorem 10 and Proposition 9.

□

In fact one can compute $l_\alpha(p)$ exactly by means of the following hierarchy:

Put $G_0(n) = 1$;

$G_{\alpha+1}(n) = G_\alpha(n+1) + 1$ for every $\alpha < \epsilon_0$;

$G_\alpha(n) = G_{(\alpha)_n}(n+1) + 1$ for every limit $\alpha < \epsilon_0$.

Basic lemma 31 (upper bound). *If $\alpha < \epsilon_0$ and $p \in \mathbb{N}$ then*

$$l_\alpha(p) = G_\alpha(p).$$

Proof. We proceed by transfinite induction on α. If $\alpha = 0$ then $l_\alpha(p) = 1 = G_\alpha(p)$. If $\alpha = \beta + 1$ is a successor ordinal then

$$l_\alpha(p) = 1 + l_\beta(p+1) = 1 + G_\beta(p+1) = G_\alpha(p).$$

If α is a limit ordinal then

$$l_\alpha(p) = 1 + l_{(\alpha)_p}(p+1) = 1 + G_{(\alpha)_p}(p+1) = G_\alpha(p).$$

□

Corollary 32. *Let T be a (rooted) tree. Then any Hercules versus Hydra game which starts at T has length at most $G_{c(T)}(1)$.*

Proof. This follows from statements (i) and (ii) in the proof of Theorem 16 and from Lemma 31.

□

Although the definitions of the Hardy Hierarchy H_α and the hierarchy G_α differ, the relations $H_\alpha(n) - n \leq G_\alpha(n), G_\alpha(m) \leq H_\alpha \circ H_\alpha(n)$ hold for $\alpha < \epsilon_0, n \geq 0, m \geq 2$ and consequently these hierarchies can play the same rôle in proofs of unprovability.

The method of "calibrating" finite sets by a well-quasi-ordering is useful not only for "unprovably" large sets. We finish this paper by giving a recent example [Lo2] related to the Ackermann hierarchy.

We will use a lexicographic ordering of nonnegative integer vectors of a fixed dimension k : $(x_1, \ldots, x_k) < (y_1, \ldots, y_k)$ iff $x_i < y_i$ for the first index i such that $x_i \neq y_i$. To obtain an upper bound on the number of elements of a given sequence coded by a descending chain of this lexicographic ordering, we define a modification of an Ackermann function with parameter f, where f is an integer function:

$$A_f(1,n) = n, \quad A_f(k,k) = 1, A_f(k,n) = 0 \text{ for } n < k,$$
$$A_f(k,n) = A_f(k,n-1) + A_f(k-1, f(A_f(k,n-1))).$$

(We define the function $A(k,n)$ for pairs of positive integers only.)

The functions $A_f(k,n)$ are often of Ackermann type. For example if $f(n) \geq n$ then $A_f(k,n) \geq A(k,n)$. As the function A_f is defined by a similar recursion to that used earlier to define the function A, for a primitive recursive function f it cannot grow too much faster than the function A. Without proof let us just state that if the function f satisfies $f(n) \leq 2^{2^{2^n}}$ for $k \leq n$ then for all $n \geq k$ holds $A_f(k,n) \leq A(k^3, n)$.

We have the following result which is perhaps of an independent interest.

Lemma 33. *Let $s_1 > s_2 > \ldots > s_n$ be $k-$dimensional positive integer vectors strictly decreasing in the lexicographic ordering. Let $f(i)$ denote the sum of all entries of all vectors $s_j, j \leq i$. Then*

$$n \leq A_f(k, f(1)).$$

Proof. It is easy to see that if the functions f, f' are monotone increasing and $f(k) \leq f'(k)$ for every k then also $A_f(k,n) \leq A_{f'}(k,n)$ for every k,n. It also follows from the definition that for a given k the function $A_f(k,n)$ is monotone increasing in n. We shall prove the Lemma by double induction on $(k, f(1))$. Obviously $f(1) \geq k$. If $f(1) = k$ then $n = 1 = A_f(k, f(1))$. If $k = 1$ then again $n \leq f(1) = A_f(k, f(1))$.

Let $k > 1$ and $f(1) > 0$. Let i_0 be the first index such that the first digit of s_{i_0} is equal to 0. Then from the induction assumption for $(k, f(1) - 1)$ it follows that $i_0 < A_f(k, f(1) - 1)$. Moreover all vectors $s_{i_0}, s_{i_0+1}, \ldots, s_n$ have the first digit equal to 0: hence they actually form $(k-1)$–dimensional vectors and thus from the induction assumption for $(k-1, f(i_0))$ it follows that $n - i_0 \leq A_f(k-1, f(i_0))$. Summarizing,

$$\begin{aligned} n &\leq A_f(k, f(1)-1) + A_f(k-1, f(i_0)) \leq \\ &\leq A_f(k, f(1)-1) + A_f(k-1, f(A_f(k, f(1)-1))) = \\ &= A_f(k, f(1)). \end{aligned}$$

□

The application we have in mind concerns the compressions of trees mentioned already in section III. Let us recall some basic definitions. All our trees will be rooted. *End-vertices* are vertices of degree 1 different from the root. Given a (rooted) tree, a *compression* is an operation which consists in following a path from an end-vertex of the tree in the direction of the root and in collapsing the tree by deleting the edges of the path and the end-vertex at which it started, and by making all other visited vertices sons of the end of the path. This is illustrated in Fig.6. Note that the path concerned may terminate before it reaches the root. A tree is called *balanced* if for every its vertex x all the branches at x (formed by all descendants of a son of x) are root-isomorphic.

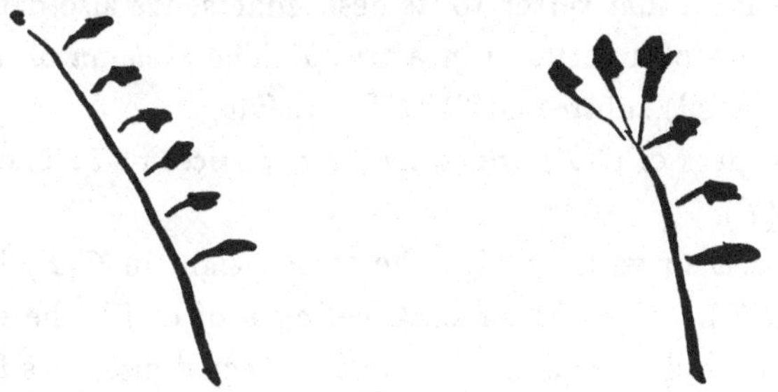

Fig.6

Usually we do not distinguish between a compression and the corresponding path.

A system $S = (C_1, C_2, \ldots, C_n)$ is called a *compression system* on a tree T if each C_i is a compression of the tree T_i obtained from T by performing compressions $C_1, \ldots, C_{i-1}$ $(T = T_1)$. Moreover we assume that from each end-vertex of T starts exactly one compression of S, and any compression C_i starts from an end-vertex of T. The *length* of S is $\sum_{i=1}^{n} |C_i|$, where $|C_i|$ is the number of edges of the path C_i.

Tarjan showed in [T1] that the maximum length of arbitrary compression systems on balanced trees is asymptotically equal to $n\alpha(n)$, n being the size of the tree.

Here we give a new proof of a non-linear lower bound. Moreover, we show that even compression systems constructed by a *greedy strategy* may have non-linear total length on balanced trees. A strategy which chooses a compression of a tree is called *greedy* if it chooses some compression which starts at an end-vertex at maximum distance from the root. A compression system $C_1, C_2, \ldots, C_n)$ is called *greedy* if each C_i in turn is chosen by a greedy strategy. In order to prove that there exists a system of compressions of non-linear length it suffices to prove the following:

Theorem 34. *Let k be a positive integer. Then there exists a binary balanced tree $T(k)$ and a greedy compression system $S(k)$ on $T(k)$ such that $|T(k)| \le A(k^3, 2^{k+1})$ and any compression from $S(k)$ has length k.*

Proof. We start with some definitions. Let T be a rooted tree with the length of each path from an end-vertex to the root not greater than k. The *height* of a vertex is its distance from the root. The *depth* of a vertex is the maximum of the distances from that vertex to its descendants. We also denote by $s(T, v)$ the number of sons of a vertex v in a tree T. The *skeleton* of a tree T is the (uniquely determined) subtree $C(T)$ of T satisfying:

(1) The end-vertices of $C(T)$ are (some) end-vertices of T. The root of T belongs to $C(T)$.

(2) Each pair v, w of vertices with the same height in $C(T)$ have the same depth in $C(T)$. Moreover for each vertex v of $C(T)$, the end-vertices of $C(T)$ which are descendants of v are all at equal distances from v.

(3) Let the vertices v, w have the same depth in $C(T)$, say equal to i. Then

$s(C(T),v) = s(C(T),w) = s_i$ and $s(T,v) = s(T,w) = S_i$, where s_i and S_i are integers depending only on i. Further we put $s_i = S_i = 2$ for all integers $i \leq k$ such that i exceeds the depth of the root of $C(T)$.

(4) The vector $W(C(T)) = (s_1, S_1, s_2, S_2, \ldots, s_k, S_k)$ is maximum in lexicographic ordering among all vectors $W(T')$ with T' satisfying (1)-(3).

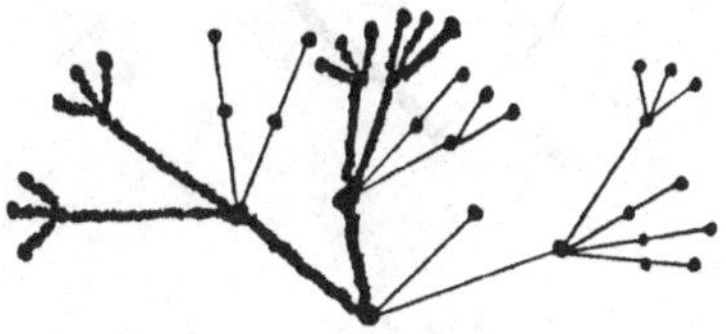

Fig.7: (tree with skeleton: $W(C(T)) = (3,3,2,4,2,4)$ if $k = 3$
$W(C(T)) = (3,3,2,4,2,4,2,2,2,2)$ if $k = 5$ (for example)

We remark that it follows from (1)-(4) that the skeleton is balanced and uniquely determined (see [LM1]). We prove the theorem by performing a construction of $S(k)$ and $T(k)$. We will construct a sequence of forests $F_0, F_1, \ldots$ with the properties:

1. Each component of F_0 is a binary balanced tree of height k,

Fig.8

2. The components of each F_i are isomorphic and each path from an end-vertex to the root of the corresponding component has length less than or equal to k,
3. F_{i+1} is obtained from F_i by performing two steps:

A. We make compressions of length k in F_i until all paths of length k disappear. Let the resulting forest after performing this operation be denoted by F'_{i+1}.

B. If all end-vertices of F_0 belong to compressions of $F_0, F_1, \ldots, F_i$ then the construction terminates. Otherwise we attach to the roots of F'_{i+1} a binary balanced forest of minimum height so that in the resulting components a path of length k appears. The resulting forest is F_{i+1}.

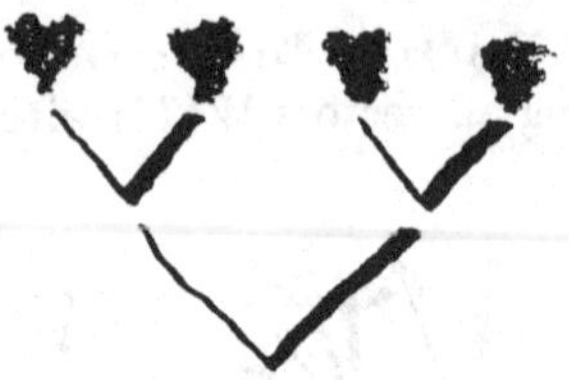

Fig.9

If this construction of $F_0, F_1, \ldots$ terminates, it constructs successively a binary balanced tree and a greedy compression system on it such that each compression has length k. Hence to prove Theorem 34 it suffices to show that the construction terminates in at most $A(k)$ steps.

Let T_i be a component of F_i. By statement B in our description of the construction of F_i from F_{i-1}, the skeleton $C(T_i)$ contains all sons of the root of T_i. From this it is not difficult to see that $W(C(T'_{i+1})) < W(C(T_i))$. Since $W(C(T_{i+1})) = W(C(T'_{i+1}))$ by the definition of W it follows that $W(C(T_{i+1})) < W(C(T_i))$.

At this moment we know that our greedy compression system has a non-linear length. In order to obtain an Ackermann type bound we have to estimate the function f in the Lemma 33. Our $f(i)$ is the sum of all entries of all vectors $W(C(T_j))$ for $j \leq i$. However the sum of entries of $W(C(T_j))$ is bounded by the sum of degrees of the trees $C(T_j)$ which in turn is bounded by 2^k times the sum of degrees of $C(T_{j-1})$. Thus $f(i) \leq 2^{2^{k+i}}$. By the above claim and Lemma 33 we have that the construction terminates in at most $A_f(k, f(1)) = A_f(k, 2^{k+1}) \leq A(k^3, 2^{k+2})$ steps.

□

Acknowledgements: We would like to thank D. Keedwell and an anonymous referee for many suggestions which improved the quality of this paper.

References

[A] W. Ackermann: Zum Hilbertschen Aufbau der reellen Zahlen, Math. Ann. 99 (1928), 118 - 133

[AKV] R. Adamec, M. Klazar, P. Valtr: Generalized Davenport-Schinzel Sequences with linear upper bound, In: Topological, algebraical and combinatorial structures (ed. J. Nešetřil), North Holland (to appear)

[ASS] P. Agarwal, M. Sharir, P. Shor: Tight bounds on the length of Davenport-Schinzel Sequences of order 4, (preprint)

[AHU] A. V. Aho, J. E. Hopcrof, J. D. Ullman: On finding lowest common ancestors in trees, SIAM j. Comp. 5, 1 (1976), 115 - 132

[AZ] M. Atallah: Dynamic computational geometry, Proc. 24th Symp. on Foundations of Computer Sci. 1983, 92 - 99

[B] R. E. Bixby: Recent algorithms for two versions of graph realization and remarks on applications two linear programming, Progress in Combinatorial Optimization / Academic press (1984), 39 - 66

[C] E. A. Cichon: A short proof of two recently discovered independence results using recursion theoretic methods, Proc. Amer. Math. Soc. 87, 4 (1983), 704 - 706

[DS] H. Davenport, A. Schinzel: A combinatorial problem connected with differential equations I and II, Amer. J. Math. 87 (1975), 683 - 694 and Acta Arithmetica 17 (1971), 363 - 372

[DDPW] D. Dolev, C. Dwork, N. Pippenger, A. Wigderson: Superconcentrators, Generalizers and Generalized Connectors with Limited Depth, STOC (1983), 42 - 51

[DL] R. A. De Millo, R. J. Lipton: The consistency of $P = NP$ and related problems with fragments of number theory, In: 12th Ann. ACM Symp. of Computing, Los Angeles (1980), 45 - 57

[E1] P. Erdős: Graphs and probability I., Canad. Math. J. (1958)

[EM] P. Erdős, G. Mills: Some bounds for the Ramsey - Paris - Harrington Numbers, J. Comb. Th. A 30 (1981), 53 - 70

[EGHPPSSS] H. Edelsbrunner, L. Guibas, J. Hershberger, J. Pach, R. Pollack, R. Seidel, M. Sharir, J. Snoeyink: On arrangements of Jordan arcs with tree intersections per pair, Discrete Comp. Geom. 4 (1989), 523 - 539

[FLN] A. S. Fraenkel, M. Loebl, J. Nešetřil: Epidemiography II., Games with a dozing yet winning player, J. Comb. Th. A 49 (1988), 129 - 144

[FN] A. S. Fraenkel, J. Nešetřil: Epidemiography, Pacific J. Math. 118 (1985), 369 - 381

[FH] Z. Füredi, P. Hajnal: Davenport-Schinzel Theory of matrices (DIMACS preprint April 1990)

[F] H. Furstenberg: Recurrence in ergodic theory and combinatorial number theory, Princeton Univ. Press, Princeton 1982

[G1] H. N. Gabow: Data structures for weighted matching and nearest common ancestors with linking (preprint July 1989)

[G2] H. N. Gabow: Two algorithms for generating weighted spanning trees in order, SIAM J. Comp. 6, 1 (1977), 139 - 150

[GT] H. N. Gabow, R. E. Tarjan: A linear time algorithm for a special case of disjoint set union, J. Computer System Sci. 30, 2 (1985), 209 - 221

[GF] B. A. Galler, M. J. Fisher: An improved equivalence algorithm, Comm. ACM 7 (1964), 301 - 303

[Go] K. Gödel: Über formal unentscheidbare Sätze der Principia Mathematica und verwandter Systeme I. Monatsh. Math. Phys. 38 (1931), 173 - 198

[Goo] R. L. Goodstein: On the restricted ordinal theorem, J. Symbol. Logic 9 (1944), 33 - 41

[GRS] R. Graham, B. Rothschild, J. Spencer: Ramsey Theory, Wiley 1979, 2nd edition 1990

[Grz] A. Grzegorczyk: Some classes of recursive functions, Rozprawy Matem. IV (1953), Warsaw

[HJ] A. W. Hales, R. I. Jewett: Regularity and positional games, Trans. Amer. math. Soc. 106 (1963), 222 - 229

[HS] S. Hart, M. Sharir: Nonlinearity of Davenport-Schinzel Sequences and of generalized path compression schemes, Combinatorica 6, 2 (1986), 151 - 177

[HU] J. E. Hopcroft, J. D. Ullman: Set merging algorithms, SIAM J. Comput. 2 (1973), 294 - 303

[HP] P. Hájek, J. Paris: Combinatorial principles concerning approximations of functions, Arch. Math. Logik 26 (1986/7), 13 - 28

[HPu] P. Hájek, P. Pudlák: Metamathematics of Peano Arithmetic, Springer Verlag (to appear)

[J] J. P. Jones: Some undecidable determined games, Int. J. of Game Theory, 11 ( ), 63 - 70

[JY] D. Joseph, P. Young: Independence results in computer science, Journal of Computer and System Sci. 23 (1981), 205 - 222

[KLPS] K. Kedem, R. Livne, J. Pach, M. Sharir: On the union of jordan regions and collision-free translation motion amidst polyhedral obstacles, Discrete Comput. Geom. 1 (1986), 59 - 71

[KP] L. Kirby, J. B. Paris: Accessible independence results for Peano Arithmetic, Bull. London Math. Soc. 14 (1982), 285 - 293

[Ko] P. Komjath: A simplified construction of nonlinear Davenport-Schinzel Sequences, J. Comb. Theory A 49 (1988), 262 - 267

[Kr] G. Kreisel: On the interpretation of non-finitistic proofs II, J. Symb. Logic 17 (1952), 43 - 58

[Kru] J. B. Kruskal: Well-quasi ordering, the tree theorem and Vaszonyi's conjecture, Trans. Amer. Math. Soc. 95 (1960), 210 - 225

[Kř] I. Kříž: A hypergraph-free construction of highly chromatic graphs without short cycles, Combinatorica 9, 2 (1989), 227 - 229

[KOR] S. Kurtz, M. J. O'Donnell, J. S. Royer: How to prove representation-independent independence results, Inf. Processing Letters 24 (1987), 5 -10

[Lo1] M. Loebl: Hercules and Hydra, Comment. Math. Univ. Carol 26, 2 (1985), 259 - 267

[Lo2] M. Loebl: Greedy compression systems (to appear in Lecture Notes in Computer Sci. - Proceedings of YMICS)

[Lo3] M. Loebl: Unprovable Combinatorial Statements. In: Topological, Algebraical and Combinatorial Strucutres (ed. J. Nešetřil), North Holland (to appear)

[L] L. Lovaśz: On chromatic Numbers of Graphs and Set Systems, Acta Math. Acad. Sci. Hungar. 17 (1966), 61 - 99

[LM] M. Loebl, J. Matoušek: Hercules versus Hidden Hydra Helper (to appear)

[LM2] M. Loebl, J. Matoušek: On the undecidability of the weakened Kruskal Theorem, In: Logic and Combinatorics (ed. S. Simpson), AMS - Contemporary Mathematics (1985), 275 - 280

[LN1] M. Loebl, J. Nešetřil: An unprovable Ramsey type theorem, (to appear in Proc. Amer. Math. Soc.)

[LN2] M. Loebl, J. Nešetřil: Linearity and unprovability of Set union problem strategies, Proc. ACM (STOC), 1988, 360 - 366

[LN3] M. Loebl, J. Nešetřil: Unprovability of set union problem strategies (preprint 1990)

[LN4] M. Loebl, J. Nešetřil: Linearity of On Line Postorder, (preprint 1989)

[LPS] A. Lubotzky, R. Phillips, P. Sarnak: Ramanujan Graphs, Combinatorica 8, 3 (1988), 261 - 277

[M] D. Miseroque: Le plus long combat d'Hercule (manuscript 1990, IIF - IMC Bruxelles)

[N] J. Nešetřil: Some non-standard Ramsey-like applications, Theor. Comp. Sci. 34 (1984), 3 - 15

[NT] J. Nešetřil, R. Thomas: Well-quasi ordering long games and a combinatorial study of undecidability, In: Logic and Combinatorics (S. Simpson, ed.), AMS (1987), 281 - 293

[NR] J. Nešetřil, V. Rödl: A short Proof of the Existence of Highly Chromatic Graphs without Short Cycles, J. Comb. Th. B 27 (1979), 225 - 227

[O] M. J. O'Donnell: A programming language theorem which is independent of Peano Arithmetic, 11th Ann. ACM Symp. on Theory of Computing, Atlanta (1979), 176 - 188

[P1] J. Paris: Combinatorial statements independent of arithmetic, In: Mathematics of ramsey Theory (ed. J. Nešetřil, V. Rödl), Springer (1990), 232 - 245

[P2] J. Paris: Some independence results for Peano Arithmetic, J. Symb. Logic 43 (1978), 725 - 731

[PH] J. Paris, L. Harrison: A mathematical incompleteness in Peano

Arithmetic, In: (J. Barwise, ed.) Handbook of Mathematical Logic (North Holland), 1132 - 1142

[PS] P. Pudlák, P. Savický: On Shifting Networks (preprint 1990)

[Sch] H. Schwichtenberg: Eine Klassifikation der ϵ_0–rekursiven Funktionen, Zeitschr. für Math. Logik 17 (1971), 61 - 74

[SCKLP] M. Sharir, R. Cole, K. Kedem, D. Leven, R. Pollack, S. Sifrony: Geometric applications of Davenport-Schinzel Sequences, In: Proc. SIGACT 1986, IEEE, 77 - 86

[Sh] S. Shelah: Primitive recursive bounds for van der Waerden numbers, Journal AMS, 1, 3 (1988), 683 - 697

[SoK] R. Solovay, J. Ketonen: Rapidly growing Ramsey functions, Ann. Math. 2, 113 (1981), 267 - 314

[SV] L. Spišiak, P. Vojtáš: Dependences between definitions of Finiteness, Czech. Math. J. 38 (1988), 389 - 397

[S] E. Szemerédi: On a problem by Davenport and Schinzel, Acta Arithmetica 15 (1974), 213 - 224

[T1] R. E. Tarjan: Efficiency of a good but not linear set union algorithm, J. ACM 22, 2 (1975), 215 - 225

[T2] R. E. Tarjan: Applications of path compression on balanced trees, J. ACM 26, 4 (1979), 690 - 715

[T3] R. E. Tarjan: Amortized computational complexity, SIAM J. Alg. Disc. Meth. 6, 2 (1985), 306 - 317

[T4] R. E. Tarjan: Data structures and network algorithms, SIAM, Philadelphia, 1983

[TL] R. E. Tarjan, J. van Leeuwen: Worst case analysis of set union algorithms, J. ACM 31,2 (1984), 245 - 281

[vW] B. L. Van der Waerden: Beweis einer Baudetschen Vermutung, Nienw. Arch. Wisk. 15 (1927), 212 - 216

[W] D. K. Wagner: An almost linear-time algorithm for graph realization (PhD Thesis), Northwestern University, 1983

[Wa1] S.S. Wainer: Ordinal recursion, and a refinement of the extended Grzegorczyk Hierarchy, J. Symb. Logic 37 (1972), 281 - 292

[Wa2] S.S. Wainer: A classification of the ordinal recursive functions, Archiv für math. Logik 13 (1970), 136 - 153

[Wi] A. Wiernick: Planar realization of non-linear Davenport-Schinzel Sequences by segments, Proc. FOCS-IEEE (1986), 97 - 106

[WS] A. Wiernick, M. Sharir: Planar realization of nonlinear Davenport Schinzel Sequences by segments, Discrete Comp. Geom. 3 (1988), 15 - 47

Orientations and Edge Functions on Graphs

OLIVER PRETZEL
Imperial College

The intention of this paper is to provide an introduction to an area of graph theory that has in the past been somewhat neglected, although it provides the natural setting for many important problems. As the title of the paper indicates, the area is the study of the orientations of a given graph. Thus our study bridges the gap between ordinary graph theory and the theory of directed graphs.

The problem that first aroused my interest in this subject was considered by Ore in his book (1962). It is to characterize those (unoriented) graphs that can be oriented as (Hasse-) *diagrams* of finite partially ordered sets.

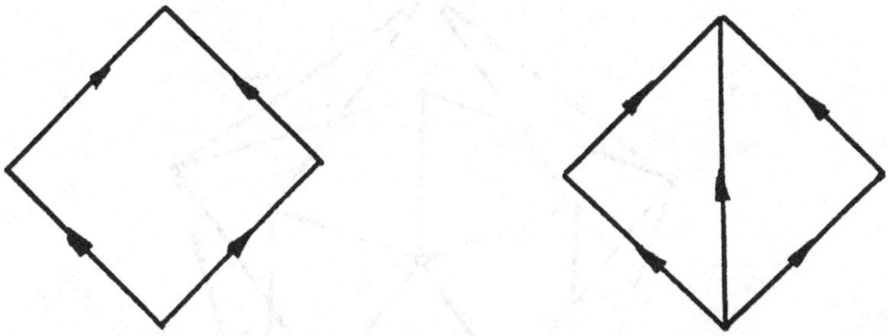

Figure 1

A diagram of a partially ordered set contains no *superfluous edges*. So the corresponding orientations can be described in the following way. Define a relation $<$ between the vertices of the graph G, by putting $u < v$ if there is a directed path from u to v. Then for the orientation to correspond to a diagram, the relation $<$ must be a strict partial ordering and if u and v are adjacent in G, there must be no w with $u < w < v$. Obviously any orientation of a triangle must produce a directed path of length 2, so it is impossible to orient a triangle in this way.

Ore's problem was to characterize those triangle-free graphs that cannot be oriented in this manner. Constructions of such graphs are the topic of section 3 of this paper. The smallest example is illustrated below. It is the the Grötzsch or Mycielski Graph M_5 (Grötzsch (1958), Mycielski (1955)) that was constructed as an example of a triangle-free 4-chromatic graph. Indeed, the whole subject of graph orientations is closely related to questions of graph-colouring.

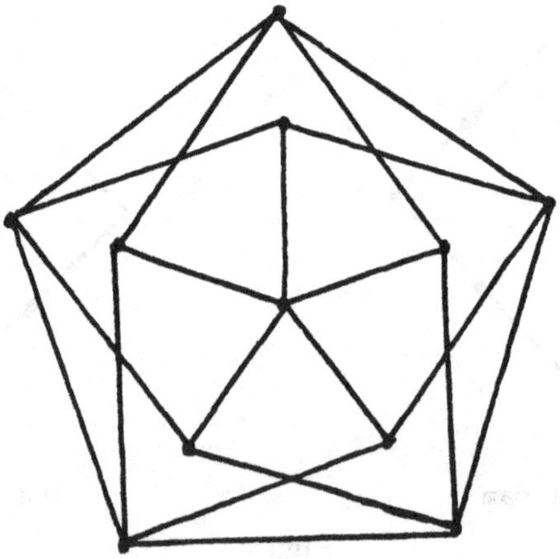

M_5

Figure 2

Our approach to the study of the orientations of a graph is to consider those properties of an orientation that remain invariant under certain transformations, or conversely to consider those transformations that preserve certain properties of orientations. It turns out that the most important tool for this study is the theory of chain groups which is described in some detail in Tutte's book (1984). Commonly, the theory is restricted to unoriented cycles, which in effect corresponds to taking $\mathbb{Z}/2$ as our coefficient field, though Tutte presents it in full generality. For our purposes it is essential to use the theory with coefficients in $\mathbb{Z}$ itself. As this version is not widely known, the paper begins with a short exposition of the theory, whose only non-standard features are the identification of orientations with certain chains, and the introduction of a *dot product* between chains. The easy proofs are, in the main, omitted and we refer the reader to Tutte's book for a fuller treatment.

The first operation that was studied for its effect on graph orientations was *pushing down* maximal vertices, which was introduced by Mosesian in the early 1970's. He observed that if you take a diagram of a partially ordered set, select a maximal vertex, and reverse all the edges incident with it, the result is a diagram of a new partially ordered set. He proved that with this operation it is possible to change a connected diagram to make any chosen vertex into the unique maximum, and using that fact he showed that the graph M_5 illustrated above cannot be oriented as a diagram (this proof is sketched in section 2).

It turns out that there is an invariant, the *flow-difference* (Pretzel (1986a)), such that two orientations have the same flow-difference if and only if one can be obtained from the other by Mosesian's pushing down operation. For each orientation of a graph, the flow-difference is a function defined on the cycles of the graph. In the language of chains, both orientations and cycles are 1-chains, and the flow-difference is just the dot product. Liu and Rival (1988) coin the term *inversion* for a re-orientation of a graph that preserves the flow-difference and discuss alternative characterizations of inversions. The obvious question that arises, is to characterize those functions on the cycles of a graph that can be flow-differences. This is achieved in Pretzel and Youngs (1990b). The characterization given there depends heavily on the length function on cycles, but requires no other information about the graph. In particular, it reduces the problem of determining whether a graph has a diagram orientation to the question of the existence of a linear function on cycles satisfying certain inequalities involving the length. It seems unlikely that a characterization requiring much less information will be possible, because of an important result of Nešetřil and Rödl (1987). This states that the decision-problem whether a given graph can be oriented as a diagram or not is *NP*-complete, and therefore sets bounds on the possibility of finding an "efficient" characterization of non-diagrams.

That leads one to ask how far the knowledge of the lengths of cycles (without any knowledge of their incidences with edges) determines the structure of the underlying graph. As this information determines all possible orientations of the graph, it must fix the structure to a large extent. This question was answered by Woodall (1991). He shows that the cycle lengths of a 3-connected graph G determine the circuit matroid of G up to isomorphism. It then follows from Whitney's Theorem ((1932); see also Aigner (1979 p352) and Welsh (1976)) that the information determines G itself up to isomorphism. In section 4 we describe a proof due to Youngs and Pretzel (1991). It is longer than Woodall's, but also answers another question about orientations. The idea of our proof is to reconstruct the edges of the graph by determining in terms of flow-differences precisely when two orientations differ by reversing a single edge.

Another approach to questions of orientability has also been actively pursued. That is to try and construct examples of graphs that cannot be oriented as diagrams, and to characterize the graphs yielded by the constructions. This approach shows very strong parallels with graph colouring theory. Indeed, the first condition that a non-diagram must satisfy is that its chromatic number must be at least equal to its girth. Thus non-diagrams of large girth must have large chromatic number. It is therefore natural to try to extend the techniques used to investigate the chromatic number of a graph to orientations. That has been done in three ways.

The most direct approach, initiated by Aigner and Prins in (1980), is to modify the chromatic number itself, by noticing that every graph colouring can be made to produce an orientation, merely by ordering the colours. If we restrict our colourings to those that produce diagrams the minimal number of colours required is denoted by χ_2. Aigner and Prins (1980) and Pretzel (1985) contain some results on this invariant.

This method has not proved very fruitful, but the idea of using probability theory in the manner of Erdös (1961,1962) has produced valuable results. The first that specifically deals with orientations is due to Nešetřil and Rödl (1978). They use a clever estimate of the number of colourings that produce forbidden subgraph orientations to show that there must exist non-diagrams of arbitrarily large girth. In a later paper Bollobás, Brightwell and Nešetřil (1986) produce a slightly simpler proof, closer to Erdös' original argument.

A parallel "semi-constructive" technique for graph colouring is characterized by the use of diagonal arguments and Ramsey Theory. This approach to colouring problems was inaugurated by the paper of Lovász (1968) which gave a constructive proof of the existence of graphs of large girth and chromatic number. The construction is somewhat simplified by Nešetřil and Rödl (1979). These papers construct gigantic graphs inductively and the induction step requires as its hypothesis a generalization of the property to be proved to hypergraphs. Křiž (1989) removes the necessity for using hypergraphs, and simplifies the arguments, but his graphs are still enormous. Furthermore, all the graphs constructed in these papers can be oriented as diagrams. In Pretzel (1987) it is shown how to extend a graph of large girth and chromatic number to a non-diagram of girth 6. The extension is quite specific, but the underlying graph needs such a high girth and chromatic number, that it is inconceivable that this method could produce a small example.

In an important paper (1988) Nešetřil and Rödl extend their "partite construction" techniques to solve the Ramsey problem for arbitrary graphs. The application of this solution to a bypass yields a proof of the existence of non-diagrams of arbitrary girth. Again the graphs are enormous, and the inductive proof requires the generalization to hypergraphs of the property to be proved. I have called these examples "semi-constructive" because the size of the graphs increases so rapidly that they cannot even be used to exhibit a specific graph of modest girth g that has chromatic number g or is not a diagram. These large constructions and the probabilistic method express the philosophy that a sufficiently large general graph must contain an obstruction to n-colouring, or a diagram orientation. At the opposite extreme there have been a number of constructions of specific small *edge-critical* non-diagrams. For such small graphs, the structure is not general, but specifically designed to be an obstruction.

Notably, the ideas of Youngs have led to a large number of families of non-diagrams. These are constructed by identifying certain vertices of *balanced graphs*. Balanced graphs were first considered by Bandelt, who stated, but never published, a theorem characterizing planar balanced graphs. One consequence of the characterization of flow-differences by Pretzel and Youngs is an easy proof of his theorem. In particular, all planar balanced graphs have girth 4. Youngs' constructions do not require the underlying graph to be fully balanced, so it may be possible to use his techniques to produce examples of slightly higher girth. Nevertheless, there remains a large gap between the theoretical existence statement and practical constructions of non-diagrams, and the question of the minimum size of a non-diagram of given girth $g > 4$ remains unanswered.

The paper is divided into the following sections.

1. *Chains, Cycles, Orientations* introduces the formal terminology of chain groups. The only new material in this section is the treatment of orientations as chains, and the description of the flow-difference function as a dot product.

2. *Pushing Down Maximal Vertices* introduces Mosesian's fundamental operation, states a theorem linking pushing down and the flow-difference, and the characterization of inversions due to Liu and Rival.

3. *Balanced Graphs and Non-Diagrams* introduces balanced graphs and explains how to use them to construct non-diagrams.

4. *Characterizing Flow Differences* states the theorem of Youngs and myself that characterizes flow-differences, deduces a theorem of Bandelt from it and discusses Woodall's Theorem, on the length function and the graph matroid.

1 CHAINS, CYCLES, ORIENTATIONS

Throughout this paper we are given a fixed connected finite graph G with vertex set V and edge set E. Loops and multiple edges are permitted. For book-keeping purposes we choose an orientation of the edges. This orientation is needed in order to distinguish the two possible ways an edge can be directed, but it disappears from all calculations in the theory and so its choice has no effect. In particular, it need not have any nice properties. The reference orientation allows us to define the head and tail of each edge e, which we shall denote by $h(e)$ and $t(e)$.

Elements of the free groups (or $\mathbb{Z}$-modules) $\langle E\rangle$ and $\langle V\rangle$ over E and V are called 1-*chains* and 0-*chains* respectively. As we shall almost exclusively be concerned with 1-chains, we adopt the convention that unless otherwise stated a *chain* is a 1-chain.

Let C be a chain[1] $C = \sum a_e e$. The *length* $|C|$ of C is $|C| = \sum |a_e|$, its *support* $Sp(C)$ is the set of edges for which $a_e \neq 0$, and C is *primitive* if $|a_e| \le 1$ for all edges e.

There is a natural homomorphism δ from $\langle E\rangle$ to $\langle V\rangle$ defined by $\delta(e) = h(e) - t(e)$. Of course $\delta(-e) = t(e) - h(e)$. Conversely, there is a natural homomorphism ∂ from $\langle V\rangle$ to $\langle E\rangle$ defined in the following way. Let D be a 0-chain, $D = \sum a_v v$. Then its image, $\partial(D)$ is the 1-chain $C = \sum (a_{h(e)} - a_{t(e)})e$.

Subgroups of $\langle E\rangle$ are called *chain groups*. The most important chain group is the *circuit group*, which is defined as the kernel of δ. Geometrically this group can be viewed as being generated by walks around cycles of the graph. We shall also need an algebraic definition of a *walk*. It is a 1-chain W such that $\delta(W) = v - u$ for two (possibly equal) vertices u and v. We then say that W is a walk from u to v, and extend the terminology for edges to walks: $u = t(W)$ is the tail of W, and $v = h(W)$ is its head. The sum of two walks is a walk if the head of one is the tail of the second.

A further important chain group is the set of *cuts*, which is defined as the image of ∂. The cut corresponding to a single vertex v can be viewed as the set of edges incident with v all directed towards v. If $C = \sum a_e e$ and $D = \sum b_e e$ are chains then their *scalar product* $C{\cdot}D$ is defined by

$$C{\cdot}D = \sum a_e b_e.$$

The product is independent of the reference orientation. For if the direction of an edge e changes, then both a_e and b_e change signs. An analogous product is defined for 0-chains.

---

[1] In the summations that follow, we adopt the convention that e and f run through E, while u, v, and w run through V. The coefficients a_e etc are integers.

Lemma

If $C = \sum a_e e$ *is a 1-chain and* $D = \sum b_v v$ *is a 0-chain, then* $\delta C \cdot D = C \cdot \partial D$.

Proof

$$\delta C = \sum a_e h(e) - a_e t(e).$$

So $$\delta C \cdot D = \sum a_e b_{h(e)} - a_e b_{t(e)}.$$

$$\partial D = \sum (b_{h(e)} - b_{t(e)}) e.$$

So $$C \cdot \partial D = \sum a_e b_{h(e)} - a_e b_{t(e)}.$$ □

Corollary

The circuit and cut groups of G are the orthogonal complements of one another. □

This corollary reflects a duality between circuits and cuts that has been studied in some detail by Youngs (1990), who proves the duals of some of the theorems presented in this paper. In our presentation cuts play a subordinate rôle.

We can retrieve the usual unoriented theory of cycles, paths and cuts by reducing everything modulo 2. The chain groups become vector spaces of dimensions $|E|$ and $|V|$ respectively, and the circuit group becomes the cycle-space. Given any connected cycle C, there are two natural circuits K mapping onto C. These are obtained by walking round C one way or the other. Now, as a subgroup of a free abelian group, the circuit group is itself free and its rank is the same as the dimension of the cycle-space, the *cyclomatic number*. Most natural cycle bases can be converted into circuit bases just by choosing for each cycle one of the two circuits that map onto it, but it is possible to construct cycle bases for which that is false.

An *orientation* R of G assigns a direction to each edge of G. We can associate a 1-chain (which we shall also denote by R) with the orientation, by assigning the coefficient 1 to those edges e for which R agrees with the reference orientation and -1 to the rest. Clearly this 1-chain depends on the reference orientation as well as R. The 1-chains that are induced by or represent orientations are precisely those that are primitive and have support *E*.

In this paper we shall be interested in those properties of *R* that can be read off from the 1-chain associated with *R* and are independent of the reference orientation. The most important of these is obtained from the dot-product. Given that *R* is an orientation, what information is given by the product $R \cdot C$ for a chain C? As the coefficients of *R* are just ±1 the sum $R \cdot C$ can be written as a difference

$$\sum_{e \in C_R^+} |a_e| - \sum_{e \in C_R^-} |a_e|,$$

where C_R^+ and C_R^- are respectively the edges where *C* and *R* *conform* or not. We also call these the *R-forward* and *R-backward* edges of *C*. For this reason the function $R \cdot C$ is called the *flow-difference* of *C* with respect to *R*. The condition that *R* is a diagram of a partially ordered set can be neatly phrased in terms of flow-differences of circuits *C*. For if *R* violates the condition then one of two types of primitive circuit must occur, which we call *monotone* circuits or *bypasses*. A monotone circuit has all its edges oriented coherently by *R*, and a bypass has all of them except one oriented in this fashion.

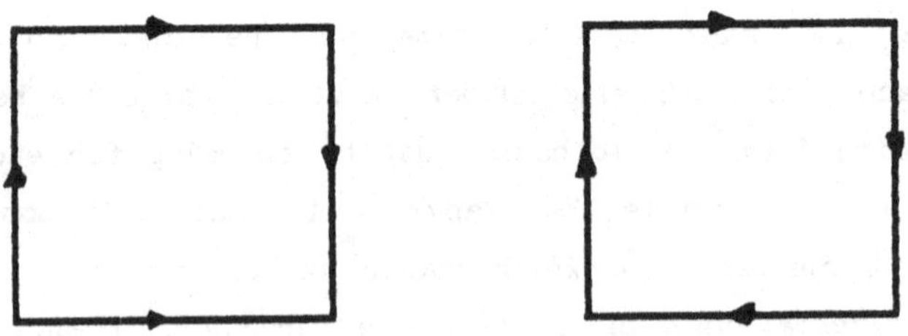

A bypass A monotone circuit

Figure 3

If an orientation contains a monotone circuit, then it does not represent a partial order at all. If it contains a bypass with exceptional edge e, then neither of the endpoints of e covers the other, so that in the diagram of the ordered set e would be removed. It can be shown (Mosesian (1972d), Pretzel (1989)) that bypasses are the essential obstacles to diagram orientation. That is, if a graph has an orientation without bypasses, then it has a diagram orientation. In terms of the dot product monotone circuits are those for which

$$|R \cdot C| = |C|,$$

and bypasses are those for which

$$|R \cdot C| = |C| - 2 .$$

That suggests the obvious definition, due to Aigner and Prins (1980). An orientation R is *k-good* if, for all circuits C,

$$|R \cdot C| \leq |C| - 2k .$$

Thus every orientation is 0-good, an orientation is 1-good or *acyclic* if it has no monotone circuits, and 2-good if it is a diagram orientation. We can also define the *orientation number* $\phi(G)$ as the maximum k for which G has a k-good orientation.

2 PUSHING DOWN MAXIMAL VERTICES

As already mentioned in the introduction, the *reorientation* operation that was first studied was Mosesian's operation of *pushing down* maximal vertices. A vertex v is *maximal* with respect to R, if, for any edge e for incident with v,

$$v = t(e) \Rightarrow R \cdot e = -1 \text{ and } v = h(v) \Rightarrow R \cdot e = 1.$$

In other words, if we take R as our reference orientation v is the head of all edges incident with it (and does not lie on any loop). Mosesian proved the following easy theorem.

Theorem (Mosesian).

Given a vertex v and an acyclic orientation R of G, it is possible to convert R into an orientation S in which v is the unique maximal vertex, by successively pushing down maximal vertices. □

He also noted that pushing down does not destroy the property of being a diagram and that enabled him to prove that M_5 cannot be oriented as a diagram. For, if it could, it would be possible to modify the orientation using pushing down so that the central vertex is the unique maximum. As the resulting orientation would still be a diagram orientation, all the edges of the graph except those on the outer 5-cycle must be directed towards the central point. Furthermore, two adjacent edges of the outer 5-cycle must be directed coherently. In Figure 2 you can see that these edges form part of a dart-shaped four-cycle, which must necessarily be a bypass, contradicting the assumption that the original orientation is a diagram.

Here is the main theorem of Pretzel (1986a).

Theorem.

Let R and S be two acyclic orientations of G. Then S can be obtained from R by pushing down maximal vertices if and only if for all circuits C of G, $R \cdot C = S \cdot C$.

Sketch of proof.

The proof contains three elements. The first of these is Mosesian's Theorem. The second is the fact that if S can be obtained from R by pushing down, then R can be obtained from S in the same manner. Indeed the effect of pushing down a maximal vertex can be undone by pushing down all the other vertices once each, as they become maximal. The final ingredient of the proof is the fact that for an orientation R with a unique maximum v, the flow-differences $R \cdot C$ for circuits determine R. □

In the same paper I also extended the result to arbitrary orientations, by allowing parts of G to be contracted before pushing down and re-expanded afterwards. In fact one can often eliminate pushing down from proofs and argue directly from flow-differences. For instance, in M_5, there are five cyclically arranged 5-cycles consisting of two radial arms and a single edge of the outer 5-cycle. If we orient these coherently as circuits, then the sum of any adjacent pair is a 6-circuit, that can be decomposed as the sum of two 4-circuits (see the diagram below). In a diagram orientation the flow-difference of a 4-circuit is 0. It follows that in a diagram orientation of M_5 any two adjacent 5-cycles must have opposite flow-differences, but since they are arranged cyclically that is impossible.

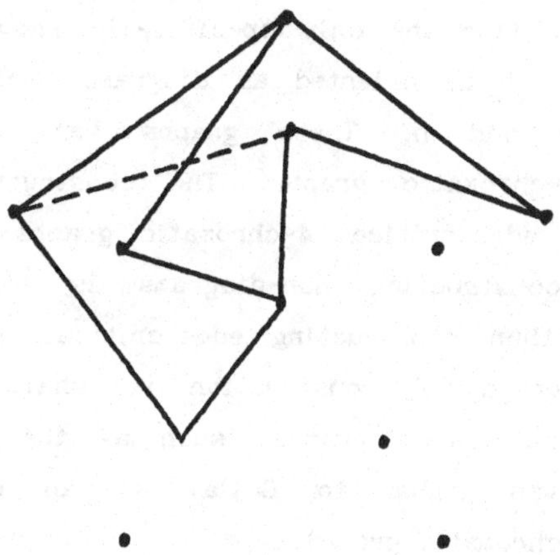

Figure 4

Liu and Rival (1988) introduce the term *inversion* for an orientation S that has the same flow-difference function as a given orientation R and give an alternative characterization of the inversions of an R. That involves the operation of reversing *monotone cuts*. Just as with a circuit, a cut C is monotone if $|R \cdot C| = |C|$. Now obviously the set of edges incident with a maximal vertex is a monotone cut. So pushing down is a special case of reversing a monotone cut. However, Rival and Liu prove a little more.

Theorem (Rival and Liu)

Let R and S be acyclic orientations of G. Then S is an inversion of R if and only if there exists a set of disjoint R-monotone cuts, such that S is obtained from R by reversing just the edges of these cuts. □

3 BALANCED GRAPHS AND NON-DIAGRAMS

For a long time the only specifically known triangle-free graphs that cannot be oriented as diagrams were the Mycielski Graphs M_n for odd n. These graphs were constructed as edge-critical 4-chromatic graphs. The constructions presented here also yield edge-critical 4-chromatic graphs and indeed, it appears that constructing non-diagrams is a slightly more stringent task than constructing edge-critical highly chromatic graphs. The method of construction we shall describe here produces many more such graphs, such as the one illustrated below (which was known to Gallai as an example of an edge-critical 4-chromatic graph).

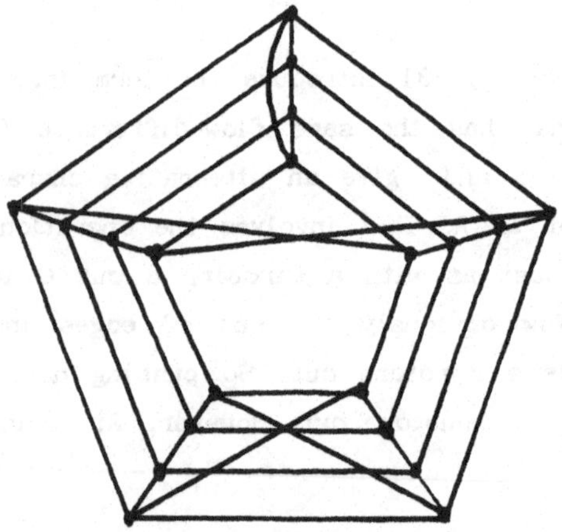

Figure 5

The idea is to begin with a graph that is such that the diagram-orientations of its circuits are very restricted. The most stringent restriction is to require that all circuits should have flow-difference 0. We call such graphs *balanced*, because any traversal that begins and ends at the same vertex must encounter the same number of forward- and backward-oriented edges. It is easy to see that a graph with a circuit basis of 4-circuits must be balanced and Bandelt stated in an unpublished theorem that for planar graphs the converse is also true. A proof of this theorem will be sketched later. I should perhaps mention that the converse is not true for non-planar graphs and examples of balanced graphs without a basis of 4-circuits are given in Youngs' thesis (1990).

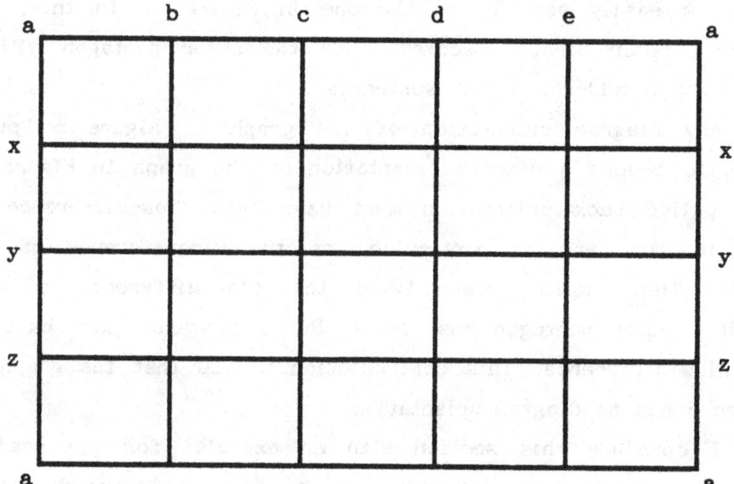

Figure 6

The example we describe is a special case of the construction of a non-diagram explained in Pretzel and Youngs (1990a), which serves to show the technique and also gives a proof that the example of Figure 5 is a non-diagram. As that graph has girth four, it also proves that it must have chromatic number at least 4. We begin with a plane 4×5 grid (Figure 6). This obviously has a cycle-basis of 4-cycles. It is staring us in the face. In any diagram orientation a 4-cycle must have two edges oriented each way. For otherwise it is a bypass or a monotone circuit. Thus it must have flow-difference 0. Our 4-cycles form a natural basis for the circuit group of the graph and so every circuit must also have flow-difference 0 in any diagram orientation. In particular, that must hold for the outer circuit. We now identify vertices on this outer circuit to produce the graph of Figure 5. To do this we first identify the vertices at opposite ends of the horizontal rows. Then we identify the vertices of the top row from left to right with those of the bottom row from right to left. In the Figure identified vertices are given the same labels. The resulting graph is easily seen to be the one in Figure 5. In this, the original outer circuit becomes twice the outer pentagon with a four circuit added and then subtracted.

Any diagram orientation of the graph of Figure 5 "pulls back" to become a diagram orientation of the graph in Figure 6. This pulled back orientation must have zero flow-difference on all circuits, and in particular, on the circumference of the graph. That implies that twice the flow-difference of the original outer pentagon must be 0. But a pentagon must have an odd flow-difference. This contradiction proves that the graph in Figure 5 has no diagram orientation.

I conclude this section with an exercise for the reader. Find an analogous construction of M_5 from a balanced graph, thereby producing a third proof that it cannot be oriented as a diagram.

4 CHARACTERIZING FLOW-DIFFERENCES

Having thoroughly characterized inversions, the next step is to ask what inversion classes a graph can have. There is a natural answer to this question, albeit one that does not produce an efficient decision algorithm. It turns out that if we write down the most obvious properties that a flow-difference must have, then these characterize flow-difference functions. To be specific, the function must be *linear*, because

$$R\cdot(aC + bD) = a(R\cdot C) + b(R\cdot D),$$

it must be *bounded* (by the length), that is

$$|R\cdot C| \le |C|,$$

and it must be *even*[2], that is

$$R\cdot C \equiv |C| \pmod 2.$$

Theorem (Pretzel and Youngs (1990b))

Let f be a function defined on the circuits of G. Then there exists an orientation R of G such that $R\cdot C = f(C)$ for all circuits C, if and only if f is linear, bounded, and even.

The proof of this theorem is rather intricate. It relies on determining how $S\cdot W$ can differ from $R\cdot W$ for a walk W when S is an inversion of R. That is done in the following lemma, which can be viewed as a grand extension of Mosesian's Theorem.

Lemma

Let W be a walk from u to v in a graph G and let R be an orientation of G. Then exactly one of the two following possibilities holds.

(a) *There exists an inversion S of R with $S\cdot W = R\cdot W + 2$.*

(b) *There exists a walk V from u to v with $R\cdot V = |V|$.* □

We omit the proof of the lemma but observe that both cases cannot occur simultaneously, because the circuit $W - V$ must have equal flow-differences, $R\cdot(W - V) = S\cdot(W - V)$. If $u = v$ then case (b) holds, as we can take $V = 0$.

---

[2] In the following, the words *bounded* and *even* when applied to functions on circuits will be taken to indicate the validity of these two formulas.

Outline proof of theorem

The necessity of the conditions is obvious. Sufficiency is proved by induction on the number of edges of G, the statement being trivial on a tree. It is sufficient to verify that $R \cdot C = f(C)$ for all primitive circuits of G. For the induction step we select an edge e lying on at least one primitive circuit and delete it. The resulting graph G' has as its circuits a subset of the circuits of G. So f is defined for G'. By induction hypothesis G' has an orientation R' such that for circuits C avoiding e,

$$R' \cdot C = f(C).$$

Now for primitive circuits $C = e + U$ we define the deficiency

$$d(C) = f(C) - R' \cdot U.$$

Note that $d(C)$ is necessarily odd, because f is even. We can extend R' to an orientation R of G so that $R \cdot C = f(C)$ if and only if $d(C) = \pm 1$. The first, straightforward, step of the proof proper is to verify that all primitive circuits C as above have the same deficiency. Next we fix a primitive circuit $C = e + U$ and show that, if $d(C) > 1$, then statement (b) of the lemma cannot hold for U and R' in G'. Thus we can find an inversion S' of R' with $S' \cdot U = R' \cdot U + 2$, reducing $d(C)$. The same argument works if $d(C) < -1$ with $-U$ in place of U. Thus in either case we can bring $d(C)$ closer to ± 1 by replacing R' by an inversion. Eventually $d(C) = \pm 1$ and R' can be extended to G. □

It is not difficult to use this theorem to construct an unbalanced flow for a plane graph G with two faces of circumference greater than 4. If a face has odd circumference it cannot have flow-difference 0 and there are no balanced orientations. If all faces are even, give one of the two large faces flow-difference 2 and the other flow-difference -2 and all other faces flow-difference 0. It is then easy to verify that the conditions of the theorem are satisfied, provided that all 4-cycles of G are face cycles. So in that case, these flow-differences produce an unbalanced diagram orientation. If there is a four cycle that is not a face cycle, we split G into its interior and exterior and proceed by induction. That proves the following corollary.

Corollary (Bandelt (unpublished), Pretzel and Youngs (1990b))

A planar balanced graph can be drawn in the plane with all face cycles except one of circumference 4. *Hence a planar graph is balanced if it has a cycle basis (in the conventional sense) consisting of 4-cycles.* □

What makes it difficult to verify whether a function f defined on the circuits of a graph G satisfies the conditions? Linearity can be forced by defining f on a circuit basis only and extending it. Evenness also will automatically follow if it is satisfied on a basis. The only difficulty is involved in verifying boundedness. As the length is not a linear function on chains, this must be verified on all *simple* circuits (that is, primitive circuits with minimal support). That leads to a question I asked at the previous British Combinatorial Conference in Norwich: how far does a knowledge of the rank and length function of the circuit group of G determine G? To be precise, given two connected graphs G and H and an isometric isomorphism of the circuit groups of G and H, does it follow that G and H themselves are isomorphic? The easy example of Figure 7 shows that this is not always true.

However, the question is answered in the affirmative for 3-connected graphs by Woodall (1991). A later proof of Woodall's result is given by Pretzel and Youngs (1991). It is longer than Woodall's own proof but gives some additional information because it uses the following lemma characterizing the flow-differences of orientations that differ by reversing a single edge.

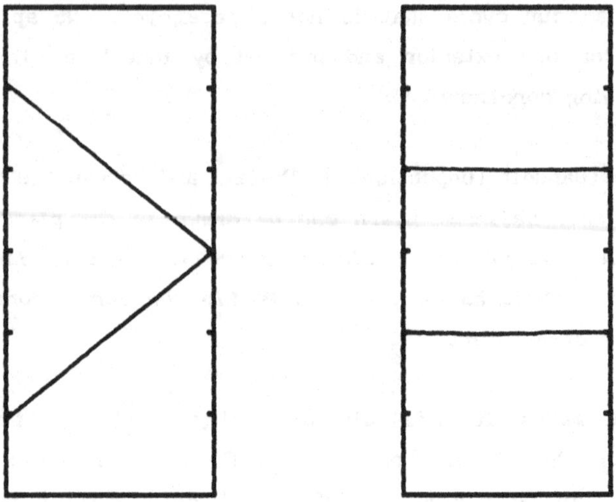

Figure 7

Lemma

Suppose that R and S are orientations of the graph G. Then there exist inversions R′ and S′ of R and S such that S′ can be obtained from R′ by reversing a single edge if and only if for all primitive circuits C of G we have

$$|S\cdot C - R\cdot C| \le 2. \qquad \square$$

The proof of this lemma depends on showing that every circuit can be covered by the right number of *conforming* primitive circuits (recall that a circuit is primitive if it has no coefficients a_e with $|a_e| > 1$, and it conforms to K if on all edges the signs of the coefficients agree). To do that, we show that there exists a primitive circuit C that simultaneously satisfies the following two conditions:

(*i*) $|K - C| = |K| - |C|$;

and (*ii*) If $K = \sum a_e e$ and $K - C = \sum b_e e$, then for the maximum $a = \max\{ |a_e| \}$, we have

$b_e < a$ for all e.

That means that C conforms to K, and it goes through all the *maximal edges* of K. The circuit C is the first member of the covering set for K. The others are found inductively by finding analogous circuits for $K - C$.

The second ingredient in our proof states that knowledge of the length function of all circuits allows us to determine the primitive circuits.

Lemma

A circuit C of a graph G is primitive if and only if for any circuits A and B satisfying

$$A + B = C \quad \text{and} \quad |A| + |B| = |C|,$$

we also have

$$|A - B| = |A| - |B|. \qquad \square$$

With these tools we can prove the following theorem, from which it immediately follows that for 3-edge connected graphs, the length function on circuits determines the circuit matroid. For the purpose of the theorem we choose an orientation R of our graph G and call a set $\mathcal{S}$ of primitive circuits of G *admissible* if there exists an orientation S such that

$$S\cdot C = R\cdot C + 2 \quad \text{for} \quad C \in \mathcal{S};$$
$$S\cdot C = R\cdot C - 2 \quad \text{for} \quad -C \in \mathcal{S};$$
$$S\cdot C = R\cdot C \quad \text{for all other primitive circuits.}$$

Theorem

Let G be a 3-edge connected graph and choose an orientation R of G. Then the admissible sets of primitive circuits are in one-to-one correspondence with the edges of G. $\square$

From the knowledge of the edges and their incidences with primitive circuits, it is easy to construct the circuit matroid of the graph. It should be stressed that this method is longer than the original proof of Woodall, but perhaps the intermediate results are of independent interest.

BIBLIOGRAPHY

In addition to the papers referred to in the text, this bibliography contains some relevant papers on graph orientations and graph colouring.

Textbooks

Aigner, M.

(1979) Combinatorial Theory, Springer, Berlin.

Nešetřil, J. and Rödl, V (editors)

(1990) The Mathematics of Ramsey Theory, Springer, Berlin.

Ore, O.

(1962) Theory of Graphs, AMS Colloquium Publications **34**, Providence RI.

Tutte, W. T.

(1984) Graph Theory, Addison-Wesley.

Welsh, D. J. A.

(1976) Matroid Theory, Academic Press, London, New York & San Francisco.

Research Papers

Aigner, M. and Prins, G.

(1980) k-Orientable Graphs, unpublished. Mathematics Department, Free University Berlin.

Bandelt, H-J. and Rival, I.

(1988) Diagrams, Orientations and Varieties, Preprint, Department of Computer Science, University of Ottawa.

Bollobás, B., Brightwell, G. and Nešetřil, J.

(1986) Random Graphs and Covering Graphs of Posets, Order **3**, 245-257.

Bondy, J. A.

(1976) Diconnected Orientations and a Conjecture of Las Vergnas, J. London Math. Soc. **14**, 277-282.

Chvátal, V.

(1970) The Smallest Triangle-Free 4-Chromatic 4-Regular Graph, J. Combin. Theory **9**, 93-94.

Erdös, P.

(1962) Graph Theory and Probability I, Canad. J. Math. **5**, 346-352.

(1962) Graph Theory and Probability II, Canad. J. Math. **6**, 122-127.

Gallai, T.

(1963a) Kritische Graphen I, Publ. Math. Inst. Hungar. Acad. Sci. **8**, 165-192.

(1963b) Kritische Graphen II, Publ. Math. Inst. Hungar. Acad. Sci. **8**, 373-395.

(1968) On Directed Paths and Circuits, in "Theory of Graphs", P. Erdös, G. Katona Eds. Academic Press, New York, 115-118.

Grötzsch, H.

(1958) Ein Dreifarbensatz für Dreikreisfreie Netze auf der Kugel, Wiss. Zeitschr. Martin-Luther-Univ. Halle -Wittenberg. Math.-Nat. **8**, 109-119.

Grünbaum, B.

(1970) A Problem in Graph Coloring, Amer. Math. Monthly **77**, 1088-1092.

Hajós, G.

(1961) Über eine Konstruktion nicht *n*-färbbarer Graphen, Wiss. Zeitschr. Martin-Luther-Univ. Halle-Wittenberg Math.-Nat. **10**, 116-117.

Jensen, T. and Royle, G. F.

(1989) Two Triangle-Free Graphs With Chromatic Number Five, Preprint, Department of Combinatorics and Optimisation, Univ. of Waterloo, Canada.

Križ, I.

(1989) Hypergraph-free Construction of Highly Chromatic Graphs without Short Cycles, Combinatorica **9**, 227-229.

Liu, W.P. and Rival, I.

(1988) Inversions, Cuts and Orientations, Preprint, Department of Computer Science, University of Ottawa.

Lovász, L.

(1968) On Chromatic Number of Finite Set Systems, Acta Math. Acad. Sci. Hung. **19**, 59-67.

Minty, G. S.

(1962) A Theorem on n-Colouring the Points of a Linear Graph, Amer. Math. Monthly **69**, 623-624.

Mosesian, K. M.

(1972a) A Minimal Graph that is not Strongly Basable [Russian], Akad. Nauk. Armian. SSR Dokl. **54**, 8-12.

(1972b) Strongly Basable Graphs [Russian], Akad. Nauk. Armian. SSR Dokl. **54**, 134-138.

(1972c) Certain Theorems on Strongly Basable Graphs [Russian], Akad. Nauk. Armian. SSR Dokl. **54**, 241-245.

(1972d) Basable and Strongly Basable Graphs [Russian], Akad. Nauk. Armian. SSR Dokl. **55**, 83-86.

(1973a) Saturated Graphs [Russian], Akad. Nauk. Armian. SSR Dokl. **56**, 257-262.

(1973b) Basis Graphs of Certain Orderings [Russian], Akad. Nauk. Armian. SSR Dokl. **57**, 264-270.

Mycielski, J.

(1955) Sur le Colorage des Graphes, Colloq. Math. **3**, 161-162.

Nešetřil, J. and Rödl, V.

(1978) On a Probabilistic Graph Theoretical Method, Proc. AMS **72**, 417-421.

(1979) A Short Proof of the Existence of Highly Chromatic Graphs without Short Cycles, J. Combin. Theory. B **27**, 225-227.

(1987) Complexity of Diagrams, Order **3**, 321-330.

(1988) Ramsey Theory for Set Structures. Preprint, Charles University, Prague.

Nielsen, F. and Toft, B.

(1975) On a Class of Planar 4-Chromatic Graphs due to T Gallai, Recent Adv. in Graph Theory, Proc. Symp. Prague, Academia Praha **74**, 425-430.

Pretzel, O. R. L.

(1985) On Graphs that can be Oriented as Diagrams of Ordered Sets, Order **2**, 25-40.

(1986a) On Reorienting Graphs by Pushing Down Maximal Vertices, Order **3**, 135-153.

(1986b) Orientations and Reorientations of Graphs, Contemporary Math. **57**, 103 -125.

(1987) A non-diagram of girth 6, Discrete Math. **63**, 241-244.

(1989) Removing Monotone Cycles From Orientations, Annals of Discrete Math. **41**, 355-362.

Pretzel, O. R. L. and Youngs, D. A.

(1990a) Balanced Graphs and Non-Covering Graphs, to appear in Discrete Math. (special ed. on ordered sets).

(1990b) Cycle lengths and Graph Orientations, to appear in SIAM J. Discrete Math. **3**, 544-553.

(1991) Cycle lengths and Graph Determination, Preprint.

Roy, B.

(1967) Nombre Chromatique et plus longs Chemins d'un Graphe, Revue AFIRO **1**, 127-132.

Toft, B.

(1974) On Critical Subgraphs of Colour Critical Graphs, Discrete Math. **7**, 377-392.

Whitney, H.

(1932) Congruent Graphs and the Connectivity of Graphs, Amer. J. Math. **54**, 150-168.

(1933) A Set of Topological Invariants for Graphs, Amer. J. Math. **55**, 231-235.

Woodall, D. R.

(1991) Cycle Lengths and Circuit Matroids of Graphs, to appear in Discrete Math.

Youngs, D. A.

(1990) Orientations of Graphs, PhD Thesis, London.

Graph Perturbations

P. ROWLINSON

Department of Mathematics
University of Stirling
Scotland FK9 4LA

1. INTRODUCTION

The topic of graph perturbations is, like the classical perturbation theory of linear operators [29], concerned primarily with changes in eigenvalues which result from various perturbations. The eigenvalues are those of an adjacency matrix of a graph G, and a perturbation of G is to be thought of as a local modification such as the addition or deletion of a vertex or edge. Here G is a finite undirected graph without loops or multiple edges, and if its vertices are labelled $1, 2, \ldots, n$ then the corresponding adjacency matrix A is (a_{ij}) where $a_{ij} = 1$ if vertices i and j are adjacent, and $a_{ij} = 0$ otherwise. The matrix A is regarded as a matrix with real entries, and since A is symmetric, its eigenvalues are real. These eigenvalues are independent of the ordering of the vertices of G, and so we refer to them as the eigenvalues of G. The n eigenvalues together comprise the *spectrum* of G.

We shall be concerned with three related questions: (1) What algebraic information about G is sufficient to determine the eigenvalues of a given perturbation of G? (2) Does a given eigenvalue increase or decrease under a given perturbation? (3) How can we compare the effects on eigenvalues of two different perturbations? Such questions were raised in 1979 by Li and Feng [31] in relation to the largest eigenvalue of a graph, and in effect they considered graphs perturbed by the relocation of certain pendant edges. Previously Hoffman and Smith [28] had answered question (2) in respect of the largest eigenvalue of a graph perturbed by the subdivision of an edge (see §3.4). Further observations on the behaviour of the largest eigenvalue were provided by Simić [48] in 1987. Many more references are given in subsequent sections, and indeed our survey reflects the emphasis in the literature on the largest eigenvalue. For background reading on the spectra of graphs the reader is referred to the monographs [14] and [13], both of which have appendices containing many examples of graph spectra. Some of the results discussed below were first conjectured on the basis of further numerica evidence furnished by the computer package "Graph" [15].

The spectrum of a graph is of interest to chemists pursuing Hückel's theory of molecular orbitals [35]; to physicists modelling a vibrating membrane [14, §8.4]; and of course to mathematicians working in algebraic graph theory. In discussing the above questions the aim is to gain a better understanding of the relationship between the structure of a graph and its algebraic invariants. Clearly two graphs having different spectra are non-isomorphic, but a graph is not determined by its spectrum: Figure 1 shows a pair of non-isomorphic cospectral graphs (in fact with fewest possible vertices if both graphs are connected). Nevertheless the members of a restricted class of graphs may

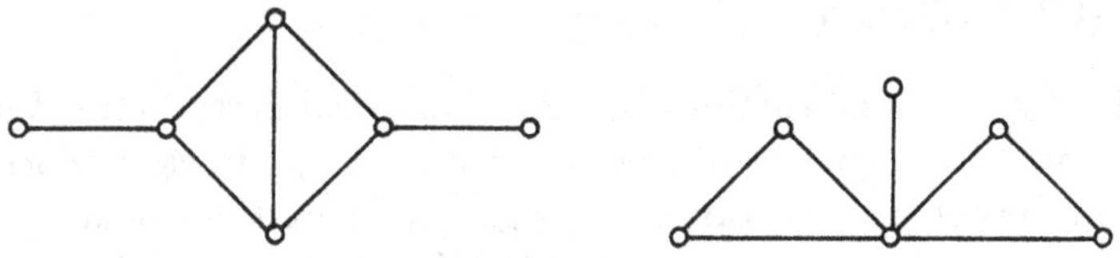

Figure 1: a pair of non-isomorphic cospectral graphs.

be characterized by their spectra: indeed the class of bicyclic Hamiltonian graphs on n vertices is an example of a class of graphs characterized by the largest eigenvalue. Such a graph consists of an n-cycle $123\dots n1$ and a chord $1k$ $(3 \leq k \leq 1+[n/2])$, and it turns out that the largest eigenvalue decreases strictly as k increases [37, 50].

If a class of graphs contains non-isomorphic cospectral graphs then we may attempt to differentiate them by considering further algebraic invariants. Here so-called graph angles have received attention of late: although in general they do not serve to distinguish cospectral graphs [11, §5] they do prove to be pertinent to the questions raised above. In order to define them we introduce some notation which will be used throughout the paper. Always A will denote the adjacency matrix of a graph G with vertices $1, 2, \dots, n$. Thus for each integer $k \geq 0$ the (u, v)-entry of A^k, denoted by $a_{uv}^{(k)}$, is the number of $u-v$ walks of length k in G. The *distinct* eigenvalues of G are denoted by $\mu_1, \dots, \mu_m$ in decreasing order, and the largest eigenvalue μ_1 is called the *index* of G, written $\mu_1(G)$, or the *spectral radius* of A, written $\mu_1(A)$. It follows from the Perron-Frobenius theory of non-negative matrices [36, §1.4] that $\mu_m \geq -\mu_1$ and A has a non-negative unit eigenvector $\mathbf{x} = (x_1, x_2, \dots, x_n)^T$ corresponding to μ_1. Moreover if G has a unique non-trivial component then μ_1 is a simple eigenvalue, $\mathbf{x}$ is determined uniquely, and $x_j = 0$ if and only if vertex j is isolated: in this case we call $\mathbf{x}$ the *principal eigenvector* of G. Here, if π is an automorphism of G then $(x_{\pi(1)}, \dots, x_{\pi(n)})^T$ remains an eigenvector corresponding to μ_1 and uniqueness ensures that $x_{\pi(u)} = x_u$ $(u = 1, \ldots, n)$. Thus if vertices u, v lie in the same orbit of $\mathrm{Aut}(G)$ then $x_u = x_v$. The notation $A', \mu_1', \mathbf{x}'$ will refer to a perturbation G' of G.

Let $\mathbf{e}_1 = (1,0,\ldots,0,0)^T, \ldots, \mathbf{e}_n = (0,0,\ldots,1)^T$, and let $\cos^{-1}(\alpha_{ij})$ be the (acute) angle between $\mathbf{e}_j$ and the i-th eigenspace $\mathcal{E}(\mu_i)$. Note that $\alpha_{1j} = x_j$ $(j = 1, \ldots, n)$ when G has a unique non-trivial component. The α_{ij} themselves have usually been referred to as the *angles* of G, abusing terminology, but it is often convenient to regard them as lengths, as we now explain. The symmetric matrix A has a spectral decomposition $A = \sum_{i=1}^{m} \mu_i P_i$, where P_i represents the orthogonal projection onto $\mathcal{E}(\mu_i)$ [26, §79], and so $\alpha_{ij} = |P_i \mathbf{e}_j|$. A *eutactic star*, in a subspace U of an inner product space V, is defined in [46] essentially as the orthogonal projection onto U of an orthonormal set of vectors in V. Thus $P_i\mathbf{e}_1, \ldots, P_i\mathbf{e}_n$ comprise a eutactic star $\mathcal{S}_i$ in $\mathcal{E}(\mu_i)$, and $\alpha_{i1}, \ldots, \alpha_{in}$ are the lengths of its arms. We shall sometimes be concerned with the angle between two given arms (in addition to their lengths), and in this case it is convenient to work with the scalar product $P_i\mathbf{e}_u . P_i\mathbf{e}_v$ because of the possibility that one or other of $P_i\mathbf{e}_u, P_i\mathbf{e}_v$ is zero. Note that $P_i\mathbf{e}_u . P_i\mathbf{e}_v$ is just the (u,v)-entry of P_i, denoted by $p_{uv}^{[i]}$. (Thus $a_{uv} = \sum_{i=1}^{m} \mu_i p_{uv}^{[i]}$, and G may be reconstructed from $\mu_1, \mu_2, \ldots, \mu_m$ and all of the $p_{uv}^{[i]}$.)

Since $A^k = \sum_{i=1}^{m} \mu_i^k P_i$, and since α_{ij}^2 is the (j,j)-entry of P_i, we have $a_{jj}^{(k)} = \sum_{i=1}^{m} \mu_i^k \alpha_{ij}^2$. It follows that knowledge of the $2m$ invariants μ_i, α_{ij} $(i = 1, 2, \ldots, m)$ is equivalent to knowledge of the number of $j-j$ walks of length k for all integers $k \geq 0$. On the other hand, knowledge of just the spectrum of G is equivalent to knowledge of the number of (rooted) closed walks of length k for all integers $k \geq 0$. To see this note first that if G has spectrum $\lambda_1, \lambda_2, \ldots, \lambda_n$ then the number of such walks of length k is $\sum_{i=1}^{n} \lambda_i^k$, the trace of A^k. Conversely, given $\sum_{i=1}^{n} \lambda_i^k$ $(k = 1, 2, \ldots, n)$ we can use symmetric functions to construct a monic polynomial of degree n whose roots are $\lambda_1, \lambda_2, \ldots, \lambda_n$. This polynomial is of course $\det(xI - A)$, which we call the characteristic polynomial of G, denoted by $\phi_G(x)$. For an interpretation of the coefficients of $\phi_G(x)$ in terms of the structure of G, see [14, §1.4].

We conclude this section with a useful theorem of Rayleigh's from matrix theory [30, §§8.4, 8.5]: it has implications for the characteristic polynomial of the perturbation $G - u$ of G which results from the deletion of vertex u and all edges containing u.

Theorem 1.1 (The Interlacing Theorem).
Let A be a Hermitian matrix with eigenvalues $\lambda_1, \lambda_2, \ldots, \lambda_n$ in non-decreasing order. If $\nu_1, \nu_2, \ldots, \nu_r$ are the eigenvalues of a principal $r \times r$ submatrix of A, in non-decreasing order, then

$$\lambda_i \leq \nu_i \leq \lambda_{n-r+i} \qquad (i = 1, \ldots, r).$$

□

Taking $r = n-1$, it follows that for any vertex u of G, the roots of $\phi_{G-u}(x)$ interlace those of $\phi_G(x)$. This conclusion may also be drawn from the very first result of the next section, which establishes an explicit relationship between $\phi_{G-u}(x)$ and $\phi_G(x)$ involving the angles of G at u.

2. PERTURBATIONS AND CHARACTERISTIC POLYNOMIALS

2.1 Deletion of a Vertex

Godsil and McKay [24] pointed out that one can obtain an expression for $\phi_{G-u}(x)$ by expressing in two ways the (u,u)-entry of the matrix generating function $\sum_{k=0}^{\infty} x^{-k}A^k$. First, for $|x| > \mu_1$, this function is $(I - x^{-1}A)^{-1}$, with (u,u)-entry $x\phi_{G-u}(x)/\phi_G(x)$ because $\phi_{G-u}(x)$ is the (u,u)-cofactor of $xI - A$. Secondly, the (u,u)-entry of $\sum_{k=0}^{\infty} x^{-k}A^k$ is $\sum_{i=1}^{m}\sum_{k=0}^{\infty} x^{-k}\mu_i^k\alpha_{iu}^2$, which is expressible as $\sum_{i=1}^{m} \frac{\alpha_{iu}^2}{1 - x^{-1}\mu_i}$. Accordingly we have:

Theorem 2.1. $$\phi_{G-u}(x) = \phi_G(x)\sum_{i=1}^{m} \frac{\alpha_{iu}^2}{x - \mu_i}.$$ □

2.2 Addition of a Pendant Edge.

Let G_u denote the graph obtained from G by adding a pendant edge at u. A straightforward determinantal expansion yields

$$\phi_{G_u}(x) = x\phi_G(x) - \phi_{G-u}(x) \tag{2.1}$$

and on combining this with Theorem 2.1 we obtain:

Theorem 2.2. $$\phi_{G_u}(x) = \phi_G(x)\{x - \sum_{i=1}^{m} \frac{\alpha_{iu}^2}{x - \mu_i}\}.$$ □

2.3 Addition of a Bridge.

Theorem 2.2 may be seen as dealing with the introduction of a bridge between G and an isolated vertex. In order to describe the general situation let H, K be disjoint graphs, and let $HuvK$ denote the graph obtained from $H \cup K$ by adding a bridge joining vertex u of H to vertex v of K. Heilbronner [27] used an appropriate Laplacian determinantal expansion to show:

$$\phi_{HuvK}(x) = \phi_H(x)\phi_K(x) - \phi_{H-u}(x)\phi_{K-v}(x). \tag{2.2}$$

Thus if $G = H \cup K$ and $G' = HuvK$ then in view of Theorem 2.1, $\phi_{G'}(x)$ is determined by $\phi_G(x)$ $(= \phi_H(x)\phi_K(x))$ together with the angles of H at u and the angles of K at v.

In order to discuss the characteristic polynomials of other perturbations, we need the following algorithm, which is a recursive formula for the characteristic polynomial of an arbitrary graph G: it expresses $\phi_G(x)$ in terms of characteristic polynomials of perturbations of G which are local modifications having fewer edges or vertices.

2.4 The Deletion-Contraction Algorithm.

This algorithm was established in [37] in a multigraph setting. For graphs, the result is as follows.

Theorem 2.3. Let $G - uv$ be the graph obtained from G by deleting the edge uv, and let $(G-uv)^*$ denote the multigraph obtained from $G - uv$ by amalgamating vertices u and v. Then

$$\phi_G(x) = \phi_{G-uv}(x) + \phi_{(G-uv)^*}(x) + (x-1)\phi_{G-u-v}(x) - \phi_{G-u}(x) - \phi_{G-v}(x). \tag{2.3}$$

□

For a graph G with non-adjacent vertices u and v, the algorithm may be restated in the following form, where $G + uv$ denotes the perturbation of G obtained by adding the edge uv:

$$\phi_{G+uv}(x) = \phi_G(x) + \phi_{G^*}(x) + (x-1)\phi_{G-u-v}(x) - \phi_{G-u}(x) - \phi_{G-v}(x). \tag{2.4}$$

The proof of (2.4) consists essentially of two straightforward determinantal expansions, each exploiting the fact that $\det(\mathbf{y}+\mathbf{z}|M) = \det(\mathbf{y}|M)+\det(\mathbf{z}|M)$ for any $n \times (n-1)$ matrix M. First one shows that

$$\phi_{G+uv}(x) = \phi_G(x) - \phi_{G-u-v}(x) - 2\psi_{uv}(x) \tag{2.5}$$

where $\psi_{uv}(x)$ denotes the (u, v)-cofactor of $xI - A$. Secondly one shows similarly that

$$\phi_{G^*}(x) = \phi_{G-u}(x) + \phi_{G-v}(x) - x\phi_{G-u-v}(x) - 2\psi_{uv}(x). \tag{2.6}$$

The result follows by eliminating $\psi_{uv}(x)$ from equations (2.5) and (2.6). As a preliminary application of Theorem 2.3 we prove the following result [37, Proposition 1.7].

Theorem 2.4. Let u, v be adjacent vertices of G, and let G_{uv} denote the graph obtained from G by subdividing the edge uv. Then

$$\phi_{G_{uv}}(x) = \phi_G(x) + (x-1)\phi_{G-uv}(x) - \phi_{G-u}(x) - \phi_{G-v}(x) + \phi_{G-u-v}(x). \tag{2.7}$$

Proof. Let $H = G_{uv}$, and let w be the vertex which subdivides uv. On applying Theorem 2.3 to H and the edge uv, we obtain

$$\phi_H(x) = \phi_{H-uw}(x) + \phi_G(x) + (x-1)\phi_{H-u-w}(x) - \phi_{H-u}(x) - \phi_{H-w}(x).$$

Now $H - uw$ has w as a pendant vertex adjacent to v, and so $\phi_{H-uw}(x) = x\phi_{G-uv}(x) - \phi_{G-v}(x)$ by equation (2.1). Similarly, $\phi_{H-u}(x) = x\phi_{G-u}(x) - \phi_{G-u-v}(x)$ and the result follows once we note that $H - u - w = G - u$ and $H - w = G - uv$. □

Although $\phi_{G-uv}(x)$ features in equations (2.3) and (2.7), and equation (2.7) provides a formula for $\phi_{G_{uv}}(x)$, Theorems 2.3 and 2.4 do not suffice to prove analogues of Theorems 2.1 and 2.2 for perturbations resulting from the deletion or subdivision of an edge. We can however use the results of this subsection to help us deal with the addition of an edge (§2.5) and the addition of a bridging vertex (§2.6).

2.5 Addition of an Edge

We begin by extending the argument of §2.1 to the non-diagonal entries of the matrix generating function $\sum_{k=0}^{\infty} x^{-k}A^k$. On expressing the (u,v)-entry in two ways, we obtain

$$\sum_{i=1}^{m} \frac{p_{uv}^{[i]}}{1 - x^{-1}\mu_i} = x\psi_{uv}(x)/\phi_G(x), \tag{2.8}$$

where $\psi_{uv}(x)$ is the (u,v)-cofactor of $xI - A$. If u, v are non-adjacent vertices and we eliminate $\psi_{uv}(x)$ from equations (2.5) and (2.8) we obtain

$$\phi_{G+uv}(x) + \phi_{G-u-v}(x) = \phi_G(x)\{1 - 2\sum_{i=1}^{m} \frac{p_{uv}^{[i]}}{x - \mu_i}\}. \tag{2.9}$$

In order to derive an analogous expression for $\phi_{G+uv}(x) - \phi_{G-u-v}(x)$, we make use of two more recursion formulae: each differs from Theorem 2.3 in that global rather than local properties of G are involved. The first formula was obtained by Schwenk [45, Theorem 3] by interpreting the coefficients of the characteristic polynomials involved (cf. [14, Theorem 1.3])

Theorem 2.5 If u, v are adjacent vertices of G then

$$\phi_G(x) = \phi_{G-uv}(x) - \phi_{G-u-v}(x) - 2\sum_{C\in\mathcal{C}} \phi_{G-C}(x), \tag{2.10}$$

where $\mathcal{C}$ denotes the set of all cycles containing the edge uv and $G - C$ denotes the graph obtained from G by deleting all the vertices of C. □

The second formula was derived by Gutman [25] from a classical theorem of Jacobi on minors of an adjugate [1, §42] using a more recent graph-theoretic interpretation of cofactors by Coates (see [9, p.47]).

Theorem 2.6. For any two vertices u, v of G,

$$\phi_{G-u}(x)\phi_{G-v}(x) - \phi_G(x)_{G-u-v}(x) = \{\sum_{P\in\mathcal{P}} \phi_{G-P}(x)\}^2, \tag{2.11}$$

where $\mathcal{P}$ denotes the set of all $u - v$ paths in G. □

As pointed out in [13, p.144], if u and v are adjacent vertices we have

$$\sum_{P\in\mathcal{P}} \phi_{G-P}(x) = \phi_{G-u-v}(x) + \sum_{C\in\mathcal{C}} \phi_{G-C}(x), \tag{2.12}$$

because uv is one of the $u - v$ paths in G while any other $u - v$ path forms a cycle with uv. On eliminating $\sum_{P\in\mathcal{P}} \phi_{G-P}(x)$ and $\sum_{C\in\mathcal{C}} \phi_{G-C}(x)$ from equations (2.10), (2.11) and (2.12) we obtain

$$4\{\phi_{G-u}(x)\phi_{G-v}(x) - \phi_G(x)\phi_{G-u-v}(x)\} = \{\phi_G(x) - \phi_{G-uv}(x) - \phi_{G-u-v}(x)\}^2 \tag{2.13}$$

Equation (2.13) may be rearranged to provide a second recursive formula for $\phi_G(x)$ involving only local modifications of G (cf. Theorem 2.3). This was exploited by Cvetković in a paper [12] on the reconstruction of graphs from eigenvalues and angles. Cvetković subsequently observed that when u and v are non-adjacent it follows from equations (2.9) and (2.13) that $\phi_G(x)$ together with the invariants $\alpha_{iu}, \alpha_{iv}, p_{uv}^{[i]}$ $(i = 1, \ldots, m)$ always suffices to determine the spectrum of $G + uv$, and not just the index of $G + uv$ under certain conditions [40, §1]. To see how this comes about, we first apply equation (2.13) to $G + uv$ and rewrite it in the form

$$2\phi_G(x)\{\phi_{G+uv}(x) - \phi_{G-u-v}(x)\} = \\ \{\phi_{G+uv}(x) + \phi_{G-u-v}(x)\}^2 + \phi_G(x)^2 - 4\phi_{G-u}(x)\phi_{G-v}(x). \tag{2.14}$$

It is now clear that we can use Theorem 2.1 and equations (2.9) and (2.14) to express both $\phi_{G+uv}(x)$ and $\phi_{G-u-v}(x)$ in terms of $\phi_G(x)$ and $\alpha_{iu}, \alpha_{iv}, p_{uv}^{[i]}$ $(i = 1, \ldots, m)$. The resulting expressions are not very illuminating and so we state the conclusions as follows.

Theorem 2.7 If u and v are non-adjacent vertices of G then both $\phi_{G+uv}(x)$ and $\phi_{G-u-v}(x)$ are determined by $\phi_G(x)$ and the invariants

$$|P_i\mathbf{e}_u|, |P_i\mathbf{e}_v|, |P_i\mathbf{e}_u + P_i\mathbf{e}_v| \ (i = 1, \ldots m).$$

□

2.6 Addition of a Bridging Vertex

This perturbation, which is of interest to chemists in relation to isospectral molecules [20, 33], consists of adding a new vertex adjacent to two non-adjacent vertices u and v. We denote the resulting graph by $G(u, v)$. Thus

$G(u,v)$ is obtained from G by adding the edge uv and then subdividing this edge. If we apply Theorem 2.4 to $G+uv$, equation (2.7) becomes

$$\phi_{G(u,v)}(x) = \phi_{G+uv}(x)+(x-1)\phi_G(x)-\phi_{G-u}(x)-\phi_{G-v}(x)+\phi_{G-u-v}(x) \quad (2.15)$$

We can now use equation (2.9) to express $\phi_{G+uv}(x)+\phi_{G-u-v}(x)$ in terms of $\phi_G(x)$ and $p_{uv}^{[i]}$ $(i=1,\ldots,m)$; and Theorem 2.1 to express $\phi_{G-u}(x), \phi_{G-v}(x)$ in terms of $\phi_G(x)$ and $\alpha_{iu}, \alpha_{iv} (i=1,\ldots,m)$. We obtain the following neat expression for $\phi_{G(u,v)}(x)$ [43, equation (2.5)].

Theorem 2.8. Let u, v be non-adjacent vertices of G and let $G(u,v)$ be obtained from G by adding a bridging vertex between u and v. Then

$$\phi_{G(u,v)}(x) = x\phi_G(x) - \phi_G(x)\sum_{i=1}^{m}\frac{|P_i\mathbf{e}_u + P_i\mathbf{e}_v|^2}{x-\mu_i} \quad (2.16)$$

□

2.7 Remarks

Theorems 2.7 and 2.8 show that the invariants $|P_i\mathbf{e}_u|$, $|P_i\mathbf{e}_v|$, $|P_i\mathbf{e}_u+P_i\mathbf{e}_v|$ $(i=1,\ldots,m)$, together with the spectrum of G, suffice to determine the spectra of $G+uv, G-u-v$ and $G(u,v)$ when u,v are non-adjacent. At the time of writing, it is not known whether the same is true of $G-uv, G-u-v$ and G_{uv} when u,v are adjacent. The invariants in question are essentially the dimensions of m parallelograms determined by the appropriate arms of the eutactic stars $\mathcal{S}_1,\ldots,\mathcal{S}_m$.

3. PERTURBATIONS AND THE LARGEST EIGENVALUE

3.1 General Observations

In the previous section we discussed perturbations G' of a graph G for which all the eigenvalues of G' could be determined, albeit implicitly, from certain algebraic invariants of G. If we are concerned only with the simpler question of whether the largest eigenvalue increases or decreases in the transition from G to G' (as for example in connection with the maximal index problems encountered in §6) then we can expect to answer this question in respect of a wider class of perturbations of G. In this section we review what can be said in this direction without recourse to the more elaborate techniques of sections 4 and 5.

First it is clear from Theorem 2.1 that the index of a connected graph G decreases with deletion of a vertex u, because $\alpha_{1u}=x_u>0$ and this ensures that $\phi_{G-u}(x)>0$ for all $x\geq\mu_1$. It follows that the index of a connected graph increases with the addition of a pendant edge or a bridging vertex. Again if G' is obtained from the connected graph G by adding an edge then $\mu_1' > \mu_1$ because $\mu_1' = \sup\{\mathbf{y}^TA'\mathbf{y} : |\mathbf{y}|=1\}$, whence $\mu_1'-\mu_1 \geq \mathbf{x}^T(A'-A)\mathbf{x} > 0$.

Taken together, the foregoing remarks serve to show that if an edge is deleted from a connected graph then the index decreases.

Suppose next that G consists of two disjoint connected graphs H and K. Recall that $HuwK$ denotes the graph obtained from G by adding a bridge joining vertex u of H to vertex w of K. The remarks of the previous paragraph show that $\mu_1(HuwK) > \mu_1(G)$. The authors of [55] provide a means of comparing the indices of two such perturbations of G: they show that if $\phi_{H_u}(x) < \phi_{H_v}(x)$ for all $x > \mu_1(H_v)$ then $\mu_1(HuwK) > \mu_1(HvwK)$ for all vertices w of K. (Here v is a vertex of H and in accordance with §2.2, H_v denotes the graph obtained from H by adding a pendant edge at vertex v.)

In §3.2 we again inspect characteristic polynomials, this time to obtain an upper bound for the increase in index resulting from the addition of a pendant edge. In §§3.3 and 3.4 it is an inspection of eigenvectors which we use to determine whether the index of a connected graph increases or decreases under two further types of perturbation.

3.2 Addition of a Pendant Edge

There have been several estimates for the increase in index resulting from the attachment of a pendant edge, as in [5, Section 3], [16, Section 3] and [18, Section 6] . We shall see in Sections 4 and 5 that if $\mu_1 - \mu_2$ is large enough, reasonable upper bounds can be obtained in terms of μ_1, μ_2 and the angle of G at the vertex u of attachment. Here we observe that if $\mu_1(\mu_1 - \mu_2) > 1$ then Theorem 2.2 can be used directly to provide an upper bound for $\mu_1(G_u)$ in terms of these parameters.

Theorem 3.1 [41, Remark 5.3]. Let G be a connected graph and let G_u denote the graph obtained from G by adding a pendant edge at vertex u. If $\mu_1(\mu_1 - \mu_2) > 1$ then $\mu_1(G_u) < \mu_1(G) + \epsilon_u$ where $\epsilon_u = \dfrac{\alpha_{1u}^2}{\mu_1 - (\mu_1 - \mu_2)^{-1}}$.

Proof. Since the eigenvalues of G interlace those of G_u (by Theorem 1.1) it suffices to prove that $\phi_{G_u}(\mu_1 + \epsilon_u) > 0$. From Theorem 2.2, we have

$$\phi_{G_u}(\mu_1 + \epsilon_u) = \phi_G(\mu_1 + \epsilon_u)\{\epsilon_u + (\mu_1 - \mu_2)^{-1} - \sum_{i=2}^{m} \frac{\alpha_{iu}^2}{\mu_1 - \mu_i + \epsilon_\mu}\}.$$

Now

$$\sum_{i=2}^{m} \frac{\alpha_{iu}^2}{\mu_1 - \mu_i + \epsilon_\mu} \le \sum_{i=2}^{m} \frac{\alpha_{iu}^2}{\mu_1 - \mu_2} = \frac{1 - \alpha_{1u}^2}{\mu_1 - \mu_2},$$

and so $\phi_{G_u}(\mu_1 + \epsilon_u) \ge \phi_G(\mu_1 + \epsilon_u)\{\epsilon_u + \alpha_{1u}^2(\mu_1 - \mu_2)^{-1}\} > 0$. □

Example 1. If G is the skeleton of an icosahedron then $m = 4$ and $\mu_1 = 5, \mu_2 = \sqrt{5}, \mu_3 = -1, \mu_4 = -\sqrt{5}$. In particular, $\mu_1(\mu_1 - \mu_2) > 1$. For any vertex u, we have $\alpha_{1u}^2 = \frac{1}{12}$ and $\epsilon_u \simeq 0 \cdot 01797$. Thus Theorem 3.1 gives an upper bound of $5 \cdot 01797$ for $\mu_1(G_u)$, compared with an actual value of $5 \cdot 01728$ (to 5 decimal places). □

3.3 Splitting a Vertex

Let v be a vertex of G, let uv $(u \in U)$ be the edges containing v, and let $U_1 \dot{\cup} U_2$ be a nontrivial bipartition of U. Let G' be the graph obtained from $G - v$ by adding two (non-adjacent) vertices v_1, v_2 and edges $u_1v_1(u_1 \in U_1), u_2v_2(u_2 \in U_2)$. We say that G' is obtained from G by *splitting* the vertex v.

Theorem 3.2 (Simić [48]) If G is connected and if G' is obtained from G by splitting a vertex then $\mu_1(G') < \mu_1(G)$.

Proof. If G' is not connected then the result is already clear from the remarks in §3.1. Otherwise let G have principal eigenvector $\mathbf{x} = (x_1, \ldots, x_n)^T$, where vertex 1 is split into vertices labelled 0 and 1. Let $\mathbf{y} = \binom{x_1}{\mathbf{x}}$ and note that $A'\mathbf{y} < \mu_1(G)\mathbf{y}$; in other words the components of the vector $\mu_1(G)\mathbf{y} - A'\mathbf{y}$ are all non-negative and not all zero. On taking the scalar product of this vector with the principal eigenvector of G' we deduce that $\mu_1(G) > \mu_1(G')$. □

3.4 Subdivision of an Edge

The first thing to observe here is that the index of a connected graph may remain unchanged when an edge is subdivided. This is clearly true of any edge in a cycle C_n $(n \geq 3)$ and of any non-pendant edge in the tree W_k $(k \geq 1)$ illustrated in Figure 2. (Both C_n and W_k have index 2, and the figure shows

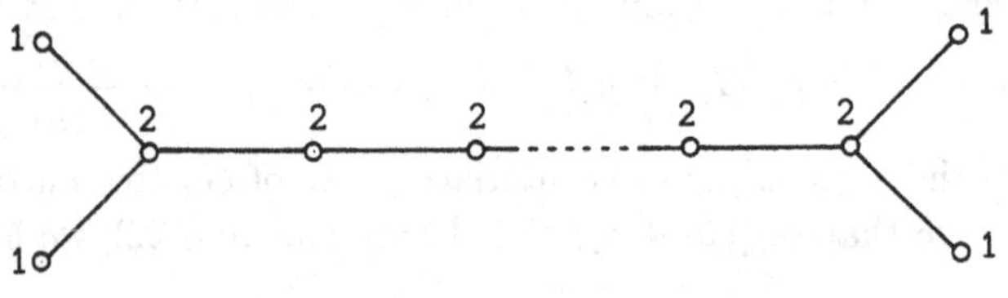

Figure 2: The graph W_k, a tree with $k + 4$ edges $(k \geq 1)$.

components of a corresponding eigenvector for W_k.) In all other instances, the index either increases strictly or decreases strictly: these cases were distinguished by Hoffman and Smith [28], whose result we now describe. An *internal path* of G is a walk $v_0v_1v_2 \ldots v_k$ $(k \geq 1)$ such that the vertices $v_1, \ldots, v_k$ are distinct, $\deg(v_0) > 2, \deg(v_k) > 2$ and $\deg(v_i) = 2$ whenever $0 < i < k$. Thus an internal path gives rise to either a subgraph W_k or a k-cycle with a pendant edge attached. Note that if G is a connected graph other than a cycle and the edge uv does not lie in an internal path then uv lies in a pendant path and so G is a proper subgraph of G_{uv}; hence $\mu_1(G_{uv}) > \mu_1(G)$ in this case. In the case that uv does lie in an internal path, one can construct from the principal eigenvector $\mathbf{x}$ of G a positive vector $\mathbf{y}$ such that $A'\mathbf{y} < \mu_1(G)\mathbf{y}$, where A' denotes the adjacency matrix of G_{uv}. It now follows as in the proof of Theorem 3.2 that $\mu_1(G) > \mu_1(G_{uv})$. We summarize the result of Hoffman and Smith as follows.

Theorem 3.2. Let uv be an edge of the connected graph G. If uv does not lie in an internal path of G and if $G \neq C_n$ then $\mu_1(G_{uv}) > \mu_1(G)$. If uv lies in an internal path and $G \neq W_k$ then $\mu_1(G_{uv}) < \mu_1(G)$. □

The only case not covered by the foregoing remarks is that in which $G = W_k$ and uv is a pendant edge; but then G is a proper subgraph of G_{uv} and so $\mu_1(G_{uv}) > \mu_1(G)$.

4. AN ANALYTICAL THEORY OF PERTURBATIONS

The various theorems from matrix theory (such as Theorem 1.1) which serve as basic tools in algebraic graph theory are listed in §0.3 of the monograph [14]. Here we show how to make use of the classical theory of matrix perturbations as described for example in [30, §§11.5, 11.6]. In general one considers a perturbation of the matrix A of the form $A(\zeta)$ where ζ is a complex variable and $A(0) = A$. Here we shall restrict ourselves to the case of a linear perturbation $A(\zeta) = A + \zeta B$, where A and B are fixed symmetric matrices. For our purposes, A and $A + B$ will be taken to be the adjacency matrices of a connected graph G and a perturbation G' of G respectively. The basic idea is to express eigenvalues and eigenvectors of $A(\zeta)$ as convergent power series in ζ. For this to be of any use in our context we need to impose conditions which ensure a radius of convergence greater than 1, so that we may obtain information about the eigenvalues of G' by setting $\zeta = 1$. For the purposes of exposition we consider changes in only the largest eigenvalue resulting from a perturbation of a connected graph G. The underlying theory is essentially the same for any simple eigenvalue. For corresponding results on multiple eigenvalues the reader is referred to [30, §11.7].

Suppose then that

$$(A + \zeta B)\mathbf{x}(\zeta) = \mu_1(\zeta)\mathbf{x}(\zeta) \tag{4.1}$$

where $\mathbf{x}(\zeta) \neq \mathbf{0}, \mathbf{x}(0) = \mathbf{x}$ (the principal eigenvector of G) and $\mu_1(0) = \mu_1$. We show first that if there exist convergent power series

$$\mu_1(\zeta) = \mu_1 + c_1\zeta + c_2\zeta^2 + \ldots, \tag{4.2}$$

$$\mathbf{x}(\zeta) = \mathbf{x} + \mathbf{x}_1\zeta + \mathbf{x}_2\zeta^2 + \ldots \tag{4.3}$$

such that

$$\mathbf{x}^T\mathbf{x}_r = 0 \text{ for each } r \in \mathbb{N} \tag{4.4}$$

then the perturbation coefficients c_r ($r \in \mathbb{N}$) and the vectors $\mathbf{x}_r$ ($r \in \mathbb{N}$) are determined by $\mathbf{x}, B$ and the spectral decomposition of A. Since $\mathbb{R}^n = \langle\mathbf{x}\rangle \oplus \langle\mathbf{x}\rangle^\perp$ and $\langle\mathbf{x}\rangle^\perp = \mathcal{E}(\mu_2) \oplus \ldots \oplus \mathcal{E}(\mu_m)$, condition (4.4) may be seen as normalizing $\mathbf{x}(\zeta)$ and constraining $\mathbf{x}(\zeta)$ to lie outside $\mathcal{E}(\mu_2) \oplus \ldots \oplus \mathcal{E}(\mu_m)$.

Let $E = \sum_{i=2}^m (\mu_1 - \mu_i)^{-1} P_i$, so that $(\mu_1 I - A)E = E(\mu_1 I - A) = I - P_1$ and $E\mathbf{x} = \mathbf{0}$. If we use equations (4.2) and (4.3) to substitute for $\mu_1(\zeta)$ and $\mathbf{x}(\zeta)$ in equation (4.1), and equate coefficients of ζ^r, then we obtain

$$\left.\begin{aligned} (\mu_1 I - A)\mathbf{x}_1 &= B\mathbf{x} - c_1\mathbf{x}, \\ (\mu_1 I - A)\mathbf{x}_r &= B\mathbf{x}_{r-1} - (c_1\mathbf{x}_{r-1} + \ldots + c_{r-1}\mathbf{x}_1) - c_r\mathbf{x} \\ & \qquad\qquad (r \geq 2). \end{aligned}\right\} \tag{4.5}$$

On taking the scalar product with $\mathbf{x}$, we obtain

$$\left.\begin{aligned} c_1 &= \mathbf{x}^T B\mathbf{x}, \\ c_r &= \mathbf{x}^T B\mathbf{x}_{r-1} \quad (r \geq 2). \end{aligned}\right\} \tag{4.6}$$

On multiplication by E, we obtain from equation (4.5):

$$\left.\begin{aligned} \mathbf{x}_1 &= EB\mathbf{x}, \\ \mathbf{x}_r &= EB\mathbf{x}_{r-1} - E(c_1\mathbf{x}_{r-1} + \ldots + c_{r-1}\mathbf{x}_1) \quad (r \geq 2). \end{aligned}\right\} \tag{4.7}$$

Equations (4.6) and (4.7) suffice to determine c_r and $\mathbf{x}_r$ recursively in terms of E, B and $\mathbf{x}$.

Let us now use equations (4.6) and (4.7) to *define* $c_r, \mathbf{x}_r$ and power series $\mu_1(\zeta) = \mu_1 + \sum_{r=1}^\infty c_r\zeta^r, \mathbf{x}(\zeta) = \mathbf{x} + \sum_{r=1}^\infty \mathbf{x}_r\zeta^r$. Then within the circle of convergence of these series, equations (4.5) and (4.1) hold. To see this, note that $(\mu_1 I - A)\mathbf{x}_1 = (I - P_1)B\mathbf{x} = B\mathbf{x} - \mathbf{x}\mathbf{x}^T B\mathbf{x} = B\mathbf{x} - c_1\mathbf{x}$; moreover $P_1\mathbf{x}_r = 0$ for all $r \in \mathbb{N}$ by induction on r, and so for $r \geq 2$, $(\mu_1 I - A)\mathbf{x}_r = (I - P_1)B\mathbf{x}_{r-1} - (c_1\mathbf{x}_{r-1} + \ldots + c_{r-1}\mathbf{x}_1)$; finally $P_1 B\mathbf{x}_{r-1} = \mathbf{x}\mathbf{x}^T B\mathbf{x}_{r-1} = c_r\mathbf{x}$. The next thing to do is to estimate R, a common radius of convergence of the series $\mu_1(\zeta), \mathbf{x}(\zeta)$. A general result [30, Theorem 11.5.1] asserts that $R > 0$, but it would be nice to have a lower bound for R in terms of $\|B\|$ and $\|E\|$, where $\|B\| = \sup\{|B\mathbf{v}| : |\mathbf{v}| = 1\}$ and $\|E\| = \sup\{|E\mathbf{v}| : |\mathbf{v}| = 1\} = (\mu_1 - \mu_2)^{-1}$. We use an argument from [5, §2] to show how R may be estimated when $\|B\| \leq 1$. We can then apply the theory in the case that B corresponds to the addition of an edge.

Suppose that $\|B\| \leq 1$, and let $b = |B\mathbf{x}|, e = (\mu_1 - \mu_2)^{-1}$, $\mathbf{x}_0 = \mathbf{x}$. Define

$V(\zeta) = \sum_{r=0}^{\infty} v_r \zeta^r$, with radius of convergence R^*, where

$$\left.\begin{array}{rcl} v_0 & = & 1,\ v_1 = eb, \\ v_r & = & e(v_0 v_{r-1} + v_1 v_{r-2} + \ldots + v_{r-1} v_0)\ (r \geq 2). \end{array}\right\} \tag{4.8}$$

By induction on r, we have $|\mathbf{x}_r| \leq v_r$ $(r \geq 0)$ and $|\mathbf{c}_r| \leq v_{r-1}$ $(r \geq 1)$; and so $R \geq R^*$. Now from equations (4.8) we have

$$V(\zeta) = e\zeta V(\zeta)^2 + 1 + \zeta(eb - e), \tag{4.9}$$

and on solving this quadratic for $e\zeta V(\zeta)$, we find that $R^* \geq 1/2e(1+\sqrt{b})$. It follows that if $\|B\| \leq 1$ and $\mu_1 - \mu_2 > 2(1 + \sqrt{|B\mathbf{x}|})$ then $R^* > 1$ and so $\mu_1(\zeta), \mathbf{x}(\zeta)$ converge at $\zeta = 1$. In particular, $\mu_1(1)$ is an eigenvalue of $A + B$. Is it the largest one? The following remark shows that the answer in our situation is 'yes'.

Remark 4.1. Let G be a connected graph and let u, v be non-adjacent vertices of G. Let A, $A+B$ be the adjacency matrices of G, $G+uv$ respectively. For each real $\zeta \in (0, R)$, $\mu_1(\zeta)$ is the largest eigenvalue of $A + \zeta B$.

Proof. The eigenvalues of $A + \zeta B$ are continuous functions of ζ. Hence if $\mu_1(\zeta)$ does not remain the largest eigenvalue of $A + \zeta B$ as ζ increases from 0 to R then for some $\xi \in (0, R)$, $\mu_1(\xi)$ coincides with a second eigenvalue which is equal largest. But $A + \xi B$ is irreducible because A is irreducible and all entries of A and ξB are non-negative. Hence the largest eigenvalue of $A + \xi B$ is simple, a contradiction. □

We can now sharpen Theorem 2.7 in respect of the index of $G + uv$.

Theorem 4.2. Let u, v be non-adjacent vertices of the connected graph G. If $\mu_1 - \mu_2 > 2(1 + \sqrt[4]{\alpha_{1u}^2 + \alpha_{1v}^2})$, in particular if $\mu_1 - \mu_2 > 4$, then $\mu_1(G + uv) = \mu_1 + \sum_{r=1}^{\infty} c_r$ where the c_r are recursively defined functions of $\mu_i, \alpha_{iu}, \alpha_{iv}, p_{uv}^{[i]}$ $(i = 1, 2, \ldots, m)$.

Proof. Here $\|B\| = 1$ because the only non-zero elements of B are ones in positions (u, v) and (v, u). By Remark 4.1 and the observations preceding it, we know that $\mu_1(G + uv) = \mu_1(1) = \mu_1 + \sum_{r=1}^{\infty} c_r$ provided that $\mu_1 - \mu_2 > 2(1 + \sqrt{|B\mathbf{x}|})$. Here $|B\mathbf{x}| = \sqrt{\alpha_{1u}^2 + \alpha_{1v}^2}$, and so it remains to show that the c_r are recursively defined functions of $\mu_i, \alpha_{iu}, \alpha_{iv}$ and $p_{uv}^{[i]}$ $(i = 1, \ldots, m)$. We describe a scalar as *known* if it is determined by these $4m$ invariants, and we take $u = 1, v = 2$ without loss of generality. We first show by induction on t $(t \geq 0)$ that if $\mathbf{w}$ is a vector of the form

$$\mathbf{w} = E^{s_t} B E^{s_{t-1}} B \ldots E^{s_2} B E^{s_1} B\mathbf{x}, \tag{4.10}$$

where each $s_i \in I\!N$, then the first two components of $\mathbf{w}$ are known. This is clear if $t = 0$ because $B\mathbf{x} = (x_2, x_1, 0, \ldots, 0)^T$ and $x_j = |P_1 \mathbf{e}_j|$ $(j = 1, 2)$. Let

$E^s = (e_{ij}^{(s)})$, so that if $\mathbf{v} = (v_1, v_2, \ldots, v_n)^T$ then the first two components of $E^s B\mathbf{v}$ are $e_{11}^{(s)}v_2 + e_{12}^{(s)}v_1$ and $e_{21}^{(s)}v_2 + e_{22}^{(s)}v_1$. Now $e_{jk}^{(s)} = \sum_{i=2}^m (\mu_1 - \mu_i)^{-s} p_{jk}^{[i]}$ and so if the first two components of $\mathbf{v}$ are known then so are the first two components of $E^s B\mathbf{v}$. In particular, the first two components of $\mathbf{w}$ are known.

We can now show by induction on r ($r \geq 1$) that c_r is known and $\mathbf{x}_r$ is a known linear combination of vectors of the form (4.10). This is clear for $r = 1$ because $c_1 = \mathbf{x}^T B\mathbf{x} = 2x_1x_2$, while $\mathbf{x}_1 = EB\mathbf{x}$. Now suppose that $r > 1$ and consider $c_r = \mathbf{x}^T B\mathbf{x}_{r-1}$, where we may suppose that $\mathbf{x}_{r-1}$ is a known linear combination of vectors of the form (4.10). Then the first two components $x_{r-1,1}, x_{r-1,2}$ of $\mathbf{x}_{r-1}$ are known, and c_r is known because $c_r = x_2x_{r-1,1} + x_1x_{r-1,2}$. Next consider the expression in (4.7) for $\mathbf{x}_r$ ($r \geq 2$). In the right-hand side, each of $c_1, \ldots, c_{r-1}$ is known and each of $\mathbf{x}_1, \ldots, \mathbf{x}_{r-1}$ is a known linear combination of vectors of the form (4.10). If we premultiply a vector of this form by either EB or E then we obtain a vector of the same form: hence $\mathbf{x}_r$ is a known linear combination of vectors of the form (4.10). Thus the vectors $\mathbf{x}_r$, and hence the perturbation coefficients c_r, are determined recursively in terms of $\mu_i, \alpha_{iu}, \alpha_{iv}, p_{uv}^{[i]}$ ($i = 1, 2, \ldots, m$). □

In the situation of Theorem 4.2, we have by way of illustration, $c_1 = 2x_1x_2$ and $c_2 = \mathbf{x}^T B\mathbf{x}_1 = \mathbf{x}^T BEB\mathbf{x} = e_{11}x_2^2 + e_{12}x_2x_1 + e_{21}x_1x_2 + e_{22}x_1^2$. Thus in terms of u and v,

$$c_1 = 2\alpha_{1u}\alpha_{1v},\ c_2 = \sum_{i=2}^{m} (\mu_1 - \mu_i)^{-1}(\alpha_{1v}^2\alpha_{iu}^2 + 2\alpha_{1v}\alpha_{1u}p_{uv}^{[i]} + \alpha_{1u}^2\alpha_{iv}^2). \quad (4.11)$$

For an explicit example, we take a graph whose symmetry makes it straightforward to calculate the invariants involved.

Example 2. Let G be the graph whose complement $\overline{G}$ is the skeleton of an icosahedron. Here $m = 4$ and $\mu_1 = 6, \mu_2 = \sqrt{5} - 1, \mu_3 = 0, \mu_4 = -\sqrt{5} - 1$. In particular, $\mu_1 - \mu_2 > 4$ and so our condition for convergence (Theorem 4.2) is satisfied. In order to find the matrices P_i ($i = 1, 2, 3, 4$) it is convenient to consider the adjacency matrix $\overline{A}$ of $\overline{G}$, which (since G is regular) is given by

$$\overline{A} = J - I - A = \overline{\mu}_1 P_1 + \overline{\mu}_2 P_2 + \overline{\mu}_3 P_3 + \overline{\mu}_4 P_4$$

where $\overline{\mu}_1 = 5, \overline{\mu}_2 = -\sqrt{5}, \overline{\mu}_3 = -1$ and $\overline{\mu}_4 = \sqrt{5}$. Since $\overline{G}$ is distance-transitive we can describe the (u, v) entries of $I, \overline{A}, \overline{A}^2, \overline{A}^3, P_1, P_2, P_3, P_4$ as follows:

$d_{\overline{G}}(u,v)$	I	$\overline{A}$	$\overline{A}^2$	$\overline{A}^3$	P_1	P_2	P_3	P_4
0	1	0	5	10	$\frac{1}{12}$	$\frac{1}{4}$	$\frac{5}{12}$	$\frac{1}{4}$
1	0	1	2	13	$\frac{1}{12}$	$\frac{\sqrt{5}}{20}$	$\frac{-1}{12}$	$\frac{-\sqrt{5}}{20}$
2	0	0	2	8	$\frac{1}{12}$	$\frac{-\sqrt{5}}{20}$	$\frac{-1}{12}$	$\frac{\sqrt{5}}{20}$
3	0	0	0	10	$\frac{1}{12}$	$\frac{-1}{4}$	$\frac{5}{12}$	$\frac{-1}{4}$

Here the entries of $\overline{A}^2$ and $\overline{A}^3$ are obtained directly from the icosahedron, while those of P_2, P_3, P_4 are calculated from the cubic expression

$$P_i = \prod_{j \neq i} (\overline{\mu}_i - \overline{\mu}_j)^{-1} (\overline{A} - \overline{\mu}_j I).$$

[We know that $P_1 = \frac{1}{12}J = I - P_2 - P_3 - P_4$.] Now to within isomorphism, there is only one graph $G + uv$, which arises when $d_{\overline{G}}(u,v) = 1$. If we use the second row of the foregoing table to substitute the appropriate values in equation (4.10), we find that $c_1 = \frac{1}{6}$ and $c_2 = \frac{49}{2376}$. Thus our first approximation to $\mu_1(G+uv)$ is 6.1667 and our second approximation is 6.1873, compared with an actual value of 6.1894 (to four decimal places).

We remark that the condition $\mu_1 - \mu_2 > 2(1 + \sqrt[4]{\alpha_{1u}^2 + \alpha_{1v}^2})$ fails for the icosahedral graph $\overline{G}$, although it is possible that the series $\sum_r c_r$ converges to $\mu_1(\overline{G} + uv) - \mu_1(\overline{G})$ nevertheless. Here there are two possibilities for $\overline{G} + uv$, according as $d_{\overline{G}}(u,v)$ is 2 or 3. For the record, the numerical results (obtained using the third and fourth rows of our table) are as follows:

if $d_{\overline{G}}(u,v) = 2$ then

$$\mu_1(\overline{G} + uv) \simeq 5.1959,\ \overline{\mu}_1 + c_1 \simeq 5.1667,\ \overline{\mu}_1 + c_1 + c_2 \simeq 5.1926;$$

if $d_{\overline{G}}(u,v) = 3$ then

$$\mu_1(\overline{G} + uv) \simeq 5.1926, \overline{\mu}_1 + c_1 \simeq 5.1667, \overline{\mu}_1 + c_1 + c_2 \simeq 5.1898.$$

□

The arguments used to prove Theorem 4.2 may be extended to deal with the attachment of a pendant edge to a connected graph G. One replaces A with A_0, the adjacency matrix of the graph G_0 which consists of G and an isolated vertex labelled 0. Although G_0 is not connected, μ_1 remains a simple eigenvalue of A_0, and the vector $\mathbf{x}$ becomes $(0, \alpha_{11}, \alpha_{12}, \ldots, \alpha_{1n})^T$. If, without loss of generality, the pendant edge is attached at vertex 1 then the entries of E which we need are:

$$e_{01}^{(1)} = e_{10}^{(1)} = 0,\ e_{00}^{(1)} = \mu_1^{-1} \text{ and } e_{11}^{(1)} = \sum_{i=2}^{m} (\mu_1 - \mu_i)^{-1} \alpha_{i1}^2.$$

If now we describe a scalar as known if it is determined by the $2m$ invariants μ_i, α_{i1} $(i = 1, \ldots, m)$ then we can show that for odd r, we have $c_r = 0, \mathbf{x}_r = (a_r, 0, \ldots, 0)^T$ where a_r is known; and for even r, c_r is known while $\mathbf{x}_r = f_r(E)(0, 1, 0, \ldots, 0)^T$ where f_r is a polynomial with known coefficients. Since $|B\mathbf{x}| = \alpha_{11}$, Remark 4.1 yields the following analogue of Theorem 4.2.

Theorem 4.3. Let G_u be the graph obtained from the connected graph G by adding a pendant edge at vertex u. If $\mu_1 - \mu_2 > 2(1 + \sqrt{\alpha_{1u}})$ then $\mu_1(G_u) = \mu_1 + \sum_{r=1}^{\infty} c_r$ where the c_r are recursively defined functions of μ_i, α_{iu} $(i = 1, 2, \ldots, m)$. □

The perturbation coefficients which arise in Theorem 4.3 were obtained in [5] by examining power series solutions $\mu_1 + \sum_{r=1}^{\infty} c_r\zeta^r$ of the equation $\det(xI - A_0 - \zeta B) = 0$. Another condition for convergence which is established in that paper is: $\mu_2(\mu_1 - \mu_2) > 4$. For even r, the coefficient c_r takes the form $\sum_{h=r/2}^{r-1} \frac{d_{hr}}{\mu_1^h}$, where each d_{hr} is a function of

$$\alpha_{1u}, \ldots, \alpha_{mu}, \quad \mu_1 - \mu_2, \quad \mu_1 - \mu_3, \quad \ldots, \quad \mu_1 - \mu_m.$$

It remains to be seen whether the convergent series $\mu_1 + \sum_{k=1}^{\infty} \sum_{h=k}^{2k-1} \frac{d_{h,2k}}{\mu_1^h}$ may be rearranged to express $\mu_1(G_u)$ as a convergent power series of the form $\mu_1 + \sum_{h=1}^{\infty} \frac{d'_h}{\mu_1^h}$, where each d'_h is a function of $\alpha_{1u}, \ldots, \alpha_{mu}, \mu_1 - \mu_2, \ldots, \mu_1 - \mu_m$. The second approximation to $\mu_1(G_u)$ afforded by Theorem 4.3 is $\mu_1 + c_2 + c_4$, which turns out to be $\mu_1 + \frac{\alpha_{1u}^2}{\mu_1} + \frac{\alpha_{1u}^2 \tau_u}{\mu_1^2} - \frac{\alpha_{1u}^4}{\mu_1^3}$, where $\tau_u = \sum_{i=2}^{m} (\mu_1 - \mu_i)^{-1} \alpha_{iu}^2$. For the icosahedral graph of example 1, a graph for which $\mu_2(\mu_1 - \mu_2) > 4$, this approximation is 5.01726, compared with $\mu_1(G_u) \simeq 5.01728$.

5. AN ALGEBRAIC THEORY OF PERTURBATIONS

5.1 Introduction

The techniques of the previous section enable the indices of certain perturbed graphs to be computed to any degree of accuracy provided that (i) the spectrum and appropriate angles of the original graph are known, and (ii) $\mu_1 - \mu_2$ is large enough. If less information is given, typically the spectrum and relevant components $x_j (= \alpha_{1j})$ of the principal eigenvector, then the methods of §4 can still be used to obtain bounds on the index of the perturbed graph. For example if G' is obtained from G by adding the edge uv then (for large enough $\mu_1 - \mu_2$) we have from equation (4.9),

$$\mu_1(G') < \mu_1 + \frac{1}{2e} - \frac{1}{2}\sqrt{(\frac{1}{2} - e)^2 - e^2 b} \qquad (5.1)$$

where $e = (\mu_1 - \mu_2)^{-1}$ and $b = \sqrt{x_u^2 + x_v^2}$. This however is a very crude bound and we can do better by invoking the algebraic theory of matrix perturbations exploited by Maas in [34]. In this paper he shows that for $G' = G + uv$ as above, we have the following result.

Theorem 5.1 If u, v are non-adjacent vertices of the connected graph G, then $\mu_1(G + uv) < \mu_1 + 1 + \delta - \gamma$ where $\delta > 0$ and

$$\gamma = \frac{\delta(1+\delta)(2+\delta)}{(x_u + x_v)^2 + \delta(2 + \delta + 2x_u x_v)} = \mu_1 - \mu_2. \tag{5.2}$$

□

(To see that Theorem 5.1 improves the inequality (5.1) for large enough $\mu_1 - \mu_2$, let $\mu_1 - \mu_2 \to \infty$.)

In §5.2 we outline the theory required to obtain Theorem 5.1 and similar results. Various applications and examples are given in §§5.3 and 5.4. The techniques may be used to obtain an upper bound on any eigenvalue of a perturbed matrix $A + B$, not just the largest. We note here that analogous lower bounds for the index of a perturbed graph are obtained easily from the fact that $\mu_1(G') = \sup\{\mathbf{y}^T(A+B)\mathbf{y} : |\mathbf{y}| = 1\}$: on taking $\mathbf{y}$ to be the principal eigenvector of G, we obtain

$$\mu_1(G') \geq \mu_1 + \mathbf{x}^T B\mathbf{x}. \tag{5.3}$$

This lower bound is just $\mu_1 + c_1$ in the notation of §4. In the case that $G' = G + uv$, the inequality (5.3) becomes $\mu_1(G') \geq \mu_1 + 2x_u x_v$. In §5.4 we shall compare the effects of two different perturbations on the index of a graph by showing that the lower limit for the index of one perturbed graph exceeds the upper limit for the index of the other.

5.2 Intermediate Eigenvalue Problems of the Second Type

Here we outline the results needed from [53] in terms of an n-dimensional Euclidean space V in which $(\mathbf{u}, \mathbf{v})$ denotes the inner product of vectors $\mathbf{u}$ and $\mathbf{v}$. The eigenvalues of a symmetric linear transformation T of V are denoted by $\lambda_1(T), \ldots, \lambda_n(T)$ where $\lambda_1(T) \leq \lambda_2(T) \leq \ldots \leq \lambda_n(T)$. Let $\tilde{A}$ be a symmetric linear transformation of V and let $\tilde{B}$ be a positive linear transformation of V: the general problem is to find lower bounds for the eigenvalues $\lambda_i(\tilde{A} + \tilde{B})$, given $\tilde{B}$ and appropriate invariants of $\tilde{A}$.

Another inner product may be defined on V by: $[\mathbf{u}, \mathbf{v}] = (\tilde{B}\mathbf{u}, \mathbf{v})$. Choose any basis $\{\mathbf{v}_1, \ldots, \mathbf{v}_n\}$ for V and, using this second inner product, let P_r be the orthogonal projection onto the subspace of V spanned by $\mathbf{v}_1, \ldots, \mathbf{v}_r$ ($r = 1, \ldots, n$). Thus $P_n = I$ and if we define P_0 to be the zero transformation of V then $[P_{r-1}\mathbf{v}, \mathbf{v}] \leq [P_r\mathbf{v}, \mathbf{v}]$ for all $\mathbf{v} \in V$ and all $r \in \{1, \ldots, n\}$. Hence

$((\tilde{A}+\tilde{B}P_{r-1})\mathbf{v},\mathbf{v}) \leq ((\tilde{A}+\tilde{B}P_r)\mathbf{v},\mathbf{v})$ for all $\mathbf{v} \in V$ and all $r \in \{1,\ldots,n\}$. It is straightforward to check that each $\tilde{B}P_r$ is a symmetric transformation of the original inner product space V. Now for any symmetric transformation T of $V, \lambda_i(T)$ is the minimum of $\max\{(T\mathbf{v},\mathbf{v}) : |\mathbf{v}| = 1, \mathbf{v} \in U\}$ taken over all i-dimensional subspaces U of V [30, Theorem 8.2.2], and it follows that $\lambda_i(\tilde{A}) \leq \lambda_i(\tilde{A}+\tilde{B}P_1) \leq \lambda_i(\tilde{A}+\tilde{B}P_2) \leq \ldots \leq \lambda_i(\tilde{A}+\tilde{B}P_{n-1}) \leq \lambda_i(\tilde{A}+\tilde{B})$ $(i = 1,\ldots,n)$.

The problem of determining the eigenvalues of $\tilde{A} + \tilde{B}P_r$ for some integer $r \in \{1,\ldots,n-1\}$ is called an intermediate eigenvalue problem of the second type. In practice we choose the basis $\{\mathbf{v}_1,\ldots,\mathbf{v}_n\}$ as follows, so that $\tilde{A}+\tilde{B}P_r$ can be represented by a matrix of simple form. Let $\tilde{A}\mathbf{u}_i = \tilde{\lambda}_i\mathbf{u}_i$ $(i = 1,\ldots,n)$ where $\tilde{\lambda}_i = \lambda_i(\tilde{A})$ $(i = 1,\ldots,n)$ and $\mathbf{u}_1\ldots,\mathbf{u}_n$ are orthonormal. If we now choose $\mathbf{v}_i = \tilde{B}^{-1}\mathbf{u}_i$ $(i = 1,\ldots,n)$ then $P_r\mathbf{u}_j = \sum_{i=1}^{r}\gamma_{ij}\mathbf{v}_i$, where $(\gamma_{ij})^{-1}$ is the $r \times r$ Gram matrix T_r with (i,j)-entry $[\mathbf{v}_i,\mathbf{v}_j]$ $(i,j = 1,\ldots,r)$; moreover the matrix of $\tilde{A} + \tilde{B}P_r$ with respect to the basis $\{\mathbf{u}_1,\ldots,\mathbf{u}_r\}$ is

$$\begin{pmatrix} \tilde{\lambda}_1 & & & \\ & \tilde{\lambda}_2 & & \\ & & \ddots & \\ & & & \tilde{\lambda}_n \end{pmatrix} + \begin{pmatrix} T_r^{-1} & 0 \\ 0 & 0 \end{pmatrix}. \tag{5.4}$$

In our applications, we take $V = \mathbb{R}^n, (\mathbf{u},\mathbf{v}) = \mathbf{u}^T\mathbf{v}$ and we identify a linear transformation of $\mathbb{R}^n$ with its matrix with respect to the standard basis of $\mathbb{R}^n$. If, as usual, A and $A + B$ are the adjacency matrices of a graph G and its perturbation G' then we take

$$\tilde{A} = -A - (\lambda_n(B) + \delta)I \text{ and } \tilde{B} = (\lambda_n(B) + \delta)I - B, \text{ where } \delta > 0.$$

Thus $\tilde{B}$ is positive, $\tilde{A} + \tilde{B} = -A - B$ and $\mu_1(G') = -\lambda_1(\tilde{A}+\tilde{B}) \leq -\lambda_1(\tilde{A}+\tilde{B}P_r)$. For a given value of r, the parameter δ is chosen to optimize this upper bound for $\mu_1(G')$. Results for various perturbations are given in the literature [34, 41] without details of the calculations involved. In §5.3 we give at least an outline of the calculations required to deal with a perturbation which (to the best of the author's knowledge) is not treated elsewhere. We shall then feel justified in omitting the calculations required for the generally simpler types of perturbation which are applied to specific examples in §5.4.

5.3 A Perturbation which Preserves Degrees

Let G be a connected graph with vertices t, u, v, w such that t is adjacent to w but not to u, and v is adjacent to u but not to w (see Figure 3). Here we discuss the perturbation G' obtained from G by replacing the edges tw, uv with edges tu, vw, thereby preserving the degrees of all vertices of

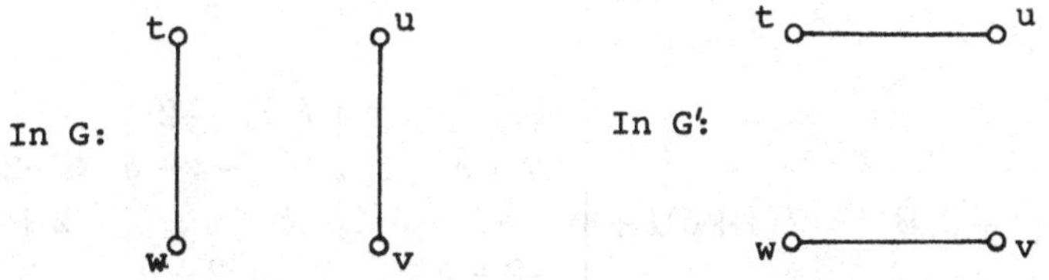

Figure 3: A perturbation $G \to G'$ which preserves degrees

G. Thus $\mathbf{x}^T B\mathbf{x} = 2(x_t - x_v)(x_u - x_w)$ and it follows from (5.2) that if $(x_t - x_v)(x_u - x_w) > 0$ then $\mu_1(G') > \mu_1(G)$; and if $(x_t - x_v)(x_u - x_w) = 0$ then $\mu_1(G') \geq \mu_1(G)$. We use the results of §5.2 to see what can be said when $(x_t - x_v)(x_u - x_w) < 0$. Our complete result is the following.

Theorem 5.2. Let G' be the graph obtained from a connected graph G by exchanging edges tw, uv for non-edges tu, vw. If $(x_t - x_v)(x_u - x_w) > 0$ then $\mu_1(G') > \mu_1$. If $(x_t - x_v)(x_u - x_w) = 0$ then $\mu_1(G') \geq \mu_1$. If $(x_t - x_v)(x_u - x_w) < 0$ and $\mu_1 - \mu_2 > \dfrac{(x_t - x_v)^2 + (x_w - x_u)^2}{(x_t - x_v)(x_w - x_u)}$ then $\mu_1(G') < \mu_1$.

Proof. We take $t = 1, u = 2, v = 3, w = 4$ without loss of generality, so that

$$B = \begin{bmatrix} 0 & 1 & 0 & -1 & \\ 1 & 0 & -1 & 0 & \\ 0 & -1 & 0 & 1 & \\ -1 & 0 & 1 & 0 & \\ & & & & O \end{bmatrix}.$$

Now $Q^T BQ = \mathrm{diag}(2, -2, 0, 0, \ldots, 0)$, where

$$Q = \begin{bmatrix} \frac{1}{2} & \frac{1}{2} & \frac{1}{2} & \frac{1}{2} & \\ \frac{1}{2} & -\frac{1}{2} & \frac{1}{2} & -\frac{1}{2} & \\ -\frac{1}{2} & -\frac{1}{2} & \frac{1}{2} & \frac{1}{2} & \\ -\frac{1}{2} & \frac{1}{2} & \frac{1}{2} & -\frac{1}{2} & \\ & & & & I \end{bmatrix}.$$

Hence $\lambda_n(B) = 2, \tilde{B} = (\delta + 2)I - B$ and

$$\tilde{B}^{-1} = (2 + \delta)^{-1} I + Q \begin{bmatrix} \delta^{-1} - (2 + \delta)^{-1} & \\ & O \end{bmatrix} Q^T.$$

Thus

$$\delta(2+\delta)(4+\delta)\tilde{B}^{-1} = \delta(4+\delta)I + \begin{bmatrix} 2 & 2+\delta & -2 & -2-\delta & \\ 2+\delta & 2 & -2-\delta & -2 & \\ -2 & -2-\delta & 2 & 2+\delta & \\ -2-\delta & -2 & 2+\delta & 2 & \\ & & & & O \end{bmatrix}.$$

We now apply the results of §5.2 with $r = 1$ and $\mathbf{u}_1 = \mathbf{x}$. Since $\tilde{\lambda}_i = \lambda_i(-A-(2+\delta)I)$ $(i = 1, 2, \ldots, n)$, we have $\tilde{\lambda}_1 = -\mu_1 - 2 - \delta$ and $\tilde{\lambda}_2 = -\mu_2 - 2 - \delta$. The matrix $\tilde{A} + \tilde{B}P_1$ is similar to $\mathrm{diag}(\tilde{\lambda}_1 + \gamma, \tilde{\lambda}_2, \tilde{\lambda}_3, \ldots, \tilde{\lambda}_n)$, where $\gamma^{-1} = \gamma_{11}^{-1} = [\mathbf{v}_1, \mathbf{v}_1] = [\tilde{B}^{-1}\mathbf{u}_1, \tilde{B}^{-1}\mathbf{u}_1] = \mathbf{x}^T\tilde{B}^{-1}\mathbf{x}$. Having calculated $\tilde{B}^{-1}$ above, we know that

$$\gamma = \frac{\delta(2+\delta)(4+\delta)}{\delta(4+\delta) + 2(x_1+x_2-x_3-x_4)^2 + 2\delta(x_1-x_3)(x_2-x_4)}.$$

Now

$$\lambda_1(\tilde{A} + \tilde{B}P_1) = \begin{cases} \tilde{\lambda}_1 + \gamma & \text{if } \gamma \le \tilde{\lambda}_2 - \tilde{\lambda}_1 \\ \tilde{\lambda}_2 & \text{if } \gamma > \tilde{\lambda}_2 - \tilde{\lambda}_1 \end{cases}$$

and so

$$\mu_1(G') \le \begin{cases} \mu_1 + 2 + \delta - \gamma & \text{if } \gamma \le \mu_1 - \mu_2 \\ \mu_2 + 2 + \delta & \text{if } \gamma > \mu_1 - \mu_2 \end{cases}.$$

As a function of δ $(\delta > 0)$, γ has range $(0, \infty)$ and so we may choose $\delta > 0$ such that $\gamma = \mu_1 - \mu_2$. This determines our upper bound for $\mu_1(G')$, namely $\mu_2+2+\delta$, and it remains to obtain conditions under which this upper bound is less than μ_1. Let us write α for $(x_1+x_2-x_3-x_4)^2$ and β for $(x_1-x_3)(x_4-x_2)$: thus $\alpha \ge 0$ and $\beta > 0$. Now $\mu_2 + 2 + \delta < \mu_1$ if and only if $\gamma - (2+\delta) > 0$; and $\gamma - (2+\delta) = \dfrac{-2(\alpha-\delta\beta)(2+\delta)}{\delta(4+\delta)+2(\alpha-\delta\beta)} = \dfrac{-2(\alpha-\delta\beta)\gamma}{\delta(4+\delta)}$. If $\gamma > 2 + \alpha\beta^{-1}$ then we have

$$\begin{aligned} -2(\alpha-\delta\beta) &= \gamma^{-1}\delta(4+\delta)[\gamma-(2+\delta)] > \gamma^{-1}\delta(4+\delta)(\alpha\beta^{-1}-\delta) \\ &= \gamma^{-1}\delta(4+\delta)\beta^{-1}(\alpha-\delta\beta), \end{aligned}$$

whence $(\alpha-\delta\beta)[2+\gamma^{-1}\delta(4+\delta)\beta^{-1}] < 0$. Hence if $\gamma > 2 + \alpha\beta^{-1}$, then $\alpha - \delta\beta < 0$ and so $\gamma - (2+\delta) > 0$, $\mu_1(G') < \mu_1$. Since $2 + \alpha\beta^{-1} = \dfrac{(x_1-x_3)^2 + (x_4-x_2)^2}{(x_1-x_3)(x_4-x_2)}$, our theorem is proved. □

5.4 Further Applications and Examples

The techniques of §5.3 may be used to prove Theorem 5.1, which concerns the addition of an edge, and the following result (stated in [41]) which deals with the deletion of an edge.

Theorem 5.3. Let G be a connected graph with principal eigenvector $(x_1, \ldots, x_n)^T$. If the graph G' is obtained from G by deleting the edge hi then $\mu_1(G') < \mu_1 + 1 + \delta - \gamma$ where

$$\gamma = \frac{\delta(1+\delta)(2+\delta)}{(x_h - x_i)^2 + \delta(2 + \delta - 2x_h x_i)} = \mu_1 - \mu_2.$$

□

Of course, in the situation of Theorem 5.3, we already know that $\mu_1(G') < \mu_1$ and so the result provides new information only when $\gamma > 1+\delta$, equivalently when $2x_h x_i > (x_h - x_i)^2$. This condition is certainly satisfied in the following example, taken from [41], because there vertices h and i lie in the same orbit of Aut(G).

Example 3. Let G_n $(n \geq 3)$ be the $3n$-vertex graph consisting of a $2n$-cycle C_{2n} with vertices $1, 2, \ldots, 2n$ (in order), an n-cycle C_n with vertices $2n+1, \ldots, 3n$ (in order) and edges from $2n+i$ to $2i-1$ and $2i$ $(i = 1, \ldots, n)$. The graph G_6 is shown in Figure 4. The orbits of Aut(G_n) are $\{1, 2, \ldots, 2n\}$

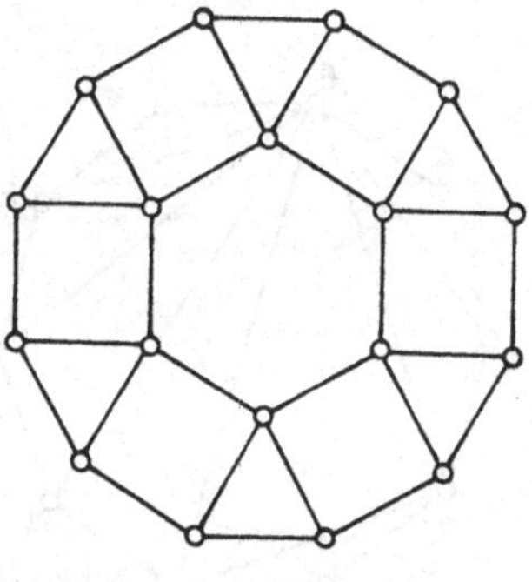

Figure 4: the graph G_6 of Example 3.

and $\{2n+1, \ldots, 3n\}$ and we have $x_1 = \ldots = x_{2n} = 1/2\sqrt{n}$, $x_{2n+1} = \ldots = x_{3n} = 1/\sqrt{2n}$. If we remove an edge of C_n we obtain a graph G_n' such that $\mu_1(G_n') < \mu_1(G_n) + 1 + \delta - \gamma$ where δ, γ are given by Theorem 5.3. If we remove an edge of C_{2n} then there are two possibilities for the resulting graph G_n'': in either case, the inequality (5.3) yields $\mu_1(G_n'') > \mu_1(G_n) - \frac{1}{2n}$. In order to conclude that $\mu_1(G_n'') > \mu_1(G_n')$ it suffices to show that $\gamma - (1+\delta) > \frac{1}{2n}$; but this is clear since $\gamma - (1+\delta) = \dfrac{1+\delta}{(2n + \delta n - 1)}$. □

The remaining results presented in this section relate to the relocation of an edge. The advantage in regarding the deletion of an edge hi and the addition of an edge jk as a single perturbation of G is that the bounds obtained depend only on the appropriate invariants of G, and not on those of $G-hi$ or $G+jk$. In §5.3, *two* edges were relocated and the same methods serve to prove the next two results, where again $x_1, \ldots, x_n$ are the components of the principal eigenvector of the connected graph G.

Theorem 5.4. [41] Let G be a connected graph with distinct vertices h, i, j, k such that h is adjacent to i, j is not adjacent to k. Let G' be the graph obtained from G by replacing edge hi with jk. If $x_j x_k \geq x_h x_i$ then $\mu_1(G') > \mu_1$. If $x_j x_k < x_h x_i$ and $\mu_1 - \mu_2 > \dfrac{x_h^2 + x_i^2 + x_j^2 + x_k^2}{2(x_h x_i - x_j x_k)}$ then $\mu_1(G') < \mu_1$. □

One detail which remains to be checked here is the case $x_j x_k = x_i x_h$, where inequality (5.3) yields only $\mu_1(G') \geq \mu_1(G)$. If however $\mu_1(G') = \mu_1(G)$ in this situation, we have $A'\mathbf{x} = \mu_1 \mathbf{x} = A\mathbf{x}$ where A' is the adjacency matrix of G'; but this is impossible as can be seen by comparing the h-th components of $A'\mathbf{x}$ and $A\mathbf{x}$.

We note in passing that if $x_j x_k < x_h x_i$ then the index may increase or decrease when hi is replaced by jk, as illustrated by the following example.

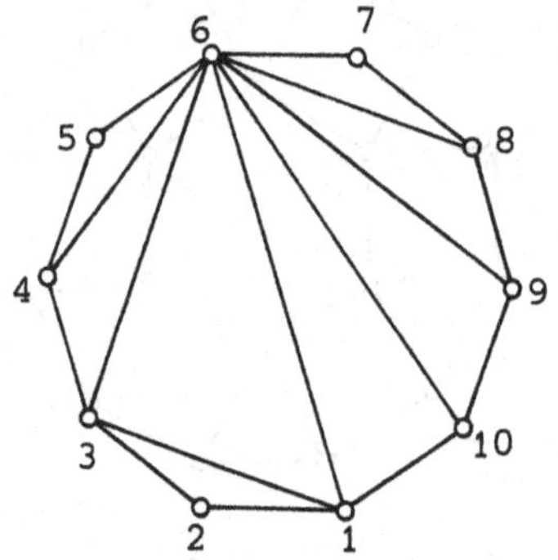

Figure 5: the graph G of Example 4.

Example 4. Let G be the labelled graph shown in Figure 5. We have $\mu_1(G) = 3{\cdot}9781, x_2 x_6 < x_1 x_3$ and $x_2 x_7 < x_1 x_3$. If edge 13 is replaced by 26 the index increases to 4·0465, while if edge 13 is replaced by 27 the index decreases to 3·8791. □

The last result deals with the situation in which the 'old' edge and the 'new' edge have a vertex in common.

Theorem 5.5.[41] Let G be a connected graph with distinct vertices h, i, k such that h is adjacent to i but not to k. Let G' be the graph obtained from

G by replacing edge hi with hk. If $x_k \geq x_i$ then $\mu_1(G') > \mu_1$. If $x_k < x_i$ and $\mu_1 - \mu_2 > \dfrac{2x_h^2 + (x_i - x_k)^2}{2x_h(x_i - x_k)}$ then $\mu_1(G') < \mu_1$. □

We illustrate Theorems 5.4 and 5.5 with an example from [41].

Example 5. Let G be the wheel W_{n+1} ($n \geq 4$) with vertex 0 adjacent to each vertex of the cycle $123...n1$. The eigenvalues of G are $1 \pm \sqrt{n+1}$ and $2\cos\frac{2\pi i}{n}$ ($i = 1, \ldots, n-1$) [14, Theorem 2.8]. Thus $\mu_1 = 1 + \sqrt{n+1}$ and $\mu_1 - \mu_2 > \mu_1 - 2$. The principal eigenvector of G is $(\alpha, \beta, \ldots \beta)^T$ where $n\beta = \mu_1 \alpha$ and $\alpha^2 + n\beta^2 = 1$. Let G' be the graph obtained from G by replacing the spoke 02 with the chord 13. To apply Theorem 5.4 with $h = 0, i = 2, j = 1, k = 3$ we require $\mu_1 - \mu_2 > \dfrac{\alpha^2 + 3\beta^2}{2(\alpha\beta - \beta^2)}$. It is sufficient to show that $\mu_1 - 2 > \dfrac{n^2 + 3\mu_1^2}{2\mu_1(n - \mu_1)}$; but this holds for large enough n (certainly for $n \geq 15$) because $\dfrac{n^2 + 3\mu_1^2}{2\mu_1(n - \mu_1)(\mu_1 - 2)} \to \frac{1}{2}$ as $n \to \infty$. Then $\mu_1(G') < \mu_1(G)$ by Theorem 5.4. Now let G'' be the graph obtained from G by replacing the spoke 02 with the chord 24. To apply Theorem 5.5 with $h = 2, i = 0, k = 4$ we require $\mu_1 - \mu_2 > \dfrac{2\beta^2 + (\alpha - \beta)^2}{2\beta(\alpha - \beta)}$, a condition which holds for $n \geq 15$ as a consequence of the corresponding inequality for G'. Then $\mu_1(G'') < \mu_1(G)$ by Theorem 5.5. □

In Example 5, we have $x_1 = x_2 = \ldots = x_n$. Accordingly the arguments show that if the spoke of a (large enough) wheel is replaced by any chord then the index decreases. On the other hand (applying the first parts of Theorems 5.4 and 5.5) if any other edge is replaced by a chord then the index increases. Which graphs have the property that relocation of *any* edge results in a decrease in index? As a first step towards answering this question, we can find the graphs with the property that the relocation of any number of edges results in a decrease in index. This problem, together with analogous questions for particular classes of graphs, is the subject of the next section.

6. MAXIMAL INDEX PROBLEMS

6.1 Introduction

By a 'maximal index problem' we mean a problem of the following sort: given a class $\mathcal{G}$ of graphs, find the graph(s) in $\mathcal{G}$ with maximal index. The case in which $\mathcal{G}$ consists of all trees with a prescribed number of vertices has received attention from a number of authors [10, 32, 52]. Lovász and Pelikán [32] suggested that the larger the index the more 'dense' the tree: the n-vertex tree with largest index (namely $\sqrt{n-1}$) is the star $K_{1,n-1}$ and that with smallest index (namely $2\cos\frac{\pi}{n+1}$) is the path P_n.

The graphs in $\mathcal{G}$ with maximal index are of course extremal graphs in the context of ordering $\mathcal{G}$ by index (or lexicographically by spectrum); and for graphs with prescribed numbers of vertices and edges, ordering by μ_1 is identical to ordering by $\mu_1 - \overline{d}$, where $\overline{d}$ denotes mean degree. Collatz and Sinogowitz [10] observed that $\mu_1 \geq \overline{d}$ for any graph G, with equality if and only if G is regular. This led them to propose $\mu_1 - \overline{d}$ as a measure of irregularity: for a discussion of $\mu_1 - \overline{d}$ in this context see [3]. The topic of μ_1-ordering was recently surveyed in §3 of [19] and we do not propose to cover the same ground here: we simply record in Table 1 several classes of graphs $\mathcal{G}$ for which the maximal index problem has been solved. In most cases the analogous minimal index problem has also been solved: details may be found in [19]. The 'other references' in Table 1 include alternative proofs, special cases, generalizations and minimal index problems. Note that an *internal triangle* of an outerplanar graph G is a 3-cycle which has no edge in common with the unique Hamiltonian cycle in G.

The class $\mathcal{G}$	Solution of the maximal index problem	Other references
Trees with n vertices	Lovász & Pelikán [32]	[10, 48, 52]
Unicyclic with n vertices	Brualdi & Solheid [8]	[16, 31, 47, 48]
Bicyclic with n vertices	Brualdi & Solheid [8]	[48, 49]
Tricyclic with n vertices	Brualdi & Solheid [8]	
Bicyclic Hamiltonian with n vertices	Simić& Kocić [50]	[37] [44, p.211]
Tricyclic Hamiltonian with n vertices	Bell & Rowlinson [4]	
Maximal outerplanar with n vertices and no internal triangles	Rowlinson [42]	
Graphs with e edges	Rowlinson [38]	[7, 21, 22, 23, 51]
Hamiltonian graphs with e edges	Rowlinson [39]	
Connected graphs with n vertices and e edges, $e = n - 1 + \binom{d-1}{2}$, $4 < d < n$	Bell [2]	[8, 17]

Table 1: References to maximal index problems and related topics.

For all of the classes listed in Table 1, the number of edges is prescribed (in some cases as a function of the number of vertices). Thus if G is a graph in $\mathcal{G}$ with maximal index, if G' is obtained from G by relocating an edge and if $G' \in \mathcal{G}$ then $\mu_1(G') \leq \mu_1(G)$, with strict inequality if G is unique. The methods of Section 5 are therefore of particular relevance in identifying candidates for G. In practice one often begins by using the elementary inequality (5.3) to rule out certain possibilities for G. In §6.2 we extend our range of techniques by replacing the inequality (5.3) with an equality (equation 6.2) and showing how it may be used to tackle one of the less tractable maximal index problems.

6.2 Graphs with a Prescribed Number of Edges

Let $\mathcal{S}(e)$ denote the class of all graphs with precisely e edges, and let $f(e) = \max\{\mu_1(G) : G \in \mathcal{S}(e)\}$. We use an argument of Brualdi and Hoffman [7] to show first that $f(e)$ is attained by a graph having an adjacency matrix (a_{ij}) which satisfies the condition

$$(*) \text{ if } i < j \text{ and } a_{ij} = 1 \text{ then } a_{hk} = 1 \text{ whenever } h < k \leq j \text{ and } h \leq i.$$

Such a matrix is called a *stepwise* matrix in view of the profile of the boundary between the zero and non-zero entries a_{ij} $(i \neq j)$. Since all of our matrices are symmetric with zero diagonal *we shall henceforth describe them only in terms of the triangle of entries* a_{ij} $(i < j)$.

Let G be a graph in $\mathcal{S}(e)$ with adjacency matrix $A = (a_{ij})$, where the n vertices of G are labelled so that $x_1 \geq x_2 \geq \ldots \geq x_n \geq 0$. If $a_{pq} = 0$ and $a_{p,q+1} = 1$ then let A' be the matrix obtained from A by interchanging the (p, q) and $(p, q+1)$-entries. Then $\mu_1(A') - \mu_1(A) \geq \mathbf{x}^T(A' - A)\mathbf{x} = 2x_p(x_q - x_{q+1}) \geq 0$. Thus the spectral radius does not decrease when all ones are brought to the front of a given row. The same is true when all ones are brought to the top of a given column. Accordingly we can always obtain a stepwise matrix B such that $\mu_1(B) \geq \mu_1(A)$. Since B is the adjacency matrix of a graph in $\mathcal{S}(e)$, it follows that if $\mu_1(A) = f(e)$ then $f(e)$ is also attained by the stepwise matrix B. In fact, Brualdi and Hoffman [7] go on to show that if $\mu_1(A) = f(e)$ then A itself is already a stepwise matrix, and so $f(e)$ is attained *only* by graphs having a stepwise adjacency matrix. They showed that if $e = \binom{d}{2}$, $G \in \mathcal{S}(e)$ and $\mu_1(G) = f(e)$ then, to within isolated vertices, G is the complete graph K_d. (Thus $f(\binom{d}{2}) = d-1$.) They conjectured (cf.[6, p.438]) that if $e = \binom{d}{2} + t$, $0 < t < d$, then the only non-trivial component of a graph in $\mathcal{S}(e)$ with index $f(e)$ is the graph G_e obtained from K_d by adding one new vertex of degree t. This conjecture had stood for some ten years before Friedland [22] proved it true for $t = d - 1$ and obtained the asymptotic result that there exists $K(t)$ such that the conjecture holds for $d \geq K(t)$. Subsequently Stanley [51] showed that $f(e) \leq \frac{1}{2}(-1 + \sqrt{1 + 8e})$, with equality precisely when $e = \binom{d}{2}$. Friedland [23] used a refinement of Stanley's inequality to prove that the conjecture holds when t is $1, d - 3$ or

$d-2$. Here we give an overview of how the conjecture was proved true in general in [38].

We wish to show that for any $n > d$, the matrix A^* with entries

$$
\begin{array}{ccccccc}
 & (2) & & & (d) & & \\
 & & & & & & \\
1 & 1 & 1 & \dots & 1 & 1 & \\
 & 1 & 1 & \dots & 1 & 1 & \\
 & & \vdots & \dots & \vdots & \vdots & \\
 & & 1 & \dots & 1 & 1 & (t) \\
 & & & & 1 & & \\
 & & & & \vdots & & \\
 & & & & 1 & & (d-1)
\end{array}
$$

is the unique $n \times n$ stepwise adjacency matrix with spectral radius $f(e)$. This would follow by successive relocation of edges *if* we could prove that $\mu_1(A') > \mu_1(A)$ whenever A is an $n \times n$ stepwise adjacency matrix (a_{ij}) as shown in Figure 6 with $h < p < q < k$, and A' is obtained from A

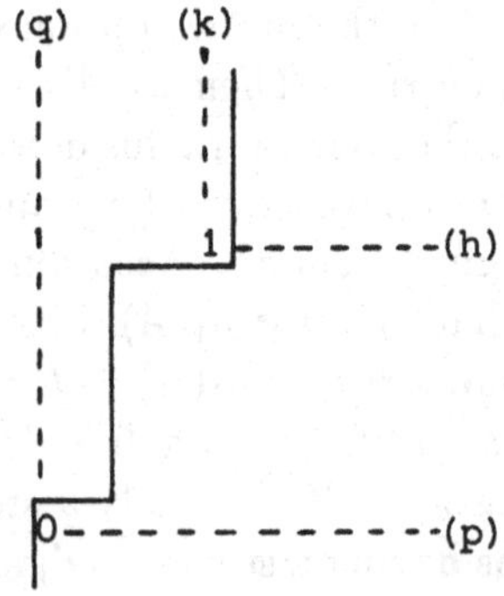

Figure 6: Part of a stepwise matrix (a_{ij}) with $a_{hk} = 1$, $a_{pq} = 0$.

by interchanging the (h, k) and (p, q) entries. This however is not true in general, but it *is* true whenever $p + q \geq h + k + 2$. This is a matter of arithmetic [38, Lemma 2], using the fact that $x_1 \geq x_2 \geq \ldots \geq x_n$ to prove that $\mathbf{x}^T(A' - A)\mathbf{x} > 0$ when $p + q \geq h + k + 2$. Accordingly it suffices to consider stepwise matrices $A = (a_{ij})$ such that

$$p + q \leq h + k + 1 \tag{6.1}$$

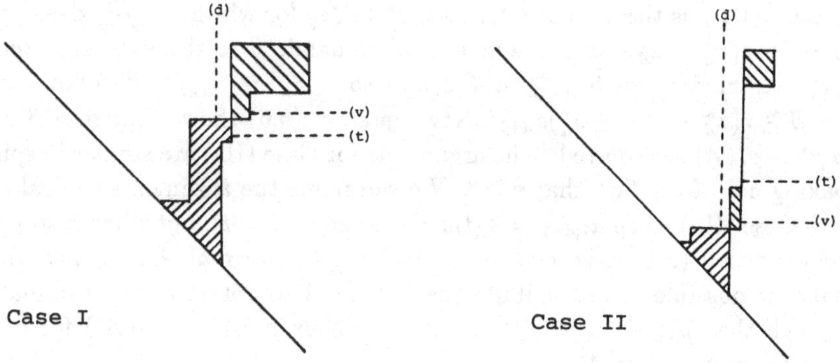

Figure 7: r entries = 1 r entries = −1

whenever $a_{hk} = 1$ and $a_{pq} = 0$ as in Figure 6 with $h < p < q < k$. We let $\mathbf{x}^*$ denote the principal eigenvector of A^* and we make use of the equality

$$\mathbf{x}^T(A^* - A)\mathbf{x}^* = (\mu_1(A^*) - \mu_1(A))\mathbf{x}^T\mathbf{x}^*. \tag{6.2}$$

Note that the right hand side of equation (6.2) has the sign of $\mu_1(A^*) - \mu_1(A)$ because $\mathbf{x}$ and $\mathbf{x}^*$ are non-negative. Accordingly we need to show that the biquadratic form $\mathbf{x}^T(A^* - A)\mathbf{x}^*$ is positive whenever $A \neq A^*$.

The argument splits into three cases, determined by the way in which the two matrices overlap, i.e. by the form of $A^* - A$: note that there exists $r > 0$ such that $A^* - A$ has $2r$ entries equal to $+1$ and $2r$ entries equal to -1. Referring to Figure 7 the three cases are (I) $v \leq t$, (II*a*) $v > t$, $r \geq t$, (II*b*) $v > t > r$. In each case, $\mathbf{x}^T(A^* - A)\mathbf{x}^*$ is expressible as $\alpha - \beta$ where each of α, β is a sum of r terms of the form $x_i x_j^* + x_i^* x_j$ $(i < j)$. Note that $x_1^* = \ldots = x_t^*$, $x_{t+1}^* = \ldots = x_d^*$ and $x_{d+2}^* = \ldots = x_n^* = 0$. The equation $A^*\mathbf{x}^* = \mu_1(A^*)\mathbf{x}^*$ is therefore equivalent to the following three equations, where we write μ_1^* for $\mu_1(A^*)$:

$$(\mu_1^* + 1)x_1^* = tx_1^* + (d-t)x_d^* + x_{d+1}^*, \tag{6.3}$$

$$(\mu_1^* + 1)x_d^* = tx_1^* + (d-t)x_d^*, \tag{6.4}$$

$$\mu_1^* x_{d+1}^* = tx_1^*. \tag{6.5}$$

On subtracting (6.4) and (6.5) from (6.3), we obtain

$$x_d^* + x_{d+1}^* = x_1^* + x_1^* t(\mu_1^* + 1)^{-1}. \tag{6.6}$$

In Case (I), α is the sum of r terms $x_i x_j^* + x_i^* x_j$ for which $i < j \le d+1$, and so $\alpha \ge r(x_d^* + x_{d+1}^*)x_{d+1}$. On the other hand, β is the sum of r terms $x_i x_j^* + x_i^* x_j$ for which $j \ge d+2$ and so $\beta \le r x_1^* x_{d+2}$. It follows that $\alpha - \beta \ge r(x_d^* + x_{d+1}^* - x_1^*)x_{d+1}$. Now equation (6.6) shows that $\alpha > \beta$, and so $\mu_1^* > \mu_1(A)$ as required. The arguments for Case (IIa) are similar in spirit, making use of the fact that $r \ge t$. We can refine the arguments to deal also with Case (IIb) when $a_{1,d+2} = 0$ (that is, when $r = v - t$); but when $a_{1,d+2} \ne 0$ the condition (6.1) is crucial in constraining the form of A in a way which makes it possible to redistribute the ones in A to obtain a stepwise matrix A' such that $\mu_1(A) < \mu_1(A') < \mu_1(A^*)$. Thus $\mu_1(A) < u_1(A^*)$ for every stepwise matrix $A \ne A^*$.

Incidentally, equation (6.2) enables us to deal immediately with the special case in which $e = \binom{d}{2}$. Here $A^* = \begin{pmatrix} J-I & O \\ O & O \end{pmatrix}$ and we obtain

$$(\mu_1(A^*) - \mu_1(A))\mathbf{x}^T\mathbf{x}^* = \mathbf{x}^T(A^* - A)\mathbf{x}^* \ge r x_1^*(2x_d - x_{d+1}) > 0.$$

6.3 Concluding Remarks

Maximal index problems closely related to the problem considered in §6.2 include those concerning (i) the Hamiltonian graphs with e edges, (ii) the connected graphs with n vertices and e edges. In order to discuss (i), note first that K_d is Hamiltonian for $d \ge 3$, and the graph G_e of §6.2 is Hamiltonian for $e = \binom{d}{2} + t$, $1 < t < d$ and $d \ge 3$. Accordingly the results of §6.2 provide the solution to the maximal index problem for Hamiltonian graphs with e edges unless $e = \binom{d}{2} + 1$. This case is dealt with in [39], where it is shown that for $d \ge 5$, the unique Hamiltonian graph with $\binom{d}{2} + 1$ edges and maximal index is the graph H_d defined as follows: if K_d^- denotes the graph obtained from K_d by deleting an edge, then H_d is obtained from K_d^- by adding one new vertex adjacent to exactly two vertices of degree $d-1$ in K_d^-. There are also unique Hamiltonian graphs with maximal index when $d = 3$ and $d = 4$, namely the 4-cycle and the unique maximal outerplanar graph with 5 vertices.

Finally we turn to the class $\mathcal{H}(n,e)$ of all connected graphs with n vertices and e edges, $n-1 \le e \le \frac{1}{2}n(n-1)$. If $e = n-1$ then such graphs are precisely the n-vertex trees discussed in §6.1. Accordingly, let $e = n+k$ where $k \ge 0$. As noted in Table 1, the maximal index problem for $\mathcal{H}(n, n+k)$ has been solved for $k = 0$ (unicyclic graphs), $k = 1$ (bicyclic graphs), $k = 2$ (tricyclic graphs) and for k of the form $\binom{d-1}{2} - 1$ $(4 < d < n)$.

Brualdi and Solheid [8, Corollary 2.2] showed that for any value of e, a graph in $\mathcal{H}(n,e)$ with maximal index again has a stepwise adjacency matrix. Then it is a consequence of connectedness that such a graph has a spanning star. In all known cases, a graph in $\mathcal{H}(n,e)$ with maximal index is one of two types, $G_{n,k}$ and $H_{n,k}$, which we now describe in terms of adjacency matrices. For $k+1=\binom{d-1}{2}+t, 0 \le t < d-1$, let $G_{n,k}$ be the graph with adjacency matrix of the form

$$
\begin{array}{cccccccccccc}
(1) & & & & & (d) & & & & & (n) & \\
 & & & & & & & & & & & \\
0 & 1 & 1 & \cdots & \cdots & 1 & 1 & 1 & \cdots & \cdots & 1 & (1) \\
 & 0 & 1 & \cdots & \cdots & 1 & 1 & & & & & \\
 & & & \cdots & \cdots & \vdots & \vdots & & & & & \\
 & & 0 & \cdots & \cdots & 1 & 1 & & & & & (t+1) \\
 & & & 0 & \cdots & 1 & & & & & & \\
 & & & & & \vdots & & & & & & \\
 & & & & 0 & 1 & & & & & & \\
 & & & & & 0 & & & & & & (d) \; .
\end{array}
$$

We can now see that the case $t=0$, i.e. the case $k=\binom{d-1}{2}-1$, corresponds to the special case $e=\binom{d}{2}$ of the maximal index problem for $\mathcal{S}(e)$ considered in §6.2.

For $k \le n-3$ let $H_{n,k}$ be the graph with adjacency matrix of the form

$$
\begin{array}{cccccccccccc}
(1) & & & & (k) & & & & & & (n) & \\
 & & & & & & & & & & & \\
0 & 1 & 1 & \cdots & 1 & 1 & 1 & 1 & 1 & \cdots & 1 & (1) \\
 & 1 & 1 & \cdots & 1 & 1 & 1 & 1 & 0 & \cdots & 0 & \\
 & & 0 & \cdots & 0 & 0 & 0 & 0 & & & & .
\end{array}
$$

In order to describe Bell's results from [2], let

$$
g(d)=\frac{1}{2}d(d+5)+7+\frac{32}{d-4}+\frac{16}{(d-4)^2} \qquad (d>4).
$$

Bell extends the techniques of §6.2 to show that if $k=\binom{d-1}{2}-1$ and $4<d<n$ then $G_{n,k}$ is the unique graph in $\mathcal{H}(n,e)$ with maximal index when $n<g(d)$, and $H_{n,k}$ is the unique graph in $\mathcal{H}(n,e)$ with maximal index when $n>g(d)$. If $n=g(d)$ then $(d,n,e) \in \{(5,60,69),(6,68,88),(8,80,85)\}$ and in these cases both $G_{n,k}$ and $H_{n,k}$ have maximal index.

For arbitrary $k \ge 3$, there is an asymptotic result [17]: there exists $N(k)>0$ such that for all $n \ge N(k)$, $H_{n,k}$ is the unique graph in $\mathcal{H}(n,n+k)$ with maximal index.

Acknowledgement. The author is indebted to C. Y. Chao for a translation of [31].

References

[1] A. C. Aitken, Determinants and Matrices (4th edn.), Oliver & Boyd, Edinburgh, 1946.

[2] F. K. Bell, On the maximal index of connected graphs, *Linear Algebra and Appl.,* 144(1991), 135-151.

[3] F. K. Bell, A note on the irregularity of graphs,*Linear Algebra and Appl.,* to appear.

[4] F. K. Bell and P. Rowlinson, On the index of tricyclic Hamiltonian graphs, *Proc. Edinburgh Math. Soc.* 33 (1990), 233-240.

[5] F. K. Bell and P. Rowlinson, The change in index of a graph resulting from the attachment of a pendant edge, *Proc. Royal Soc. Edinburgh* 108A(1988), 67-74.

[6] J.C. Bermond, J.-C. Fournier M. Las Vergnas and D. Sotteau (Eds.), Problèmes Combinatoire et Théorie des Graphes, Coll. Int. C.N.R.S., No.260 (Orsay 1976); *C.N.R.S. Publ.*, 1978.

[7] R. A. Brualdi and A. J. Hoffman, On the spectral radius of 0-1 matrices, *Linear Algebra and Appl.* 65(1985), 133-146.

[8] R. A. Brualdi and E. S. Solheid, On the spectral radius of connected graphs, *Publ. Inst. Math. (Beograd)* 39(53)(1986), 45-54.

[9] C. L. Coates, Flow-graph solutions of linear algebraic equations, *IRE Trans. Circuit Theory* CT-6 (1959), 170-187.

[10] L. Collatz and U. Sinogowitz, Spektren endlicher Grafen, *Abh. Math. Sem. Univ. Hamburg,* 21(1957), 63-77.

[11] D. Cvetković, Constructing trees with given eigenvalues and angles, *Linear Algebra and Appl.* 105(1988), 1-8.

[12] D. Cvetković, Some possibilities of constructing graphs with given eigenvalues and angles, *Ars Combinatoria* 29A(1990), 179-187.

[13] D. Cvetković, M. Doob, I. Gutman, and A. Torgašev, Recent results in the theory of graph spectra, North Holland, Amsterdam, 1988.

[14] D. Cvetković, M. Doob and H. Sachs, Spectra of graphs, Academic Press, New York, 1980.

[15] D. Cvetković, L. L. Kraus and S. K. Simić, Discussing graph theory with a computer I: implementation of graph theoretic algorithms, *Univ. Beograd Publ. Elektrotehn. Fak. Ser. Math. Fiz.* Nos. 716-734(1981), 100-104 (No. 733).

[16] D. Cvetković and P. Rowlinson, Spectra of unicyclic graphs, *Graphs and Combinatorics* 3(1987), 7-23.

[17] D. Cvetković and P. Rowlinson, On connected graphs with maximal index, *Publ. Inst. Math.(Beograd)* 44(58)(1988), 29-34.

[18] D. Cvetković and P. Rowlinson, Further properties of graph angles, *Scientia* 1(1988), 41-51.

[19] D. Cvetković and P. Rowlinson, The largest eigenvalue of a graph: a survey, *Linear and Multilinear Algebra*, 28(1990), 3-33.

[20] S. S. D'Amato, B. M. Gimarc and N. Trinajstić, Isospectral and subspectral molecules, *Croatica Chemica Acta* 54(1981), 1-52.

[21] S. Friedland, The maximal value of the spectral radius of graphs with e edges, *Linear and Multilinear Algebra* 23(1988), 91-93.

[22] S. Friedland, The maximal eigenvalue of 0-1 matrices with prescribed number of ones, *Linear Algebra and Appl.* 69(1985), 33-69.

[23] S. Friedland, Bounds on spectral radius of graphs with e edges, *Linear Algebra and Appl.* 101(1988), 81-86.

[24] C. D. Godsil and B. D. McKay, Spectral conditions for the reconstructibility of a graph, *J. Combin. Theory* Ser. B 30 (1981), 285-289.

[25] I. Gutman, Some relations for graphic polynomials, *Publ. Inst. Math. Beograd.* 39(53)(1986) 55-62.

[26] P. R. Halmos, Finite-dimensional Vector Spaces (2nd edn.), van Nostrand, Princeton, 1958.

[27] E. Heilbronner, Das Komposition-Prinzip: Eine anschauliche Methode zur elektronen-theoretischen Behandlung nicht oder niedrig symmetrischer Molekeln im Rahmen der MO-Theorie, *Helv. Chim. Acta* 36(1953), 170-188.

[28] A. J. Hoffman and J. H. Smith, On the spectral radii of topologically equivalent graphs, in: Recent Advances in Graph Theory, ed. M. Fiedler, Academia Praha, 1975, 273-281.

[29] T. Kato, Perturbation Theory for Linear Operators, Springer-Verlag, Berlin, 1966.

[30] P. Lancaster and M. Tismenetsky, Theory of Matrices (2nd. edn.), Academic Press, New York, 1985.

[31] Qiao Li and Ke Qin Feng : On the largest eigenvalue of graphs (Chinese), *Acta Math. Appl. Sinica* 2(1979), 167-175 (MR80k : 05079).

[32] L. Lovász and J. Pelikán, On the eigenvalues of trees, *Periodica Math. Hung.* 3(1973), 175-182.

[33] J. P. Lowe and M. R. Soto, Isospectral graphs, symmetry and perturbation theory, *Match* 20(1986), 21-51.

[34] Ch. Maas, Perturbation results for the adjacency spectrum of a graph, *Z. angew. Math. Mech.* 67(1987), No. 5, 428-430.

[35] R. B. Mallion, Some chemical applications of the eigenvalues and eigenvectors of certain finite planar graphs, in: Applications of Combinatorics, ed. R. J. Wilson, Shiva, Nantwich, 1982, 87-114.

[36] H. Minc, Nonnegative Matrices, Wiley, New York, 1988.

[37] P. Rowlinson, A deletion-contraction algorithm for the characteristic polynomial of a multigraph, *Proc. Royal Soc. Edinburgh* 105A(1987), 153-160.

[38] P. Rowlinson, On the maximal index of graphs with a prescribed number of edges, *Linear Algebra and Appl.* 110(1988), 43-53.

[39] P. Rowlinson, On Hamiltonian graphs with maximal index, *European J. Combinatorics* 10 (1989), 489-497.

[40] P. Rowlinson, On angles and perturbations of graphs, *Bull. London Math. Soc.* 20(1988), 193-197.

[41] P. Rowlinson, More on graph perturbations, *Bull. London Math. Soc.* 22(1990), 209-216.

[42] P. Rowlinson, On the index of certain outerplanar graphs, to appear.

[43] P. Rowlinson, Graph angles and isospectral molecules, to appear.

[44] H. Sachs (Ed.), Graphs, Hypergraphs and Applications, Teubner-Texte zur Mathematik, Band 73, Teubner, Leipzig, 1985.

[45] A. J. Schwenk, Computing the characteristic polynomial of a graph, in: Graphs and Combinatorics (Lecture Notes in Mathematics 406, ed. R. Bari & F. Harary), Springer-Verlag, New York, 1974, pp 153-172.

[46] J. J. Seidel, Eutactic stars, *Colloq. Math. Soc. János Bolyai* **18**(1976), 983-999.

[47] S. K. Simić, On the largest eigenvalue of unicyclic graphs, *Publ. Inst. Math. (Beograd)* 42(56)(1987), 13-19.

[48] S. K. Simić, Some results on the largest eigenvalue of a graph, *Ars Combinatoria*, 24A(1987), 211-219.

[49] S. K. Simić, On the largest eigenvalue of bicyclic graphs, *Publ. Inst. Math. (Beograd)* 46(60)(1989), 101-106.

[50] S. K. Simić and V. Lj. Kocić, On the largest eigenvalue of some homeomorphic graphs, *Publ. Inst. Math (Beograd)* 40(54) (1986), 3-9.

[51] R. P. Stanley, A bound on the spectral radius of graphs with e edges, *Linear Algebra and Appl.* 87(1987), 267-269.

[52] N. S. Wang, On the specific properties of the characteristic polynomial of a tree (Chinese, English summary), *J. Lanzhou Railway Coll.* 5(1986), 89-94 (MR88d: 05126).

[53] A. Weinstein and W. Stenger, Methods of Intermediate Problems for Eigenvalues, Academic Press, New York, 1972.

[54] H. Yuan, A bound on the spectral radius of graphs, *Linear Algebra and Appl.* 108(1988), 135-139.

[55] F. J. Zhang, Z. N. Zhang and Y. H. Zhang, Some theorems about the largest eigenvalue of graphs (Chinese, English summary), *J. Xinjiang Univ. Nat. Sci.* (1984) no.3, 84-90 (MR 88d: 05128).

[illegible]

[4[illegible]] [illegible] Math. [illegible]

[illegible] R[illegible] On the [illegible] of [illegible]

[illegible]

[illegible]

[illegible]

[illegible] [illegible] the [illegible] of a graph, [illegible]

[illegible]

[illegible]

[illegible]

[illegible] Chinese [illegible] [illegible] (1983) [illegible]

[illegible] [illegible] Press, New York [illegible]

[illegible] the [illegible] [illegible] (1981), [illegible]

[55] F. J. Zhang [illegible] [illegible] Zhang, some [illegible] about the [illegible] graphs [illegible] [illegible] ... 856 (1985).

A Graph Reconstructor's Manual

J.A. BONDY

University of Waterloo

1 THE RECONSTRUCTION CONJECTURE

Fifty years have elapsed since the Reconstruction Conjecture was proposed, by S.M. Ulam and P.J. Kelly, in 1941. This seems a fitting time, therefore, to report on its status. Because particular aspects of this elusive question have been amply addressed in a recent flurry of expository articles (Ellingham, 1988; Lauri, 1987; Manvel, 1988; Stockmeyer, 1988), we have chosen to present here an overview of the techniques employed, a manual to reconstruction. Each technique will be illustrated and its applications noted, accompanied by appropriate references. For clarity and simplicity, we shall concentrate on the two principal open questions of reconstruction, namely the Reconstruction Conjecture itself and its companion version for edges, the Edge Reconstruction Conjecture, formulated by F. Harary in 1964. Some related questions, amenable to the same or similar techniques, will be discussed briefly at the end of the article. We start with the basic definitions.

1.1 Definitions. A *vertex-deleted subgraph* of a graph G is a subgraph $G - v$ obtained by deleting a vertex v and its incident edges. The *deck* of a graph G is the family of (unlabelled) vertex-deleted subgraphs of G; these are the *cards* of the deck. A *reconstruction* of a graph G is a graph H with the same deck as G. A graph G is *reconstructible* if every reconstruction of G is isomorphic to G. Similar definitions apply to digraphs and hypergraphs.

1.2 Remark. If H is a reconstruction of G, there is no loss of generality in assuming that $V(H) = V(G) := V$ and $H - v \cong G - v$, for all $v \in V$; this we shall do. Moreover, with each card of the common deck of G and H, we shall associate a unique element of the set V.

1.3 Examples. The graphs $G := K_2$ and $H := 2K_1$ are reconstructions of one another. Therefore, neither of these graphs is reconstructible. The graph G of Figure 1, on the other hand, is reconstructible. While this is not immediately evident, one can check that G is indeed the only graph, up to isomorphism, with this particular deck.

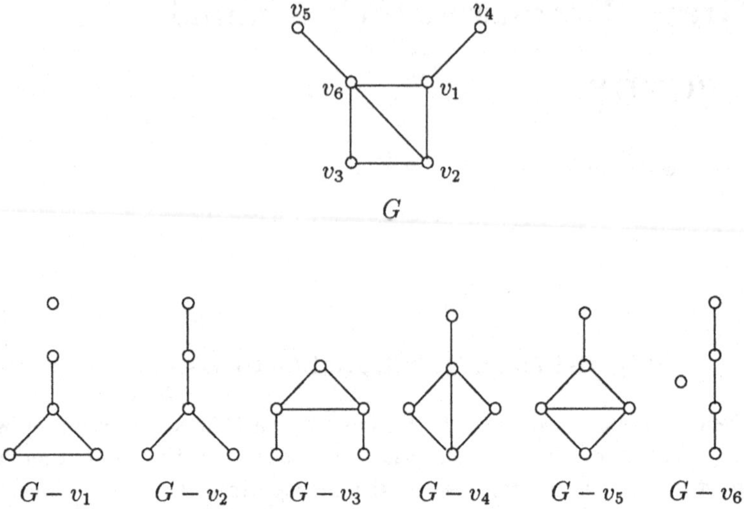

Figure 1: A reconstructible graph.

1.4 Remark. The graph G of Figure 1 is reconstructible, but only just so: there exists a graph H with vertex set $\{v_1, v_2, v_3, v_4, v_5, v_6\}$ such that $H - v_i \cong G - v_i$, $1 \leq i \leq 5$, but $H - v_6 \ncong G - v_6$ (Stockmeyer, 1976).

1.5 The Reconstruction Conjecture (P.J. Kelly and S.M. Ulam; see Kelly, 1942 or Ulam, 1960). *All finite simple graphs on at least three vertices are reconstructible.*

1.6 Remarks.

– One may go a step further and ask how many cards are needed to reconstruct a graph. The example of Figure 1 is deceptive — even though the first five cards fail to distinguish this graph, the last three clearly suffice. In fact, Bollobás (1990) has shown that almost all graphs (in the probabilistic sense) are reconstructible from three well-chosen cards; and a moment's reflection shows that two never suffice. It is a different matter if one is not allowed to choose the cards from the deck. In this case, certain families of trees are not even reconstructible from half their deck (Myrvold, 1988).

– It is easily seen that a simple graph is reconstructible if and only if its complement is reconstructible. Yang (1988) made use of this observation to show that the Reconstruction Conjecture can be reduced to nonseparable graphs.

– The Reconstruction Conjecture has no direct algorithmic implications. It

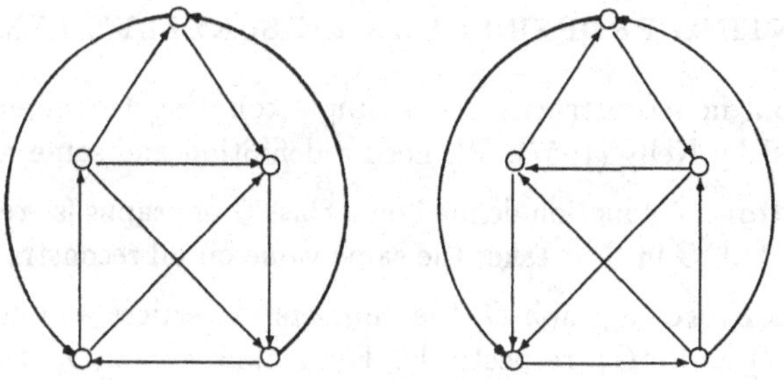

G H

Figure 2: A pair of nonreconstructible digraphs.

is concerned not with the process of reconstruction but with the end product, and asserts that this end product, the reconstructed graph, is unique up to isomorphism. The reconstructibility of any particular finite graph can, of course, be tested in finite time because the number of potential reconstructions is finite; indeed, the Reconstruction Conjecture has been verified for all graphs on nine or fewer vertices (McKay, 1977).

The natural setting for a discussion of computational complexity in this context is the *Legitimate Deck Problem.* Define a *deck* to be any multiset $\mathcal{G}$ of graphs, and call $\mathcal{G}$ *legitimate* if $\mathcal{G}$ is the deck of some graph. The Legitimate Deck Problem, posed by Harary (1969), asks for a characterization of legitimate decks. More generally, one may ask how hard it is to test a deck for legitimacy. This question has been studied by Mansfield (1982).

– Digraphs, infinite graphs and hypergraphs are not reconstructible in general. For instance, the digraphs G and H of Figure 2 are reconstructions of one another, as are the infinite graphs $G := T_\infty$ and $H := 2T_\infty$, where T_∞ denotes a regular tree of infinite degree, and the hypergraphs G and H with vertex set $V := \{1,2,3,4,5\}$ and edge sets $E(G) := \{123, 125, 135, 234, 345\}$ and $E(H) := \{123, 135, 145, 234, 235\}$. Indeed, there exist infinite families of nonreconstructible tournaments (Stockmeyer, 1977, 1981, 1988; see, also, Kocay, 1985), nonreconstructible locally-finite countable forests (Harary, Schwenk and Scott, 1972), and nonreconstructible 3-uniform hypergraphs (Kocay, 1987). On the positive side, many of the techniques to be described here apply equally well to finite digraphs and hypergraphs. The reconstruction of infinite graphs, however, relies on quite different methods.

2 COUNTING PROPER SUBGRAPHS: KELLY'S LEMMA

The basic tool in reconstruction is a simple counting technique developed and employed by Kelly (1957). We need a definition and some notation.

2.1 Definition. A function defined on a class $\mathcal{G}$ of graphs is *reconstructible* if, for each graph G in $\mathcal{G}$, it takes the same value on all reconstructions of G.

2.2 Notation. For a graph G, the numbers of vertices and edges are denoted by $v(G)$ and $e(G)$, respectively. For graphs F and G, the number of subgraphs of G isomorphic to F is denoted by $s(F, G)$.

2.3 Kelly's Lemma (P.J. Kelly, 1957). *For any two graphs F and G such that $v(F) < v(G)$, $s(F, G)$ is reconstructible.*

Proof. Each subgraph of G isomorphic to F occurs in exactly $v(G) - v(F)$ of the vertex-deleted subgraphs $G - v$. Therefore

$$s(F,G) = \frac{1}{v(G) - v(F)} \sum_{v \in V} s(F, G - v) \, .$$

Since the right-hand side of this identity is reconstructible, so too is the left-hand side. ∎

2.4 Corollary (Kelly, 1957). *For any two graphs F and G such that $v(F) < v(G)$, the number of subgraphs of G that are isomorphic to F and include a given vertex v is reconstructible.*

Proof. This number is $s(F,G) - s(F, G - v)$. ∎

2.5 Corollary (Kelly, 1957). *The number of edges and the degree sequence are reconstructible.*

Proof. Take $F = K_2$ in Kelly's Lemma and in Corollary 2.4. ∎

2.6 Definition. A *class* of graphs is a family of graphs closed under isomorphism. A class $\mathcal{G}$ of graphs is *reconstructible* if every graph in $\mathcal{G}$ is reconstructible.

2.7 Corollary (Kelly, 1957). *Regular graphs are reconstructible.*

Proof. Let G be a k-regular graph. Since the degree sequence is reconstructible, all reconstructions of G are k-regular. Also, all k-regular reconstructions of G are isomorphic, because each can be obtained, up to isomorphism, from any $G - v$, by adding a vertex and joining it to every vertex of degree $k - 1$ in $G - v$. It follows that all reconstructions of G are isomorphic. ∎

Kelly (1957) also applied his lemma to show that disconnected graphs and trees are reconstructible. These facts can be established more easily using the techniques to be described in Sections 3 and 4.

3 COUNTING MAXIMAL PROPER SUBGRAPHS: THE GREENWELL-HEMMINGER LEMMA

3.1 Definitions. Let $\mathcal{F}$ be a class of graphs and G a graph. An *$\mathcal{F}$-graph* is a member of $\mathcal{F}$. An *$\mathcal{F}$-subgraph* of G is an $\mathcal{F}$-graph that is contained in G. A *maximal $\mathcal{F}$-subgraph* of G is an $\mathcal{F}$-subgraph that is contained in no other $\mathcal{F}$-subgraph of G.

3.2 Example. Let $\mathcal{F}$ be the class of connected graphs. Then the maximal $\mathcal{F}$-subgraphs of a graph G are the connected components of G.

3.3 Notation. The number of maximal $\mathcal{F}$-subgraphs of G isomorphic to a given $\mathcal{F}$-graph F is denoted by $m(F,G)$, the role of $\mathcal{F}$ being suppressed for notational simplicity.

In this section, we prove that, under appropriate conditions on $\mathcal{F}$ and G, $m(F,G)$ is reconstructible for every $\mathcal{F}$-subgraph F of G. To formulate these conditions, the following definitions are required.

3.4 Definitions. A class $\mathcal{G}$ of graphs is *recognizable* if, for each $\mathcal{G}$-graph G, every reconstruction of G is also in $\mathcal{G}$. A class $\mathcal{G}$ of graphs is *weakly reconstructible* if, for each $\mathcal{G}$-graph G, every reconstruction of G that belongs to $\mathcal{G}$ is isomorphic to G.

3.5 Remark. It follows from these definitions that a class of graphs is reconstructible if and only if it is both recognizable and weakly reconstructible. Observe that the proof of Corollary 2.7 proceeds along precisely these lines.

3.6 The Greenwell-Hemminger Lemma (D.L. Greenwell and R.L. Hemminger, 1973). *Let $\mathcal{F}$ be an arbitrary class of graphs and let $\mathcal{G}$ be a recognizable class of graphs such that, for every $\mathcal{G}$-graph G and every $\mathcal{F}$-subgraph F of G:*

(i) $v(F) < v(G)$;

(ii) F is contained in a unique maximal $\mathcal{F}$-subgraph of G.

Then, for every $\mathcal{F}$-graph F and every $\mathcal{G}$-graph G, $m(F,G)$ is reconstructible.

Proof. Define an *(F,G)-chain* of *length* n to be a sequence $(X_0, X_1, \ldots, X_n)$ of $\mathcal{F}$-graphs such that $X_0 \subset X_1 \subset \ldots \subset X_n$, where $X_0 \cong F$ and $s(X_n, G) > 0$. Two (F,G)-chains are *isomorphic* if they have the same length and corresponding terms are isomorphic graphs.

By (ii),

$$s(F,G) = \sum_X s(F,X)m(X,G) \tag{1}$$

where the sum extends over all isomorphism types X of $\mathcal{F}$-graph. Inverting this identity (see, for example, Bondy and Hemminger, 1977), we obtain

$$m(F,G) = \sum_{n=0}^{\infty} \sum (-1)^n s(F,X_1)s(X_1,X_2)\dots s(X_{n-1},X_n)s(X_n,G) \qquad (2)$$

where the inner sum extends over all isomorphism types $(X_0, X_1, \dots, X_n)$ of (F,G)-chain. By (i), Kelly's Lemma now implies that the right-hand side of (2) is reconstructible. Hence the left-hand side is also. ■

3.7 Remark. The Greenwell-Hemminger Lemma includes Kelly's Lemma: choose $\mathcal{F}$ to be the class of all graphs isomorphic to F and $\mathcal{G}$ to be the class of all reconstructions of G.

3.8 Corollary (Kelly, 1957). *Disconnected graphs are reconstructible.*

Proof. A graph is disconnected if and only if at most one vertex-deleted subgraph is connected. Therefore, disconnected graphs are recognizable. The Greenwell-Hemminger Lemma, with $\mathcal{F}$ as the class of connected graphs and $\mathcal{G}$ as the class of disconnected graphs, establishes weak reconstructibility. ■

3.9 Remark. The Greenwell-Hemminger Lemma can also be used to prove that trees are reconstructible (see Bondy and Hemminger, 1977).

4 COUNTING SPANNING SUBGRAPHS: KOCAY'S LEMMA

It was Tutte (1979) who first realized how to reconstruct the numbers of spanning subgraphs of certain types. His approach was based on the theory of dichromatic polynomials, but an elementary technique was subsequently found by Kocay (1981) and it is this that we describe here.

4.1 Definition. Let G be a graph and $\mathcal{F} := (F_1, F_2, \dots, F_m)$ a sequence of graphs (not necessarily distinct). A *cover* of G by $\mathcal{F}$ is a sequence $(G_1, G_2, \dots, G_m)$ of subgraphs of G such that $G_i \cong F_i$, $1 \le i \le m$, and $\bigcup_{i=1}^{m} G_i = G$. The number of covers of G by $\mathcal{F}$ is denoted by $c(\mathcal{F}, G)$.

4.2 Example. Let $\mathcal{F} := (F_1, F_2)$, where $F_1 := K_2$ and $F_2 := K_{1,2}$. Figure 3 depicts the covers of G by $\mathcal{F}$ for each graph G such that $c(\mathcal{F}, G) > 0$. The edge of F_1 is shown as a thick line.

4.3 Kocay's Lemma (W.L. Kocay, 1981). *For any graph G and any sequence $\mathcal{F} := (F_1, F_2, \dots, F_m)$ of graphs such that $v(F_i) < v(G)$, $1 \le i \le m$, the parameter*

$$\sum_X c(\mathcal{F}, X)s(X,G) \qquad (3)$$

is reconstructible, where the sum extends over all isomorphism types X of graph such that $v(X) = v(G)$.

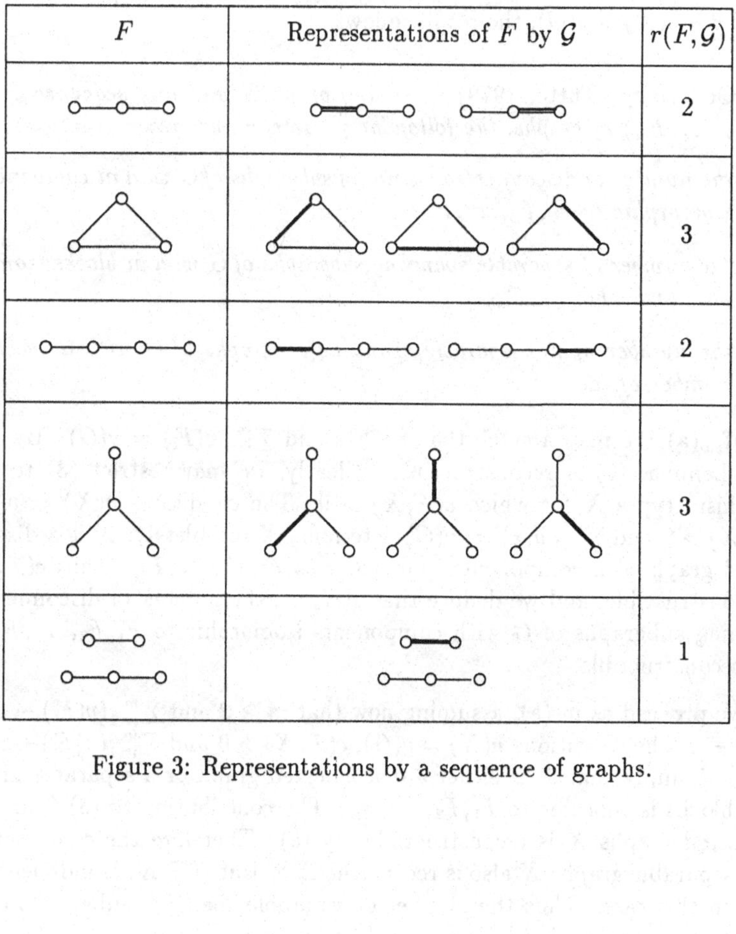

Figure 3: Representations by a sequence of graphs.

Proof. Counting in two ways the sequences $(G_1, G_2, \ldots G_m)$ of subgraphs of G such that $G_i \cong F_i$, $1 \le i \le m$, we have

$$\prod_{i=1}^{m} s(F_i, G) = \sum_X c(\mathcal{F}, X)s(X, G),$$

where the sum extends over all isomorphism types X of graph. By Kelly's Lemma, the left-hand side is reconstructible because $v(F_i) < v(G)$, $1 \le i \le m$. Thus the right-hand side is reconstructible, too. Since the summands $c(\mathcal{F}, X)s(X, G)$ are reconstructible by Kelly's Lemma if $v(X) < v(G)$ and trivially if $v(X) > v(G)$, the result follows. ■

4.4 Corollary (Tutte, 1979). *For any graph G and any sequence $\mathcal{F} := (F_1, F_2, \ldots, F_m)$ of graphs, the following parameters are reconstructible:*

(a) the number of disconnected spanning subgraphs of G with m components isomorphic to $F_1, F_2, \ldots, F_m$;

(b) the number of separable spanning subgraphs of G with m blocks isomorphic to $F_1, F_2, \ldots, F_m$;

(c) the number of nonseparable spanning subgraphs of G with a specified number of edges.

Proof. (a) We may assume that $m \ge 2$ and $\sum_{i=1}^{m} v(F_i) = v(G)$. By Kocay's Lemma, (3) is reconstructible. Clearly, we may restrict (3) to isomorphism types X for which $c(\mathcal{F}, X) > 0$. The conditions $v(X) = v(G)$, $c(\mathcal{F}, X) > 0$ and $\sum_{i=1}^{m} v(F_i) = v(G)$ determine X completely: X is a disconnected graph with components isomorphic to $F_1, F_2, \ldots, F_m$. Thus $c(\mathcal{F}, X)$ is reconstructible, and we deduce that $s(X, G)$, the number of disconnected spanning subgraphs of G with components isomorphic to $F_1, F_2, \ldots, F_m$ is also reconstructible.

(b) We proceed as in (a), assuming now that $m \ge 2$ and $\sum_{i=1}^{m}(v(F_i) - 1) = v(G) - 1$. The conditions $v(X) = v(G)$, $c(\mathcal{F}, X) > 0$ and $\sum_{i=1}^{m}(v(F_i) - 1) = v(G) - 1$ imply that X is either a disconnected graph or a separable graph with blocks isomorphic to $F_1, F_2, \ldots, F_m$. The contribution to (3) from disconnected graphs X is reconstructible, by (a). Therefore the contribution from separable graphs X also is reconstructible. But $c(\mathcal{F}, X)$ is independent of X in this case. Thus the number of separable spanning subgraphs of G with blocks isomorphic to $F_1, F_2, \ldots, F_m$ is reconstructible.

(c) Set $F_i \cong K_2$, $1 \le i \le m$, in Kocay's Lemma. The contributions to (3) from isomorphism types X that are disconnected, separable, or that have fewer vertices than G are reconstructible, by (a), (b) and Kelly's Lemma. It follows that the contribution from nonseparable spanning subgraphs of G

with at most m edges is reconstructible. Moreover, this is true for every value of m. Replacing m by $m-1$ and subtracting the latter sum from the former, we see that the contribution to (3) from nonseparable spanning subgraphs X of G with exactly m edges is reconstructible. Since $c(\mathcal{F}, X) = m!$ for each such X, the number of nonseparable spanning subgraphs of G with exactly m edges is reconstructible. ∎

4.5 Remark. In the special case where $F_i \cong K_2$, $1 \leq i \leq m$, Corollary 4.4(a) and 4.4(b) assert that the numbers of perfect matchings and spanning trees, respectively, are reconstructible. And the case $m := v(G)$ of Corollary 4.4(c) shows that the number of Hamilton cycles is reconstructible. Corollary 4.4(a) also yields an alternative proof of Corollary 3.8.

4.6 Corollary (Kelly, 1957). *Disconnected graphs are reconstructible.*

Proof. Let G be a disconnected graph. By Corollary 4.4(a), the number of spanning subgraphs of G of each isomorphism type X is reconstructible. Clearly, if $s(X, G) > 0$ and X is maximal with respect to this condition, $X \cong G$. ∎

4.7 Remark. It is reasonable to believe that Kocay's Lemma can be used to reconstruct various classes of spanning subgraphs in addition to those described in Corollary 4.4. A natural one is the class of trees.

4.8 Problem (Kocay, 1982). Prove that the number of spanning trees of each isomorphism type is reconstructible.

5 APPLICATIONS OF KOCAY'S LEMMA: THE CHARACTERISTIC AND CHROMATIC POLYNOMIALS

5.1 Definitions. The *characteristic polynomial* $\phi(G, x)$ of a graph G is defined by

$$\phi(G, x) = \det\,(x\mathbf{I} - \mathbf{A})\ ,$$

where $\mathbf{A}$ is the adjacency matrix of G. An *S-graph* is a graph in which each component is an edge or a cycle.

5.2 Notation. For a graph G, the number of components of G is denoted by $c(G)$ and the number of cycles by $z(G)$.

5.3 Proposition (Sachs, 1964). *For any graph G*

$$\phi(G, x) = \sum_{i=0}^{v} a_i x^{v-i}$$

where

$$a_i = \sum_X (-1)^{c(X)} 2^{z(X)} s(X, G) \tag{4}$$

and the sum extends over all isomorphism types X of S-graph on i vertices.

5.4 Corollary (Tutte, 1979). *The characteristic polynomial is reconstructible.*

Proof. The coefficient a_i defined in (4) is reconstructible for each i, $0 \leq i \leq v$ — by Kelly's Lemma if $0 \leq i < v$, by Corollary 4.4(a) if $i = v$ and X is disconnected, and by Corollary 4.4(c) if $i = v$ and X is connected (and hence is a Hamilton cycle). ■

5.5 Remark. The *idiosyncratic polynomial* $\psi(G, x, y)$ of a graph G is defined by

$$\psi(G, x, y) = \det\,(x\mathbf{I} - \mathbf{B}),$$

where

$$\mathbf{B} = \mathbf{A} + y(\mathbf{J} - \mathbf{A}),$$

$\mathbf{A}$ being the adjacency matrix of G and $\mathbf{J}$ the matrix in which each entry is 1. The characteristic and idiosyncratic polynomials are related by the identity

$$\phi(G, x) = \psi(G, x, 0).$$

It follows easily from Kelly's Lemma, Corollary 4.4 and an appropriate generalization of Proposition 5.3 that this more general polynomial is reconstructible (Tutte, 1979).

5.6 Definitions. Let x be a positive integer. A *proper x-vertex colouring* of a graph G is a mapping $c : V \to \{1, 2, \ldots, x\}$ such that adjacent vertices in G receive distinct values of c. The number of such colourings is denoted by $P(G, x)$ and called the *chromatic polynomial* of G.

5.7 Remark. That $P(G, x)$ is indeed a polynomial in x was proved by Birkhoff (1912).

5.8 Proposition (Whitney, 1932). *For any graph G,*

$$P(G, x) = \sum_{i=1}^{v} b_i x^i$$

where

$$b_i = \sum_X (-1)^{e(X)} s(X, G) \tag{5}$$

and the sum extends over all isomorphism types X of graph on v vertices and i components.

5.9 Corollary (Tutte, 1979). *The chromatic polynomial is reconstructible.*

Proof. The coefficient b_i defined in (5) is reconstructible for each i, $1 \leq i \leq v$: by Corollary 4.4(a) if $1 < i \leq v$, and by Corollary 4.4(b), (c) if $i = 1$. ∎

5.10 Corollary (Tutte, 1979). *The chromatic number is reconstructible.*

Proof. The chromatic number of a graph G is the least positive integer k such that $P(G, k) > 0$. By Corollary 5.9, this is reconstructible. ∎

5.11 Remarks.

– The *dichromatic polynomial* $Q(G, x, y)$ is defined by

$$Q(G,x,y) = \sum_{i=1}^{v}\sum_{j=0}^{e} c_{ij}(xy)^i y^{j-v}$$

where

$$c_{ij} = \sum_F s(F, G)$$

and the sum extends over all isomorphism types X on v vertices, i components and j edges. The chromatic and dichromatic polynomials are related by the identity

$$P(G,x) = (-1)^v Q(G, -x, -1).$$

It follows easily from Corollary 4.4 that this more general polynomial is reconstructible (Tutte, 1979).

– Although many fundamental parameters have been shown reconstructible, there remain several whose status has yet to be decided. Indeed, the problem of reconstructing the (abstract) automorphism group is just as hard as the Reconstruction Conjecture itself (Statman, 1981). Another basic parameter still to be proved reconstructible is the edge chromatic number.

5.12 Problem (Bondy and Hemminger, 1977). Prove that the edge chromatic number is reconstructible.

An alternative treatment of this material can be found in Tutte (1984).

6 THE EDGE RECONSTRUCTION CONJECTURE

6.1 Definitions. An *edge-deleted subgraph* of a graph G is a subgraph $G - e$ obtained by deleting an edge e. The *edge deck* of a graph G is the family of edge-deleted subgraphs of G. An *edge reconstruction* of a graph G is a graph H with the same edge deck as G. A graph G is *edge reconstructible* if every edge reconstruction of G is isomorphic to G.

6.2 Remark. If H is an edge reconstruction of G, it is convenient to assume that $E(H) = E(G) := E$ and $H - e \cong G - e$ for all $e \in E$.

6.3 Example. $G := 2K_2$ and $H := K_1 + K_{1,2}$ are edge reconstructions of one another, as are $G := K_{1,3}$ and $H := K_1 + K_3$. Therefore none of these graphs is edge reconstructible. Further examples may be obtained by adding a fixed number of isolated vertices to each of these pairs.

6.4 Edge Reconstruction Conjecture (F. Harary, 1964). All finite simple graphs on at least four edges are edge reconstructible.

In the sequel, we restrict our attention to *finite simple graphs on at least four edges.*

6.5 Definitions. A class $\mathcal{G}$ of graphs is *edge reconstructible* if every $\mathcal{G}$-graph is edge reconstructible. A function defined on a class $\mathcal{G}$ of graphs is *edge reconstructible* if, for each $\mathcal{G}$-graph G, it takes the same value on all edge reconstructions of G.

6.6 Kelly's Lemma (edge version). For any two graphs F and G such that $e(F) < e(G)$, $s(F, G)$ is edge reconstructible.

6.7 Corollary. *The number of isolated vertices is edge reconstructible.*

Proof. Let G be a graph and let m be the minimum number of isolated vertices in an edge-deleted subgraph of G. If G contains a path or cycle of length three, the number of isolated vertices in G is m. If G contains neither of these subgraphs, the number of isolated vertices in G is $m-1$ if G contains a path of length two, and $m - 2$ otherwise. ∎

6.8 Corollary. *The Edge Reconstruction Conjecture is valid for all graphs provided that it is valid for all graphs without isolated vertices.*

Proof. Suppose that the Edge Reconstruction Conjecture is valid for all graphs without isolated vertices. Let G be an arbitrary graph, and let H be an edge reconstruction of G. If G has n isolated vertices then so has H, by Corollary 6.7. We may therefore write $G := G' + nK_1$ and $H := H' + nK_1$, where G' and H' have no isolated vertices and thus are edge reconstructible. Since $H - e \cong G - e$ for all $e \in E$, $H' - e \cong G' - e$ for all $e \in E$, whence $H' \cong G'$ and $H \cong G$. Therefore G is edge reconstructible. ∎

6.9 Remark. In view of Corollary 6.8, we restrict our attention, henceforth, to *graphs without isolated vertices.*

6.10 Definitions. A class $\mathcal{G}$ of graphs is *edge recognizable* if, for each $\mathcal{G}$-graph G, every edge reconstruction of G is also in $\mathcal{G}$. A class $\mathcal{G}$ of graphs is *weakly edge reconstructible* if, for each $\mathcal{G}$-graph G, all edge reconstructions of G that belong to $\mathcal{G}$ are isomorphic to G.

6.11 Remark. It follows from these definitions that a class of graphs is edge reconstructible if and only if it is both edge recognizable and weakly edge reconstructible.

6.12 The Greenwell-Hemminger Lemma (edge version). *Let $\mathcal{F}$ be an arbitrary class of graphs, and let $\mathcal{G}$ be an edge recognizable class of graphs such that, for every $\mathcal{G}$-graph G and every $\mathcal{F}$-subgraph F of G:*

(i) $e(F) < e(G)$;

(ii) F is contained in a unique maximal $\mathcal{F}$-subgraph of G.

Then, for every $\mathcal{F}$-graph F and every $\mathcal{G}$-graph G, $m(F,G)$ is edge reconstructible.

6.13 Corollary (Greenwell, 1971). *Let G be a graph. Then the deck of G is edge reconstructible.*

Proof. We apply the Greenwell-Hemminger Lemma with $\mathcal{F}$ as the class of graphs on $v(G)-1$ vertices and $\mathcal{G}$ as the class of edge reconstructions of G. By Remark 6.9, G has no isolated vertices. By Corollary 6.7, each edge reconstruction of G has this property, too. It follows that condition (i) of the lemma is satisfied. Condition (ii) is also satisfied, because the maximal $\mathcal{F}$-subgraphs of G are its vertex-deleted subgraphs. By the Greenwell-Hemminger Lemma, these are edge reconstructible. ■

6.14 Corollary.

(a) The degree sequence is edge reconstructible;

(b) Regular graphs, disconnected graphs and trees are edge reconstructible.

Proof. These assertions follow directly from Corollary 6.13 and the results of Section 2. ■

6.15 Kocay's Lemma (edge version). *For any graph G and any sequence $\mathcal{F} := (F_1, F_2, \ldots, F_m)$ of graphs such that $e(F_i) < e(G)$, $1 \le i \le m$, the number of covers of G by $\mathcal{F}$ is edge reconstructible.*

The wealth of information contained in Kocay's Lemma can be judged from the following corollary, which implies that many parameters of interest are edge reconstructible.

6.16 Corollary. *For any graph G and any sequence $(F_1, F_2, \ldots, F_m)$ of graphs such that $e(F_i) < e(G)$, $1 \le i \le m$, and $\sum_{i=1}^{m} e(F_i) = e(G)$, the number of decompositions of G into m subgraphs isomorphic to F_1, $F_2, \ldots,$ F_m is edge reconstructible.*

6.17 Remark. The edge version of Kocay's Lemma is discussed in a more general setting by Avellis and Borzacchini (1988).

7 FORCED MOVES AND EXCLUDABLE CONFIGURATIONS: HOFFMAN'S LEMMA

The notion of an excludable configuration is widely used in combinatorial proofs. In order to show that an object does not exist, one assumes the contrary and establishes properties of the hypothetical object, the aim being to find properties that are mutually contradictory. Often, such properties assert that the object cannot contain specific configurations. This strategy was first used on the Four Colour Conjecture. It has also been applied with some success to the Edge Reconstruction Conjecture.

7.1 Definitions. A *configuration* is a pair $C := (F, d)$, where F is a graph and d is a function that assigns to each vertex v of F a nonnegative integer $d(v)$. An *embedding* of a configuration (F, d) in a graph G is an isomorphism from F to a subgraph of G, mapping each vertex v of F to a vertex of degree $d(v)$ in G. If such an isomorphism exists, one says that G *contains* the given configuration. A configuration C is *excludable* from an edge recognizable class $\mathcal{G}$ of graphs if every member of $\mathcal{G}$ that contains C is edge reconstructible.

7.2 Examples. Let $\mathcal{G}$ be the class of graphs of minimum degree δ, where $\delta \ge 1$. Then the configuration (F, d), where $F := K_1$ and $d(v) = \delta - 1$ for $v \in F$, is excludable from $\mathcal{G}$. Therefore the configuration (F, d), where $F := K_2$ and $d(v) = \delta$ for $v \in F$, is also excludable from $\mathcal{G}$ — if $G \in \mathcal{G}$ and e is an edge of G joining two vertices of degree δ, the edge-deleted subgraph $G - e$ has two vertices of degree $\delta - 1$, both of which are excluded. Thus e is the only edge that can be added to $G - e$ to obtain an edge reconstruction of G.

The simple idea used in the second of these examples can be formalized and extended as follows.

7.3 Definitions. Let G be a graph. An edge e of G is a *forced edge* if it is the only edge that can be added to $G - e$ to yield an edge reconstruction of G. An ordered pair $e \to f$ of edges of G is a *forced move* if f is the only edge other than e that can be added to $G - e$ to yield an edge reconstruction of G. A sequence $e_1 \to f_1,\ e_2 \to f_2, \ldots,\ e_k \to f_k$ of forced moves is *conservative* if

$$\{e_1, \ldots, e_k\} = \{f_1, \ldots, f_k\}.$$

Forced moves are the key to finding excludable configurations, and were first used systematically by Hoffman (1977).

7.4 Hoffman's Lemma (D.G. Hoffman, 1977). *Let G be a graph that admits a conservative odd-length sequence of forced moves:*

$$e_1 \to f_1,\ e_2 \to f_2, \ldots,\ e_{2k+1} \to f_{2k+1}.$$

Then G is edge reconstructible.

Proof. Suppose that G is not edge reconstructible, and let H be an edge reconstruction of G not isomorphic to G. Since $e_1 \to f_1$ is a forced move, G and H are the only edge reconstructions of G, and $G - e_1 + f_1 \cong H$. By the same reasoning, $H - e_2 + f_2 \cong G$ and, in general, $G - e_i + f_i \cong H$ for odd i and $H - e_i + f_i \cong G$ for even i. In particular, $G - e_{2k+1} + f_{2k+1} \cong H$. But $G - e_{2k+1} + f_{2k+1} = G$ because the sequence of forced moves is conservative. Thus $G \cong H$, a contradiction. ■

7.5 Corollary (Hoffman, 1977). *Let $\mathcal{G}$ be the class of graphs of minimum degree δ, where $\delta \geq 1$. Then the configuration $C_m := (K_{1,m}, d)$, where $m \geq 0$, $d(u) = \delta + m - 1$ and $d(v) = \delta$ for $v \neq u$, u being the vertex of degree m in $K_{1,m}$, is excludable from $\mathcal{G}$.*

Proof. The configurations C_0 and C_1 were shown excludable in Example 7.2. Let G be a graph of minimum degree δ and let C_2 be a configuration in G consisting of two vertices v_1 and v_2 of degree δ joined to a vertex u of degree $\delta + 1$ in G. Denote the edge uv_i by e_i, $i = 1, 2$. Then $G - e_1$ contains the excludable configuration C_1. Thus either G contains C_1 or $e_1 \to e$ is a forced move, where $e := v_1v_2$. In the former case, G is edge reconstructible, so we assume the latter. By the same reasoning, $e_2 \to e_1$ and $e \to e_2$ are now forced moves. This sequence of three forced moves results in G. Applying Hoffman's Lemma, we deduce that G is edge reconstructible and hence that C_2 is excludable.

We now establish by induction on m that C_m is excludable for $m \geq 3$. Let $e := uv$ be an edge of C_m. Then $G - e$ contains the excludable configurations C_0 and C_{m-1}. Thus either G contains at least one of C_0 and C_{m-1} or e is a forced edge. In both cases, G is edge reconstructible. ■

The excludable configurations C_m of Corollary 7.5 were used by Hoffman (1977) to deduce the following result (see Bondy, 1983).

7.6 Corollary (Hoffman, 1977). *A graph G of minimum degree δ and average degree d is edge reconstructible if*

$$d < \delta + 1 - \frac{1}{\delta + 1}.$$

Forced moves and excludable configurations have also been employed to show that two special classes of graphs are edge reconstructible.

7.7 Definitions. A *bidegreed graph* is a graph in which each vertex is of degree either δ or Δ. A *claw-free graph* is one that contains no induced subgraph isomorphic to $K_{1,3}$.

7.8 Remark. In a bidegreed graph with $\Delta > \delta + 1$, each edge is forced. Such a graph is therefore reconstructible from any edge-deleted subgraph.

Forced moves and excludable configurations were used extensively to handle the case of bidegreed graphs in which $\Delta = \delta + 1$.

7.9 Theorem (Myrvold, Ellingham and Hoffman, 1987). *Bidegreed graphs are edge reconstructible.*

These techniques were also used, in combination with the methods of the next section, to establish the edge-reconstructibility of claw-free graphs.

7.10 Theorem (Ellingham, Pyber and Yu, 1988). *Claw-free graphs are edge reconstructible.*

The expository article by Lauri (1987) describes several other applications, notably to classes of planar graphs.

8 COUNTING MAPPINGS: NASH-WILLIAMS' LEMMA

We come now to the most powerful tool developed thus far in the assault on the Edge Reconstruction Conjecture. Due to Nash-Williams (1978), it is based on a technique pioneered by Lovász (1972) and refined by Müller (1977). The idea, roughly speaking, is to search for an isomorphism between a graph G and an edge reconstruction H of G by considering all the bijections between their vertex sets.

	F		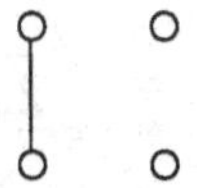	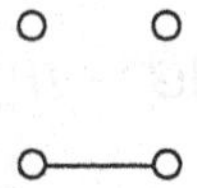	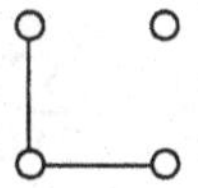
G : 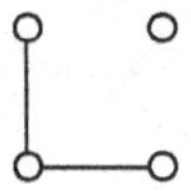	$\|G \to G\|_F$	10	6	6	2
H : 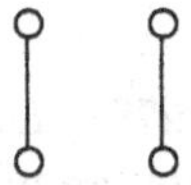	$\|G \to H\|_F$	8	8	8	0

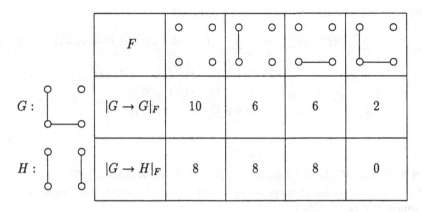

Figure 4: An illustration of Nash-Williams' Lemma

8.1 Notation. Let V be a finite set and Π the set of permutations of V. For $\pi \in \Pi$, we extend π to unordered pairs uv of elements of V by setting $\pi(uv) = \pi(u)\pi(v)$. If G is a graph with vertex set V and edge set E, we denote by $\pi(G)$ the graph with vertex set V and edge set $\{\pi(e) : e \in E\}$. Furthermore, if F is a spanning subgraph of G and H is any graph with vertex set V, we set

$$|G \to H|_F := |\{\pi \in \Pi : \pi(G) \cap H = \pi(F)\}|.$$

8.2 Example. Figure 4 gives the values of $|G \to G|_F$ and $|G \to H|_F$ for all spanning subgraphs F of G when $G := K_1 + K_{1,2}$ and $H := 2K_2$.

8.3 Notation. Let F and G be graphs on the same vertex set V. We denote by $|F \to G|$ the number of embeddings of F in G:

$$|F \to G| := |\{\pi \in \Pi : \pi(F) \subseteq G\}|,$$

and by $x(F, G)$ the number of extensions of F to a graph isomorphic to G:

$$x(F, G) := |\{X : F \subseteq X \cong G\}|.$$

8.4 Proposition. *For any two graphs G and H on the same vertex set, and any spanning subgraph F of G,*

$$\sum_{F \subseteq X \subseteq G} |G \to H|_X = |F \to H| = \text{aut}(F)s(F, H) = \text{aut}(H)x(F, H), \qquad (6)$$

where aut(G) *denotes the number of automorphisms of G.*

Proof. We have the following string of identities:

$$\begin{aligned}\sum_{F\subseteq X\subseteq G} |G\to H|_X &= \sum_{F\subseteq X\subseteq G} |\{\pi\in\Pi:\ \pi(G)\cap H = \pi(X)\}| \\ &= |\{\pi\in\Pi:\ \pi(G)\cap H \supseteq \pi(F)\}| \\ &= |\{\pi\in\Pi:\ \pi(F)\subseteq H\}| \\ &= |\{\pi\in\Pi:\ F\subseteq \pi(H)\}|.\end{aligned}$$

∎

8.5 Nash-Williams' Lemma (C.St.J.A. Nash-Williams, 1978). *Let G be a graph, F a spanning subgraph of G and H an edge reconstruction of G not isomorphic to G. Then*

$$|G\to G|_F - |G\to H|_F = (-1)^{e(G)-e(F)}\mathrm{aut}(G). \tag{7}$$

Proof. By (6),

$$\sum_{F\subseteq X\subseteq G} |G\to H|_X = \mathrm{aut}(F)s(F,H).$$

We invert this identity by inclusion-exclusion (see, for example, Liu, 1968):

$$|G\to H|_F = \sum_{F\subseteq X\subseteq G} (-1)^{e(X)-e(F)}\mathrm{aut}(X)s(X,H). \tag{8}$$

Because H is an edge reconstruction of G, $s(X,G) = s(X,H)$ for every proper spanning subgraph X of G, by Kelly's Lemma. Therefore

$$|G\to G|_F - |G\to H|_F = (-1)^{e(G)-e(F)}\{\mathrm{aut}(G)s(G,G) - \mathrm{aut}(H)s(G,H)\}.$$

But $s(G,G)=1$, whereas $s(G,H)=0$ since $e(G)=e(H)$ and $G\not\cong H$. ∎

8.6 Remark. With the aid of Corollary 4.4(a), identity (8) also yields a vertex version of Nash-Williams' Lemma: let G be a graph, F a spanning subgraph of G and H a reconstruction of G not isomorphic to G. Then

$$|G\to G|_F - |G\to H|_F = \sum_X (-1)^{e(X)-e(F)}\mathrm{aut}(X)(s(X,G)-s(X,H)),$$

where the sum is over all connected spanning subgraphs X of G that contain F. When G is a tree, this identity simplifies to (7).

8.7 Corollary (Nash-Williams, 1978). *A graph G is edge reconstructible if either of the following two conditions holds:*

(a) there exists a spanning subgraph F of G such that $|G\to H|_F$ is the same for all edge reconstructions H of G;

(b) there exists a spanning subgraph F of G satisfying any of the following equivalent inequalities:

(i) $x(F,G) < 2^{e(G)-e(F)-1}$;

(ii) $|F \to G| < 2^{e(G)-e(F)-1}\mathrm{aut}(G)$;

(iii) $\mathrm{aut}(F)s(F,G) < 2^{e(G)-e(F)-1}\mathrm{aut}(G)$.

Proof. Let H be an edge reconstruction of G. If condition (a) holds, the left-hand side of (7) is zero whereas the right-hand side is nonzero. Also, each of the three inequalities in condition (b) is equivalent, by (6), to the inequality

$$\sum_{F \subseteq X \subseteq G} |G \to G|_X < 2^{e(G)-e(F)-1}\mathrm{aut}(G).$$

But this implies that $|G \to G|_X < \mathrm{aut}(G)$ for some spanning subgraph X of G such that $e(G) - e(X)$ is even, and identity (7) is again violated (with $F := X$). Thus, in both cases, Nash-Williams' Lemma implies that H is isomorphic to G. ■

Choosing F as the empty graph in Corollary 8.7 yields two sufficient conditions for the edge reconstructibility of a graph in terms of its edge density, due to Lovász (1972) and Müller (1977), respectively.

8.8 Corollary (Lovász, 1972; Müller, 1977). *A graph G is edge reconstructible if either of the following two conditions holds:*

(a) $e(G) > \frac{1}{2}\binom{v(G)}{2}$;

(b) $2^{e(G)-1} > v(G)!$

Proof.

(a) If $e(G) > \frac{1}{2}\binom{v(G)}{2}$ and F is the empty subgraph of G, $|G \to H|_F = 0$ for every edge reconstruction H of G. By Corollary 8.7(a), G is edge reconstructible.

(b) If $2^{e(G)-1} > v(G)!$ and F is the empty subgraph of G,

$$|F \to G| = v(G)! < 2^{e(G)-1} \leq 2^{e(G)-e(F)-1}\mathrm{aut}(G).$$

By Corollary 8.7(b), G is edge reconstructible. ■

8.9 Remarks.

– Hong Yuan (1983) has observed that Corollary 8.7(a) is also a necessary condition for a graph G to be edge reconstructible. This follows directly from (8) on applying Kelly's Lemma.

– By Stirling's Formula, Corollary 8.8(b) implies that a graph G is edge reconstructible if $e(G) > v(G)\log_2 v(G)$.

– Pyber (1988) has constructed an infinite family $\mathcal{G}$ of graphs such that:

(i) $e(G) \geq v(G)\lfloor \log_2 v(G)/2 \rfloor^{1/2}$, for every $G \in \mathcal{G}$;

(ii) $|G \to G|_F \geq \mathrm{aut}(G)$, for every $G \in \mathcal{G}$ and every spanning subgraph F of G.

These examples serve to show that the lower bound on $e(G)$ given in Corollary 8.8(b) cannot be improved significantly by the use of Corollary 8.7(b) alone.

– The 'balance equations' developed by Krasikov and Roditty (1987) and used by them to derive Corollary 8.8 can be deduced easily from Nash-Williams' Lemma.

8.10 Problem (Bondy and Hemminger, 1977). Find a constructive proof of Nash-Williams' Lemma.

9 APPLICATIONS OF NASH-WILLIAMS' LEMMA

Corollary 8.8 demonstrates the effectiveness of Nash-Williams' Lemma as a tool in reconstruction. Here we describe several other interesting applications of Nash-Williams' Lemma. All are derived from Corollary 8.7(b). The strategy is simple: one seeks a spanning subgraph F of G having relatively few embeddings in G but also relatively few edges. As we shall see, a well-chosen spanning tree often achieves the right balance between these two conflicting demands. For notational clarity in this section, we denote the numbers of vertices and edges of G simply by v and e, respectively.

9.1 BIDEGREED GRAPHS

As noted earlier, bidegreed graphs were proved edge reconstructible by Myrvold, Ellingham and Hoffman (1987) using the techniques of forced moves and excludable configurations. A much shorter proof, based on Nash-Williams' Lemma (but valid only for bidegreed graphs of minimum degree at least seven) was found by Caunter and Nash-Williams (1982). It relies on the following estimate for the number of embeddings of a spanning tree in a graph of given maximum degree.

9.1 Proposition (Caunter and Nash-Williams, 1982). *Let G be a graph of maximum degree Δ, and let T be a spanning tree of G. Then*

$$|T \to G| \leq v\Delta!(\Delta - 1)^{v-\Delta-1}. \tag{9}$$

Proof. Consider an embedding π of T in G. Let $v_1, v_2, \ldots, v_v$ be an ordering of the vertices of T such that, for each i, $1 \leq i \leq v$, the subgraph of T induced by $v_1, v_2, \ldots, v_i$ is a tree. There are v choices for $\pi(v_1)$, at most Δ choices for

$\pi(v_2)$, at most $\Delta - 1$ choices for each of $\pi(v_3), \pi(v_4), \ldots, \pi(v_{v-\Delta+2})$, and at most $\Delta - i$ choices for $\pi(v_{v-\Delta+i+1})$, $2 \leq i \leq \Delta - 1$. Therefore, the number of embeddings of T in G satisfies (9). ∎

9.2 Corollary. *A graph G of maximum degree Δ is edge reconstructible if*

$$v\Delta!(\Delta - 1)^{v-\Delta-1} < 2^{e-v}. \tag{10}$$

Proof. This follows directly from Corollary 8.7(b), using (9). ∎

9.3 Corollary (Caunter and Nash-Williams, 1982). *A graph G of maximum degree Δ and average degree d is edge reconstructible if*

$$2\log_2(2\Delta) \leq d. \tag{11}$$

Proof. If $\Delta \leq 2$, G is edge reconstructible by Corollary 6.14(b). If $\Delta \geq 3$,

$$v\Delta!(\Delta - 1)^{v-\Delta-1} < v(\Delta - 1)^{v-1} < (1 + (\Delta - 1))^v = \Delta^v.$$

Therefore, by Corollary 9.2, G is edge reconstructible if $\Delta^v \leq 2^{e-v}$. This inequality is equivalent to (11). ∎

9.4 Corollary (Caunter and Nash-Williams, 1982). *A graph G of minimum degree δ and maximum degree Δ is edge reconstructible if*

$$2\log_2(2\Delta) \leq \delta + 1 - \frac{1}{\delta + 1}.$$

Proof. This follows directly from Corollary 7.6 and Corollary 9.3. ∎

9.5 Corollary (Caunter and Nash-Williams, 1982). *Let G be a bidegreed graph of minimum degree δ, where $\delta \geq 7$. Then G is edge reconstructible.*

Proof. As noted in Remark 7.8, we may assume that $\Delta = \delta + 1$. If $\delta \geq 8$, then $\Delta \geq 9$ and G is edge reconstructible by Corollary 9.4. The case $\delta = 7$ is settled by a more careful application of Corollary 7.6 and Corollary 9.2. ∎

9.6 Remark. The ideas of Caunter and Nash-Williams (1982) and the theorem of Myrvold, Ellingham and Hoffman (7.9) have been used by Krasikov and Roditty (1990a) to determine sufficient conditions in terms of the minimum and average degrees for the edge reconstructibility of graphs whose degrees take on one of four possible values.

9.2 SQUARES OF GRAPHS

In the previous section, Corollary 8.7(b) was used in conjunction with an upper bound on the number of embeddings of a spanning tree in a graph. Caro, Krasikov and Roditty (1985) derived an alternative upper bound on this number, in terms of the degree sequence of the spanning tree and the maximum and average degrees of the graph.

9.7 Proposition. *Let G be a graph with maximum degree $\Delta(G)$ and average degree $d(G)$, and let T be a spanning tree of G. Then*

$$|T \to G| \leq \frac{\Delta(T)}{\Delta(G)}(d(G))^{v(G)} \prod_{v \in V} (d_T(v) - 1)! \tag{12}$$

Proof. Let u be a vertex of maximum degree in G. The edges of any spanning tree of G can be oriented so that each vertex other than u has indegree one. Thus each spanning tree of G is determined by the choice of one incident edge at each vertex other than u, and the total number of spanning trees of G, $\tau(G)$, satisfies the inequality

$$\tau(G) \leq \frac{\prod_{v \in V} d_G(v)}{d_G(u)} \leq \frac{(d(G))^{v(G)}}{\Delta(G)}$$

Also, if $\mathrm{aut}(T, v)$ is the number of automorphisms of T fixing vertex v, then $\mathrm{aut}(T, w) = d_T(w)!$ for at most one $w \in V$, and $\mathrm{aut}(T, v) \leq (d_T(v) - 1)!$ for all $v \in V \setminus w$. It follows that

$$\mathrm{aut}(T) \leq \Delta(T) \prod_{v \in V} (d_T(v) - 1)!$$

Combining these two inequalities, we obtain

$$|T \to G| = \mathrm{aut}(T)s(T, G) \leq \mathrm{aut}(T)\tau(G) \leq \frac{\Delta(T)}{\Delta(G)}(d(G))^{v(G)} \prod_{v \in V} (d_T(v) - 1)!$$

■

Proposition 9.7 is effective when G has a spanning tree with small maximum degree and hence few automorphisms. For instance, the square of a connected graph has a spanning tree with maximum degree at most three (Caro, Krasikov and Roditty, 1985). This fact, together with inequality (12) and Corollary 8.7(b), produces the following result.

9.8 Theorem (Krasikov and Roditty, 1990b). *The square of a graph is edge reconstructible if its average degree is at least 9.5.*

9.9 Remark. Barnette (1966) has shown that any three-connected planar graph has a spanning tree with maximum degree at most three. This fact might be of use in proving that three-connected planar graphs are edge reconstructible.

9.3 HAMILTONIAN GRAPHS

The class of hamiltonian graphs is a natural candidate for the approach described in the preceding section — a Hamilton path is a spanning tree with only two automorphisms. By finding a good upper bound on the number of Hamilton paths in a graph with a given number of vertices and edges, Lovász (1983) deduced that any sufficiently large graph of average degree at least four is edge reconstructible if it contains a Hamilton path (see also Fiorini and Lauri, 1983 or Bondy, 1983). This implies that all sufficiently large four-connected planar graphs are edge reconstructible, since such graphs are necessarily hamiltonian (Tutte, 1956). It thus yields a relatively easy proof, for large graphs, of the theorem of Fiorini and Lauri (1982) that all four-connected planar graphs are edge reconstructible. The same idea was further exploited by Pyber (1990), who showed that the constraint on the average degree can be dispensed with.

9.10 Theorem (Pyber, 1990). *Let G be a graph that contains a Hamilton path. Then G is edge reconstructible provided that its order is sufficiently large.*

9.4 $K_{1,m}$-FREE GRAPHS

We have seen that claw-free graphs are edge reconstructible (Theorem 7.10). Krasikov (to appear) considered the wider class of $K_{1,m}$-free graphs, where $m \geq 3$, and showed that they too are edge reconstructible provided that the maximum degree is sufficiently large. This is yet another application of Corollary 8.7(b). However, the spanning subgraph F is not chosen to be a tree in this instance.

Given a $K_{1,m}$-free graph G, let u be a vertex of maximum degree in G whose neighbours induce a subgraph U with as few edges as possible, and let P be a *path system* of U (a spanning subgraph of U each of whose components is a path) with as few components as possible. Define F to be the spanning subgraph of G with edge set $E(G) \setminus (E(U) \setminus E(P))$.

Consider the number $x(F,G)$ of extensions of F to a graph isomorphic to G. The choice of u implies that such an extension can only be obtained by joining pairs of vertices of U; moreover, if G' is such an extension, there can be at most $1 + \Delta + \binom{\Delta}{2}$ possibilities for the isomorphism type U' of the subgraph of G' induced by the neighbours of u. The choice of P and the fact that G is $K_{1,m}$-free imply that P has fewer than m components. Thus, if $e(U) := |E(U)|$ and $e(P) := |E(P)|$, there are at most $\binom{e(U)}{e(P)} 2^{m-1}(m-1)!$ extensions of F resulting in any particular isomorphism type U'. Consequently,

$$x(F,G) \leq (1 + \Delta + \binom{\Delta}{2})\binom{e(U)}{e(P)} 2^{m-1}(m-1)!$$

Corollary 8.7(b) now yields the following theorem.

9.11 Theorem (Krasikov, to appear). *A $K_{1,m}$-free graph G is edge reconstructible if $\Delta > cm(\log m)^{1/2}$, where c is an appropriate constant.*

9.12 Remark. Krasikov and Roditty (1990b) have shown that P_4-free graphs are edge reconstructible. This can be seen easily as follows. Let G be P_4-free. If G is disconnected, then G is edge reconstructible. If, on the other hand, G is connected, then the complement of G is disconnected (Seinsche, 1974) and hence reconstructible. By Remark 1.6, G too is reconstructible, and therefore edge reconstructible.

10 ALGEBRAIC TECHNIQUES: THE FRANKL-PACH LEMMA AND THE GODSIL-KRASIKOV-RODITTY LEMMA

Since reconstruction is intimately linked to isomorphism, it is likely, if not inevitable, that group-theoretic methods will play a decisive role in the resolution of the Vertex and Edge Reconstruction Conjectures. The effectiveness, as exemplified by Nash-Williams' Lemma, of the mapping counting technique introduced by Lovász (1972) and described in Section 8 certainly supports this contention. Linear-algebraic methods were introduced by Stanley (1984), and have led, via work of Frankl and Pach (1983), Godsil, Krasikov and Roditty (1987) and Alon, Caro, Krasikov and Roditty (1989) to an alternative derivation, and generalization, of Nash-Williams' Lemma. The basis of this work is a matrix well-known to design-theorists.

10.1 Notation. Let S be an n-set, and X_i, $1 \le i \le \binom{n}{\ell}$, and Y_j, $1 \le j \le \binom{n}{m}$, the ℓ-subsets and m-subsets of S, respectively, where $\ell \le m \le n$. The matrix whose (i,j)-th entry is 1 if $X_i \subseteq Y_j$ and 0 otherwise is denoted by $\mathbf{H}(\ell, m, n)$.

10.2 Remark. For $k < \ell$, the matrices $\mathbf{H}(k,m,n)$ and $\mathbf{H}(\ell,m,n)$ are linked by the identity

$$\binom{m-k}{\ell-k}\mathbf{H}(k,m,n) = \mathbf{H}(k,\ell,n)\mathbf{H}(\ell,m,n). \tag{13}$$

10.3 The Frankl-Pach Lemma (P. Frankl and J. Pach, 1983). *Let* $\mathbf{Hx} = \mathbf{0}$, *where* $\mathbf{H} := \mathbf{H}(\ell-1,m,n)$ *and* $\ell \le m$. *Then, for all ℓ-subsets X_j of S and all subsets Z of X_j,*

$$\sum_{\{i: Y_i \cap X_j = Z\}} x_i = (-1)^{\ell-|Z|} y_j. \tag{14}$$

where x_i and y_j are the ith and jth entries of $\mathbf{x}$ *and* $\mathbf{y} := \mathbf{H}(\ell,m,n)\mathbf{x}$, *respectively.*

Proof. Because $\mathbf{Hx} = \mathbf{0}$, it follows from (13) that

$$\mathbf{H}(k,m,n)\mathbf{x} = \mathbf{0} \text{ for all } k \leq \ell - 1.$$

Also, by definition,

$$\mathbf{H}(\ell,m,n)\mathbf{x} = \mathbf{y}.$$

Thus

$$\sum_{\{i: Y_i \supseteq Y\}} x_i = \begin{cases} 0 & \text{if } Y \subset X_j, \\ y_j & \text{if } Y = X_j. \end{cases}$$

Summing over all Y, $X \subseteq Y \subseteq X_j$,

$$\sum_{X \subseteq Y \subseteq X_j} \sum_{\{i: Y_i \cap X_j \supseteq Y\}} x_i = y_j \text{ for all } X \subseteq X_j.$$

Inverting this by inclusion-exclusion,

$$\sum_{\{i: Y_i \cap X_j \supseteq Y\}} x_i = (-1)^{\ell - |Y|} y_j \text{ for all } Y \subseteq X_j.$$

Equivalently,

$$\sum_{Y \subseteq Z \subseteq X_j} \sum_{\{i: Y_i \cap X_j = Z\}} x_i = (-1)^{\ell - |Y|} y_j \text{ for all } Y \subseteq X_j.$$

A second application of inclusion-exclusion now yields (14). ∎

10.4 Remark. Our formulation of the Frankl-Pach Lemma stems from Thatte (1990), where the case $\ell = m$ is treated.

We apply the Frankl-Pach Lemma in the case where S is the set of edges of a complete graph K. Subsets of S may then be identified with spanning subgraphs of K.

10.5 Notation. The spanning subgraphs of K corresponding to the ℓ-subsets X_i and m-subsets Y_j of S are denoted by F_i and G_j, respectively, $1 \leq i \leq \binom{n}{\ell}$, $1 \leq j \leq \binom{n}{m}$. For a graph G on m edges, the vector whose jth entry is $\text{aut}(G)$ if $G_j \cong G$ and 0 otherwise is denoted by $\mathbf{x}(G)$.

10.6 The Godsil-Krasikov-Roditty Lemma (C.D. Godsil, I. Krasikov and Y. Roditty, 1987). *Let* $\mathbf{H} := \mathbf{H}(\ell - 1, m, n)$, *where* $\ell \leq m$. *Then the vector* $\mathbf{Hx}(G)$ *is edge reconstructible.*

Proof. The ith entry of $\mathbf{Hx}(G)$ is equal to $\text{aut}(G)x(F_i, G)$ and thus, by (6), to $\text{aut}(F_i)s(F_i, G)$. By Kelly's Lemma, this is edge reconstructible. ∎

Nash-Williams' Lemma follows directly from the preceding two lemmas. Let G be a graph and H an edge reconstruction of G not isomorphic to G. Set $m := e(G)$, $n := \binom{v(G)}{2}$ and $\mathbf{x} := \mathbf{x}(G) - \mathbf{x}(H)$. By the Godsil-Krasikov-Roditty Lemma, $\mathbf{Hx} = \mathbf{0}$. Now set $\ell := m$. Because $\mathbf{H}(m, m, n)\mathbf{x} = \mathbf{Ix} = \mathbf{x}$, the Frankl-Pach Lemma yields

$$\sum_{\{i:Y_i \cap Y_j = Z\}} x_i = (-1)^{m-|Z|} x_j. \tag{15}$$

Denoting the spanning subgraph of K with edge set Z by F, we see that identities (15) and (7) are one and the same.

11 RELATED QUESTIONS

Our aim in this article has been to review the status of the two principal questions of reconstruction, with particular reference to the techniques developed to attack them. Here, we touch briefly on three related problems that are susceptible to similar techniques.

11.1 SHUFFLED DECKS

11.1 Definitions. Two families of graphs $\mathcal{G}$ and $\mathcal{H}$ are *isomorphic* if they admit a bijection mapping each graph of $\mathcal{G}$ to an isomorphic graph of $\mathcal{H}$. The *shuffled deck* of a family $\mathcal{G}$ of graphs is the multiset union of the decks of the members of $\mathcal{G}$. A *reconstruction* of a family $\mathcal{G}$ of graphs from its shuffled deck is a family $\mathcal{H}$ of graphs whose shuffled deck is isomorphic to that of $\mathcal{G}$. A family $\mathcal{G}$ of graphs is *reconstructible* from its shuffled deck if every reconstruction of $\mathcal{G}$ from its shuffled deck is isomorphic to $\mathcal{G}$. The edge analogues of these concepts are defined in a similar manner.

Kocay (1979) initiated the study of shuffled deck reconstruction. He noted the obvious analogue of Kelly's Lemma and formulated the following conjecture, establishing the case $k = 2$ for families consisting either of disconnected graphs or of trees.

11.2 Conjecture (Kocay, 1979). *For $k \geq 1$, every family of graphs with k members, each of order at least $2\lceil \log_2 k \rceil + 3$, is reconstructible from its shuffled deck.*

As with graph reconstruction, the edge version of this problem is far more tractable. Indeed, the Frankl-Pach Lemma combined with a straightforward extension of the Godsil-Krasikov-Roditty Lemma (see Thatte, 1990) yields a shuffled edge deck generalization of Nash-Williams' Lemma, and this in turn gives analogues of the results of Lovász and Müller (8.8). A slight strengthening of Müller's condition is required; Lovász's condition, however, remains unchanged —every family of graphs, each with v vertices and more than $\frac{1}{2}\binom{v}{2}$ edges, is reconstructible from its shuffled edge deck.

Thatte (1990) conjectured that every family of connected graphs is reconstructible from its shuffled edge deck. This was disproved by Miklós (1991). His argument makes elegant use of elementary linear algebra.

Consider a set $\mathcal{G} := \{G_1, G_2, \ldots, G_m\}$ of pairwise nonisomorphic graphs. Let $\mathcal{F} := \{F_1, F_2, \ldots, F_n\}$ be a list of the pairwise nonisomorphic graphs in the shuffled edge deck of $\mathcal{G}$. Suppose that $m > n$; this is the case, for example, when $\mathcal{G}$ is the set of all trees on ten vertices (Miklós, 1991). Associate with each graph G_i, $1 \le i \le m$, a variable x_i and with each graph F_j, $1 \le j \le n$, the linear equation $\sum_{i=1}^{m} x_i a_{i,j} = 0$, where $a_{i,j}$ is the number of cards in the deck of G_i that are isomorphic to F_j, $1 \le i \le m$, $1 \le j \le n$. Since $m > n$, this system of homogeneous equations has a nontrivial solution $x_i = b_i$, $1 \le i \le m$. Denote by P and N the sets of indices i, $1 \le i \le m$, for which b_i is positive or negative, respectively, and let $\mathcal{G}^+$ be the family of graphs consisting of b_i copies of G_i, $i \in P$, and $\mathcal{G}^-$ the family of graphs consisting of $-b_i$ copies of G_i, $i \in N$. Then $\mathcal{G}^+$ and $\mathcal{G}^-$ are nonisomorphic families with isomorphic shuffled edge decks.

11.2 k-RECONSTRUCTION

11.3 Definitions. A *k-vertex-deleted subgraph* is a subgraph obtained by deleting a set of k vertices. The *k-deck* of a graph G is the family of k-vertex-deleted subgraphs of G. A *k-reconstruction* of a graph G is a graph H with the same k-deck as G. A graph G is *k-reconstructible* if every k-reconstruction of G is isomorphic to G. Similar definitions apply to edge deletions.

The method of proof of Kelly's Lemma shows that if $\ell > k$, the ℓ-deck of a graph is determined by its k-deck. Thus the larger k is, the more likely it is for a graph to be k-nonreconstructible. Nýdl (1991) showed that k does not need to be very large compared to the order of the graph for this to occur. Indeed, for any $c > 0$, there are arbitrarily large graphs G that are not $\lfloor cv(G) \rfloor$-reconstructible. On the other hand, a generalization of Kocay's Lemma due to Taylor (1985) shows that the number of k-component spanning subgraphs of each isomorphism type is $(k-1)$-reconstructible. In particular, the number of perfect matchings of a graph on $2n$ vertices is $(n-1)$-reconstructible, a result first obtained by Godsil (1981) using much the same sort of approach.

Nash-Williams' Lemma generalizes to k-edge reconstructibility; it suffices to set $\ell := m - k$ in the Frankl-Pach and Godsil-Krasikov-Roditty Lemmas, noting (as did Godsil, Krasikov and Roditty, 1983) that the conclusion of the latter lemma can be strengthened accordingly. Alon, Caro, Krasikov and

Roditty (1989) applied this generalized Nash-Williams' Lemma to obtain corresponding generalizations of the theorems of Lovász and Müller (8.8) and Caunter and Nash-Williams (9.2).

11.3 VERTEX SWITCHING

11.4 Definitions. Let G be a simple graph and v a vertex of G. To *switch* G at v is to replace the edges incident to v in G by the edges incident to v in the complement of G. The resulting graph is called a *vertex switching* of G and denoted by G_v. The *vertex-switching deck* of a graph G is the family of vertex switchings of G. A *vertex-switching reconstruction* of a graph G is a graph H with the same vertex-switching deck as G. A graph G is *vertex-switching reconstructible* if every vertex-switching reconstruction of G is isomorphic to G.

11.5 Remark. A graph is vertex-switching reconstructible if and only if its complement is vertex-switching reconstructible.

11.6 Examples. The graphs $G := K_4$ and $H := 2K_2$ are vertex-switching reconstructions of one another. Hence, neither of these graphs is vertex-switching reconstructible; by Remark 11.5, neither are their complements.

Algebraic methods have proved just as effective here as in edge reconstruction. They were employed by Stanley (1985) to establish the vertex-switching reconstructibility of all graphs G such that $v(G) \not\equiv 0 \pmod 4$, although N. Alon and D.W. Coppersmith (see Stanley, 1985) later deduced the same result by elementary counting arguments, as did Krasikov and Roditty (1987) using their method of 'balance equations'. Linear-algebraic techniques were also applied by Ellingham and Royle (1990) and Krasikov and Roditty (1989) to derive a vertex-switching analogue of Kelly's Lemma. By combining this lemma with the balance equations of Krasikov and Roditty (1987), Ellingham and Royle (1990) proved that triangle-free graphs on n vertices, where $n \neq 4$, are vertex-switching reconstructible. It remains an open question whether all graphs G such that $v(G) \neq 4$ are vertex-switching reconstructible.

ACKNOWLEDGEMENTS

I wish to thank Mark Ellingham for making available to me his bibliographic database on reconstruction, and Ian Goulden for a helpful remark. I also acknowledge the financial support of the Natural Sciences and Engineering Research Council of Canada under Operating Grant A7331.

REFERENCES

N. Alon, Y. Caro, I. Krasikov and Y. Roditty, Combinatorial reconstruction problems, *J. Combin. Theory Ser. B* **47** (1989), 153–161.

G. Avellis and L. Borzacchini, Reconstruction theory for g-graphs, *Ars Combin. Ser. A* **25** (1988), 173-177. MR89f:05125.

D. Barnette, Trees in polytopal graphs, *Canad. J. Math.* **18** (1966), 731–736.

G.D. Birkhoff, A determinantal formula for the number of ways of coloring a map, *Ann. Math.* (2) **14** (1912), 42–46.

B. Bollobás, Almost every graph has reconstruction number three, *J. Graph Theory* **14** (1990), 1–4.

J.A. Bondy, The reconstruction of graphs, *preprint*, University of Waterloo (1983).

J.A. Bondy and R.L. Hemminger, Graph reconstruction — a survey, *J. Graph Theory* **1** (1977), 227–268.

Y. Caro, I. Krasikov and Y. Roditty, Spanning trees and some edge-reconstructible graphs, *Ars Combin. Ser. A* **20** (1985), 109–118.

J. Caunter and C.St.J.A. Nash-Williams, Degree conditions for edge reconstruction, *manuscript*, 1982.

M.N. Ellingham, Recent progress in edge reconstruction, *Congressus Numerantium* **62** (1988), 3–20.

M.N. Ellingham, L. Pyber and Yu Xingxing, Claw-free graphs are edge reconstructible, *J. Graph Theory* **12** (1988), 445–451. MR 89g:05078.

M.N. Ellingham and G.F. Royle, Vertex-switching reconstruction of subgraph numbers and triangle-free graphs, *preprint*, Vanderbilt University, 1990.

S. Fiorini and J. Lauri, Edge-reconstruction of 4-connected planar graphs, *J. Graph Theory* **6** (1982), 33–42. MR 83g:05054.

S. Fiorini and J. Lauri, Edge-reconstruction of graphs with topological properties, in *Combinatorial Mathematics* (Marseille-Luminy, 1981), North-Holland Math. Stud. **75** North-Holland, Amsterdam-New York, 1983, 285–288.

P. Frankl and J. Pach, On the number of sets in a null t-design, *Europ. J. Combin.* **4** (1983), 21–23.

C.D. Godsil, Matchings and walks in graphs, *J. Graph Theory* **5** (1981), 285–297.

C.D. Godsil, I. Krasikov and Y. Roditty, Reconstructing graphs from their k-edge deleted subgraphs, *J. Combin. Theory Ser. B* **43** (1987), 360–363. MR 89a:05106.

D.L. Greenwell, Reconstructing graphs, *Proc. Amer. Math. Soc.* **30** (1971), 431–433. MR 44:3908.

D.L. Greenwell and R.L. Hemminger, Reconstructing the n-connected components of a graph, *Aequationes Math.* **9** (1973), 19–22. MR 52:5488.

F. Harary, The four color conjecture and other graphical diseases, in *Proof*

Techniques in Graph Theory (Proceedings of the Second Ann Arbor Graph Theory Conference Ann Arbor, Mich., 1968, F. Harary, ed.) Academic Press, New York, 1969, 1–9. MR 40:7150.

F. Harary, On the reconstruction of a graph from a collection of subgraphs, in *Theory of Graphs and its Applications* (Proceedings of the Symposium held in Prague, 1964, M. Fiedler, ed.), Czechoslovak Academy of Sciences, Prague; reprinted by Academic Press, New York, 1964, 47–52. MR 30:5296.

F. Harary, A.J. Schwenk and R.L. Scott, On the reconstruction of countable forests, *Publ. Math. Inst. (Beograd)* **13** (1972), 39–42. MR48:5894.

D.G. Hoffman, *personal communication*, 1977.

Hong Yuan, On the edge reconstruction of graphs, *Ars Combin.* **25** (1988), 185–188. MR 89e:05150.

P.J. Kelly, *On Isometric Transformations*, Ph.D. thesis, University of Wisconsin, 1942.

P.J. Kelly, A congruence theorem for trees, *Pacific J. Math.* **7** (1957), 961–968. MR 19:442.

W.L. Kocay, *k-Reconstruction of Graphs*, Ph.D. Thesis, University of Waterloo, 1979.

W.L. Kocay, On reconstructing spanning subgraphs, *Ars Combin.* **11** (1981), 301–313. MR 82m:05072.

W.L. Kocay, Some new methods in reconstruction theory, in *Combinatorial Mathematics IX* (Proc. 9th Australian Conf. on Combinatorial Mathematics, Univ. of Queensland, Brisbane), Lecture Notes in Mathematics **952**, Springer, 1982, 89–114. MR 83k:05079.

W.L. Kocay, On Stockmeyer's nonreconstructible tournaments, *J. Graph Theory* **9** (1985), 473–476.

W.L. Kocay, A family of nonreconstructible hypergraphs, *J. Combin. Theory Ser. B* **42** (1987), 46–63.

I. Krasikov, A note on the edge-reconstruction of $K_{1,m}$-free graphs, *J. Combin. Theory Ser. B*, to appear.

I. Krasikov and Y. Roditty, Balance equations for reconstruction problems, *Arch. Math. (Basel)* **48** (1987), 458–464. MR 88g:05096.

I. Krasikov and Y. Roditty, Switching reconstruction and diophantine equations, *preprint*, Tel-Aviv University, 1989.

I. Krasikov and Y. Roditty, Some applications of the Nash-Williams lemma to the edge-reconstruction conjecture, *Graphs and Combinatorics* **6** (1990a), 37–39.

I. Krasikov and Y. Roditty, Recent applications of Nash-Williams lemma to the edge-reconstruction conjecture, *Ars Combin. Ser. A* **29** (1990b), 215–224.

J. Lauri, Graph reconstruction — some techniques and new problems, Proceedings of the First Catania International Combinatorial Conference on Graphs, Steiner Systems, and their Applications, Vol. 2 (Catania, 1986),

Ars Combin. Ser. B **24** (1987), 35-61. MR 89f:05128.
C.L.Liu, *Introduction to Combinatorial Mathematics*, McGraw-Hill, New York, 1968.
L. Lovász, A note on the line reconstruction problem, *J. Combin. Theory Ser. B* **13** (1972), 309–310. MR 46:8913.
L. Lovász, Some problems of graph theory, *Matematikus Kurir* (1983).
A. Mansfield, The relationship between the computational complexities of the legitimate deck and isomorphism problems, *Quart. J. Math. (Oxford) Ser. (2)* **33** (1982), 345–347. MR 83i:68073.
B. Manvel, Reconstruction of graphs: progress and prospects, *Congressus Numerantium* **63** (1988), 177–187.
B.D. McKay, Computer reconstruction of small graphs, *J. Graph Theory* **1** (1977), 281–283.
D. Miklós, A false conjecture in the reconstruction theory, *preprint*, Hungarian Academy of Sciences, 1991.
V. Müller, The edge reconstruction hypothesis is true for graphs with more than $n\log_2 n$ edges, *J. Combin. Theory Ser. B* **22** (1977), 281–283.
W.J. Myrvold, *The Ally and Adversary Reconstruction Problems*, Ph.D. Thesis, University of Waterloo, 1988.
W.J. Myrvold, M.N. Ellingham and D.G. Hoffman, Bidegreed graphs are edge reconstructible, *J. Graph Theory* **11** (1987), 281–302. MR 88i:05137.
C.St.J.A. Nash-Williams, The reconstruction problem, in *Selected Topics in Graph Theory* (L. W. Beineke and R. J. Wilson, eds.), Academic Press, London, 1978, Chapter 8.
V. Nýdl, Finite undirected graphs which are not reconstructible from their large cardinality subgraphs, in *Topological, Algebraic and Combinatorial Structures* (J. Nešetřil, ed.), Ann. Discrete Math., 1991, to appear.
L. Pyber, *Extremal Structures and Covering Problems*, Candidate's Thesis, 1988.
L. Pyber, The edge reconstruction of hamiltonian graphs, *J. Graph Theory* **14** (1990), 173–179.
H. Sachs, Beziehungen zwischen den in einem Graphen enthaltenen Kreisen und seinem charakteristischen Polynom, *Publ. Math. Debrecen* **11** (1964), 119–134.
D. Seinsche, On a property of the class of n-colorable graphs, *J. Combin. Theory Ser. B* **16** (1974), 191–193.
R.P. Stanley, Quotients of Peck posets, *Order* **1** (1984) 29-34. MR 85m:06007.
R.P. Stanley, Reconstruction from vertex-switching, *J. Combin. Theory Ser. B* **38** (1985), 132–138. MR 86f:05096.
R. Statman, Reductions of the graph reconstruction conjecture, *Discrete Math.* **36** (1981), 103–107. MR 83a:05105.
P.K. Stockmeyer, *personal communication*, 1976.
P.K. Stockmeyer, The falsity of the reconstruction conjecture for tourna-

ments, *J. Graph Theory* **1** (1977), 19–25.
P.K. Stockmeyer, A census of non-reconstructable digraphs. I. Six related families, *J. Combin. Theory Ser. B* **31** (1981), 232–239. MR 83f:05052.
P.K. Stockmeyer, Tilting at windmills, or my quest for non-reconstructible graphs, in *Congressus Numerantium* **63** (1988), 188–200.
R. Taylor, Subgraph identities and reconstruction, *Ars Combin. Ser. A* **19** (1985), 245–256. MR 86k:05087.
B.D. Thatte, On the Nash-Williams' lemma in graph reconstruction theory, *preprint*, Indian Institute of Science, 1990.
W.T. Tutte, A theorem on planar graphs, *Trans. Amer. Math. Soc.* **82** (1956), 99–116.
W.T. Tutte, All the king's horses. A guide to reconstruction, in *Graph Theory and Related Topics* (Proceedings of the Conference held in honour of W.T. Tutte on the occasion of his 60th birthday, Waterloo, 1977, J.A. Bondy and U.S.R. Murty, eds.), Academic Press, New York, 1979, 15–33. MR 81a:05096.
W. T. Tutte, *Graph Theory*, Encyclopedia of Mathematics and its Applications **21**, Addison-Wesley Publishing Co., Reading, Mass., 1984. MR 87c:05001.
S.M. Ulam, *A Collection of Mathematical Problems*, Wiley (Interscience), New York, 1960, 29. MR 22:10884.
H. Whitney, A logical expansion in mathematics, *Bull. Amer. Math. Soc.* **38** (1932), 572–579.
Yang Yong Zhi, The reconstruction conjecture is true if all 2-connected graphs are reconstructible, *J. Graph Theory* **12** (1988), 237–243. MR 89e: 05149.

Turán Type Problems

ZOLTÁN FÜREDI

Mathematical Institute of the Hungarian Academy of Sciences,
P. O. B. 127, Budapest 1364 Hungary

Abstract. Given a graph $\mathcal{F}$, what is $\mathrm{ex}(n, \mathcal{F})$, the maximum number of edges in a graph of order n not containing $\mathcal{F}$ as a subgraph? Turán answered this question in the case of complete graphs, $\mathcal{F} = \mathcal{K}(r)$; for example, $\mathrm{ex}(n, \mathcal{K}(3)) = \lfloor n^2/4 \rfloor$. Despite the enormous progress since this result, there are a lot of unsolved problems in the area.
We shall investigate the more general hypergraph problems, where the forbidden configuration is a k-uniform hypergraph. For example, if the excluded hypergraph consists of two disjoint edges, that is the family $\mathcal{H}$ of k-sets is intersecting, then $|\mathcal{H}| \le \binom{n-1}{k-1}$ for $n \ge 2k$.
The aim here is to survey the most important methods (and results) of the field such as the application of linear algebra, random constructions, and geometry. In particular, we describe the delta system method which was developed especially for the handling of Turán type problems.

1 INTRODUCTION, UNAVOIDABLE SUBGRAPHS

One of the most important problems in every branch of mathematics is the description of the structure of its objects. What is the structure of graphs? There are several approaches, but the seemingly easiest and the most obvious one is to find the substructures, the subgraphs. In this paper we restrict ourselves only to *local* properties; e.g., the characterization of the *minors* is not a local property. We naturally get the following problem: Given the positive integers n and e what are the subgraphs of a graph with n vertices and e edges?

Our starting point is the following special case of Turán's theorem: If $\mathcal{G}$ has n vertices and more than $n^2/4$ edges then it contains a triangle, $\mathcal{C}(3)$. (This special case is due to Mantel (1907).) In general, let $\mathcal{G}$ be an arbitrary graph and let $\mathrm{ex}(n, \mathcal{G})$ denote the maximum number of edges in a graph on n vertices not containing $\mathcal{G}$ as a subgraph. $\mathcal{G}$ is called the *forbidden* subgraph. In other words, if a graph has more than $\mathrm{ex}(n, \mathcal{G})$ edges then $\mathcal{G}$

is *unavoidable*. Mantel's result can be restated as

$$\text{ex}(n, \mathcal{C}(3)) = \lfloor n^2/4 \rfloor. \tag{1.1}$$

Erdős (1938) proved already 53 years ago that

$$\text{ex}(n, \mathcal{C}(4)) = \Theta(n^{3/2}); \tag{1.2}$$

however, the exact value of $\text{ex}(n, \mathcal{C}(4))$ is only known for small n's (for $1 \leq n \leq 21$ by Clapham, Flockart and Sheehan (1989)) and for $n = q^2 + q + 1$ where q is a prime power, $q \geq 14$ (Füredi (1983a), (1991)). Turán determined $\text{ex}(n, \mathcal{K}(r))$ for all $n \geq k \geq 3$, where $\mathcal{K}(r)$ stands for the complete graph on r vertices. In his memory the determination of $\text{ex}(n, \mathcal{G})$ is called a Turán type problem and the value of $\text{ex}(n, \mathcal{G})$ is the *Turán number* of $\mathcal{G}$.

Turán type problems are often difficult and very little is known even about the simple cases when $\mathcal{G}$ is a fixed even cycle $\mathcal{C}(2k)$ or a fixed complete bipartite graph $\mathcal{K}(k, k)$.

Similar extremal problems can be investigated for k-uniform hypergraphs, i.e. collections of k-element sets, as well. The first such result is due to Erdős, Ko, and Rado (1961). If a k-uniform family $\mathcal{F}$ has n vertices, $n \geq 2k$, and it has more than $\binom{n-1}{k-1}$ members, then it contains two disjoint sets. In general, hypergraph problems are even more difficult. Despite this difficulty, so many results have emerged that this paper is devoted only to the survey of various methods.

We shall discuss different generalizations of theorems (1.1) and (1.2) for k-uniform hypergraphs. After introducing notations (Ch. 2) and a short overview of graph problems (Ch. 3), the Turán number of a hypergraph is defined in Ch. 4. Even such a short survey cannot avoid mentioning the best estimates for $T(n, \ell, k)$ which is the most frequently investigated problem of this field, (Ch. 5). After collecting all known Turán numbers of order $\Theta(n^k)$ (in Ch. 6), we return to another old problem, the maximum size of t-intersecting families.

Our primary purpose is to exhibit the delta system method, a method developed especially for the handling of Turán type problems (Sections 8–12). We also show examples of classical combinatorial methods like constructions by polynomials over finite fields (Ch. 13) and by random choice (Ch. 14), proofs by transformations (shiftings, Ch. 7), and by other discrete procedures. Finally, in Ch. 15, a number of applications are given where Turán numbers naturally emerge.

2 DEFINITIONS

A hypergraph $\mathcal{H}$ is a pair $(V(\mathcal{H}), \mathcal{E}(\mathcal{H}))$ where $V(\mathcal{H})$ is a finite set, the set of *vertices*, and $\mathcal{E}(\mathcal{H})$, the *edge set*, is a collection of subsets of $V(\mathcal{H})$. Sometimes we may allow $\mathcal{E}(\mathcal{H})$ to contain the same set more than once. If we want to emphasize that $\mathcal{H}$ contains (or might contain) multiple edges then we call it a *multihypergraph*. If $\mathcal{H}$ does not contain multiple edges then it is called a *simple* hypergraph. Usually, by 'hypergraph' we mean a simple hypergraph. The cardinality of the vertex set V is called the *order* of $\mathcal{H}$, while the cardinality of $\mathcal{E}(\mathcal{H})$ is the *size* of $\mathcal{H}$. Where no confusion will result we abbreviate $V(\mathcal{H})$ and $\mathcal{E}(\mathcal{H})$ by V and $\mathcal{H}$. So several times we identify the hypergraph $\mathcal{H}$ by its edge set and talk about the *family* $\mathcal{H}$. In this way $E \in \mathcal{H}$ means an edge from $\mathcal{E}(\mathcal{H})$.

2^S is the set of all subsets of S. $\binom{S}{k}$ denotes the set of all k-subsets of the set S $(k \geq 0)$. Obviously, $|\binom{S}{k}| = \binom{n}{k}$ for $|S| = n$. A hypergraph is a k-graph, or *k-uniform* hypergraph if all edges have k elements. The 2-graphs are called *graphs*. The *rank* of $\mathcal{H}$ is k if $\max\{|E| : E \in \mathcal{E}(\mathcal{H})\} = k$. $\binom{S}{k}$ is called the *complete k-graph* over S, and it is abbreviated to $\mathcal{K}_k(n)$. Note that this notation differs slightly from the usual one since the order of the graph stands in parentheses. So the complete graph on n vertices is denoted by $\mathcal{K}_2(n)$, or briefly by $\mathcal{K}(n)$ (instead of $\mathcal{K}_n$). In order to be consistent, we use $\mathcal{C}(n)$ for the cycle of length n (instead of the common C_n). $\mathcal{K}(A, B)$ denotes the *complete bipartite* graph with parts A and B, and $\mathcal{K}(a, b)$ stands for a complete bipartite graph with $|A| = a$, $|B| = b$.

$\mathcal{F}$ is a *subhypergraph* of $\mathcal{H}$ if $V(\mathcal{F}) \subseteq V(\mathcal{H})$ and $\mathcal{E}(\mathcal{F}) \subseteq \mathcal{E}(\mathcal{H})$. For $A \subseteq V(\mathcal{H})$, we denote by $\mathcal{H}|A$ the *induced subhypergraph* on A with edge set $\{E \in \mathcal{E}(\mathcal{H}) : E \subseteq A\}$. Moreover, let $\mathcal{H}[B]$ denote the members of $\mathcal{E}(\mathcal{H})$ containing B; i.e. $\mathcal{H}[B] = \{E \in \mathcal{E}(\mathcal{H}) : B \subseteq E\}$. The cardinality of $\mathcal{H}[\{v\}]$ for an element $v \in V$ is called the *degree* of $\mathcal{H}$ (at v), and it is denoted by $\deg_{\mathcal{H}}(v)$, or briefly by $\deg(v)$.

We use the notations $\lfloor x \rfloor$ and $\lceil x \rceil$ for the lower and upper integer part of x, respectively. For integers $a < b$, let $[a, b] := \{a, a+1, \ldots b\}$. $[1, n]$ is abbreviated to $[n]$. The vertex set V of a hypergraph always has n elements and to avoid double indices it is identified with $[n]$, unless otherwise stated. We also use the notations O, o, Ω and Θ to compare the order of magnitudes of the functions f_n and g_n; namely, $f_n = \Omega(g_n)$ means that for some constants $c > 0$ and n_0 we have $f_n \geq c g_n$ for $n > n_0$, (i.e. $g_n = O(f_n)$), while $f_n = \Theta(g_n)$ means $c'' g_n \geq f_n \geq c' g_n$ for all sufficiently large n.

3 GRAPHS

In this section we recall the basic facts of the theory of forbidden subgraphs. We shall see that the order of magnitude of $\mathrm{ex}(n, \mathcal{G})$ is determined by the chromatic number $\chi(\mathcal{G})$, unless $\mathcal{G}$ is bipartite.

Let $\mathcal{T}^r(n)$ denote the complete r-colored graph on n vertices with almost equal parts. This means that $|V(\mathcal{T}^r(n))| = n$, $V = V_1 \cup \cdots \cup V_r$, $|V_i| = \lfloor (n+i-1)/r \rfloor$, and the edge set of $\mathcal{T}^r(n)$ consists of all edges connecting distinct parts ($n \geq r > 1$). Sometimes this graph is called the r-colored *Turán graph.* Let $t^r(n)$ denote the number of edges of $\mathcal{T}^r(n)$. We have $t^r(n) = (1 - 1/r)\binom{n}{2} + O(n)$. Turán's (1941), (1954) theorem says that

$$\mathrm{ex}(n, \mathcal{K}(r)) = t^{r-1}(n) \tag{3.1}$$

for all $n \geq r \geq 3$. Moreover, $\mathcal{T}^{r-1}(n)$ is the only graph of order n and size $t^{r-1}(n)$ that does not contain a $\mathcal{K}(r)$.

Concerning the complete bipartite graph $\mathcal{K}(s,t)$, Kővári, Sós, and Turán (1954) proved that

$$\mathrm{ex}(n, \mathcal{K}(s,t)) < \frac{1}{2}(s-1)^{1/t} n^{2-1/t} + (t-1)n/2. \tag{3.2}$$

Let $\mathcal{K}^r(t)$ denote the graph $\mathcal{T}^r(rt)$. Erdős and Stone (1946) discovered that

$$\mathrm{ex}(n, \mathcal{K}^r(t)) = t^{r-1}(n) + o(n^2) \tag{3.3}$$

for fixed r and t while $n \to \infty$. The precise order of magnitude of $\mathrm{ex}(n, \mathcal{K}^r(t)) - t^{r-1}(n)$ was determined by Bollobás, Erdős, and Simonovits (1976).

Let $\mathbf{F}$ be a family of graphs, called *forbidden* subgraphs. Write $\mathrm{ex}(n, \mathbf{F})$ for the maximum size of a graph of order n not containing any forbidden subgraph. Erdős and Simonovits (1966) observed that (3.2) and (3.3) have the following immediate consequence.

$$\lim_{n \to \infty} \mathrm{ex}(n, \mathbf{F}) / \binom{n}{2} = 1 - \frac{1}{r}, \tag{3.4}$$

where $r + 1 = \min\{\chi(\mathcal{F}) : \mathcal{F} \in \mathbf{F}\}$, $r \geq 1$.

In the case $r = 1$, (i.e. in the case of bipartite graphs) the following basic result is due to Bondy and Simonovits (1974). If $\mathcal{G}$ is a graph of order n

and size at least $90kn^{(k+1)/k}$ then it contains cycles of length 2ℓ for every integer ℓ, $k \leq \ell \leq kn^{1/k}$. This implies $\mathrm{ex}(n, \mathcal{C}(2k)) \leq 90kn^{1+1/k}$.

Finally, if $\mathcal{G}$ is of order k and has no cycle (i.e. G is a tree or a forest) then the Turán number is linear: $\mathrm{ex}(n, \mathcal{G}) \leq n(k-2)$. The famous Erdős–Sós conjecture states (see, e.g., Erdős (1964a)) that

$$\mathrm{ex}(n, \mathcal{G}) \leq n(k-2)/2 \quad (?) \tag{3.5}$$

This conjecture was proved by Sidorenko (1989a) for trees having a vertex with at least $\geq (k-2)/2$ neighbouring leaves.

So the order of magnitude of $\mathrm{ex}(n, \mathcal{G})$ can be determined from the above bounds except for bipartite graphs. Here are the most important results and conjectures for that case. Erdős, Rényi, and Sós (1966) and independently Brown (1966) showed the following improvement of (1.2).

$$\mathrm{ex}(n, \mathcal{C}(4)) = \frac{1}{2}n^{3/2} + O(n^{4/3}). \tag{3.6}$$

Brown (1966) showed by an algebraic construction that $\frac{1}{2}(1+o(1))n^{5/3} \leq \mathrm{ex}(n, \mathcal{K}(3,3))$. It is not known if $\lim_{n\to\infty} \mathrm{ex}(n, \mathcal{K}(3,3))n^{-5/3}$ exists. It seems probable that

$$\mathrm{ex}(n, \mathcal{K}(t,t)) = \Omega(n^{2-1/t}) \quad (?) \tag{3.7}$$

but this has not been shown for $t > 3$. The best lower bound, obtained by probabilistic methods, is due to Erdős and Spencer (1974): $\mathrm{ex}(n, \mathcal{K}(t,t)) \geq (1/2)n^{2-2/(t+1)}$. Simonovits (1978) suggested that to give a lower bound for $\mathcal{K}(t, 2^t)$ seems simpler.

Concerning cycles $\mathcal{C}(2k)$, the Bondy–Simonovits upper bound is matched in general by $\mathrm{ex}(n, \mathcal{C}(2k)) = \Omega(n^{(2k+1)/2k})$. (Obtained by probabilistic method, too, by Erdős (1959).) Constructions giving $\Omega(n^{1+1/k})$ are known only for $k = 2, 3$, and 5, see Benson (1966). Recently Wenger (1990) gave completely new constructions but his method works only for the same values of k.

Let $\mathcal{Q}^3$ denote the graph formed by the 12 edges and 8 vertices of a 3-dimensional cube. Erdős and Simonovits (1970) proved $\mathrm{ex}(n, \mathcal{Q}^3) = O(n^{8/5})$. They conjecture (see Simonovits (1984)) that, for all rationals $1 < p/q < 2$ there exists a bipartite graph $\mathcal{G}$ with

$$\mathrm{ex}(n, \mathcal{G}) = \Theta(n^{p/q}), \quad (?) \tag{3.8}$$

and that every bipartite graph has a rational exponent r with

$$\text{(3.9)} \qquad \mathrm{ex}(n,\mathcal{G}) = \Theta(n^r) \quad (?)$$

One of the few exact results of this field is due to Erdős and Gallai (1961). For $n \geq 2t$

$$\text{(3.10)} \quad \mathrm{ex}(n, t \text{ disjoint edges}) = \max\{\binom{2t-1}{2}, \binom{t-1}{2} + (t-1)(n-t+1)\}.$$

More results on the finer structure of the extremal graphs (for $\chi(\mathcal{G}) > 2$), as well as several additional results and conjectures can be found in Bollobás' book (1978) and in the survey papers of Simonovits (1983), (1984).

4 k-PARTITE HYPERGRAPHS

Let $\mathbf{F}$ be a collection of k-uniform hypergraphs. Define $\mathrm{ex}(n,\mathbf{F})$ as $\max|\mathcal{H}|$ where $\mathcal{H} \subseteq \binom{V}{k}$ with $|V| = n$ and $\mathcal{H}$ contains no copy of any $\mathcal{F} \in \mathbf{F}$. Such an $\mathcal{H}$ is called $\mathbf{F}$-*free*. To indicate that we consider k-graphs we sometimes write $\mathrm{ex}_k(n,\mathbf{F})$. From an averaging argument it follows (Katona, Nemetz and Simonovits (1964)) that $\mathrm{ex}_k(n,\mathbf{F})/\binom{n}{k}$ is a monotone decreasing function of n. (More exactly, non-increasing.) Hence, the *coefficient of saturation* $\pi(\mathbf{F}) := \lim_{n\to\infty} \mathrm{ex}(n,\mathbf{F})/\binom{n}{k}$ always exists.

The complete k-partite hypergraph $\mathcal{K}(t_1, t_2, \ldots, t_k)$ has a partition of its vertex set $V = V_1 \cup \cdots \cup V_k$, such that $|V_i| = t_i$, and $\mathcal{E}(\mathcal{K}) = \{E \subseteq V : |E \cap V_i| = 1\}$ for all $1 \leq i \leq k$. It has $t_1 \cdot \ldots \cdot t_k$ edges. If all parts are equal we use the notation $\mathcal{K}_k(t,\ldots,t)$. If $|V_i| = \lfloor (n+i-1)/k \rfloor$, i.e. they are almost equal, then $(V, \mathcal{E}(\mathcal{K}))$ is called the *complete equipartite* k-graph (over n vertices). This hypergraph and its size are denoted by $\mathcal{P}_k(n)$ and $p_k(n)$, respectively. A k-graph is k-*partite* if it is a subhypergraph of a complete k-partite one. Erdős (1964b) proved the following theorem in an implicit form. For any positive integers k and $t_1 \leq \ldots \leq t_k$ there exist positive constants c' and c'' such that the following holds. If $\mathcal{H}$ is an arbitrary k-graph with n vertices and $|\mathcal{H}| \geq c' n^{k-\varepsilon}$ edges where $\varepsilon = 1/(t_1 \cdots t_{k-1})$, then $\mathcal{H}$ contains at least

$$\text{(4.1)} \qquad c'' \frac{|\mathcal{H}|^{t_1 \cdots t_k}}{n^{kt_1 \ldots t_k - t_1 - \ldots - t_k}}$$

copies of $\mathcal{K}(t_1, \ldots, t_k)$. This implies the following extension of (3.2).

$$\text{(4.2)} \qquad \mathrm{ex}_k\left(n, \mathcal{K}_k(t,\ldots,t)\right) \leq c_{k,t} n^{k-1/t^{k-1}},$$

where $c_{k,t}$ is a constant depending only on k and t. Erdős conjectures that this is the best exponent but no lower bound of the form $k - (C/t^{k-1})$ has yet been proved for an absolute constant C.

If $\mathbf{F}$ is k-partite (has a k-partite member) then (4.2) implies $\pi(\mathbf{F}) = 0$. Otherwise, if no member of $\mathbf{F}$ is k-partite then the example of the complete equipartite k-graph of order n shows that

$$\text{(4.3)} \qquad \mathrm{ex}(n, \mathbf{F}) \geq (1 + o(1))(n/k)^k,$$

implying $\pi(\mathbf{F}) \geq k!/k^k$.

For ordinary graphs ($k = 2$), the limit $\pi(\mathcal{G})$ is $1 - (1/r)$, by (3.4). Erdős (1981) conjectured that for every fixed k the set of $\{\pi(\mathcal{G}) : \mathcal{G} \text{ is a } k\text{-graph}\}$ forms a discrete sequence in the interval $[k!/k^k, 1)$ (the famous 'jumping constant' conjecture). This was disproved by Frankl and Rödl (1984) (of course, only for $k \geq 3$). Their result is especially interesting because for $k \geq 3$ the coefficients of saturation, $\pi(\mathcal{G})$, are known only for a very few cases and all but one was determined well after 1984. However, part of the original conjecture might be true: namely that the value $k!/k^k$ is isolated. We list all known positive $\pi(\mathcal{G})$ in Section 6.

The following simple but useful lemma, due to Erdős and Simonovits (1983) (also see Frankl, Rödl (1984)) justifies the name coefficient of saturation. Let $\mathcal{F}$ be a k-graph with vertices $\{x_1, \ldots, x_s\}$. Then for every $\delta > 0$ there exists an $\varepsilon = \varepsilon(\delta, s)$ and $n_0 = n_0(\delta, s)$ such that whenever $\mathcal{H} \subseteq \binom{V}{k}$ satisfies $|V| = n$, $n > n_0$ and $|\mathcal{H}| > (\pi(\mathcal{F}) + \delta)\binom{n}{k}$, then there exist at least

$$\text{(4.4)} \qquad \varepsilon \binom{n}{s} s!$$

sequences $v_1, \ldots, v_s$ from V with the property that $x_i \mapsto v_i$ defines an embedding of $\mathcal{F}$ into $\mathcal{H}$. In other words, there are $\mathcal{F}$-free k-graphs of size $\pi(\mathcal{F})\binom{n}{k}$ but if a k graph has a few more edges then immediately it contains almost as many copies of $\mathcal{F}$ as possible; i.e. at least a positive fraction of them.

4.1 Large k-partite Subhypergraphs

We close this section with another important lemma which enables us to simplify the case when the forbidden subgraph is k-partite, however, in doing so we lose a constant factor of $k^k/k!$. In that case the right order of magnitude of $\mathrm{ex}_k(n, \mathcal{F})$ is the first question, so we are ready to pay this cost. The lemma states that every k-uniform family $\mathcal{H}$ contains a k-partite subfamily $\mathcal{H}' \subseteq \mathcal{H}$ of size

$$|\mathcal{H}'| \geq \frac{k!}{k^k}|\mathcal{H}|. \tag{4.5}$$

This lemma is due to Erdős (1967) for $k = 2$ and Erdős and Kleitman (1968) for general k. For the proof we consider a random k-partition of the underlying set of $\mathcal{H}$ and count the expected number of multicolored edges.

5 COMPLETE k-GRAPHS, THE TURÁN CONJECTURE

Let $\text{ex}_k(n, \mathcal{K}_k(\ell))$ denote the maximum size of a k-graph of order n without a complete k-graph on ℓ vertices. In this section we overview the bounds for $k \geq 3$. The limit of the monotone non-increasing sequence $\text{ex}_k(n, \mathcal{K}_k(\ell))/\binom{n}{k}$ is denoted by $\pi_k(\ell)$. No value of $\pi_k(\ell)$ is known for $\ell > k \geq 3$.

Turán conjectured that $\text{ex}_3(n, \mathcal{K}_3(4))$ is attained in the 3-graph $\mathcal{T}_3(n, 4)$ defined as follows. Let $V = V_0 \cup V_1 \cup V_2$ be an equipartition of V, $|V| = n$, and let the hypergraph $\mathcal{T}_3(n, 4)$ have as edges those $E \in \binom{V}{3}$ which either intersect all V_i's or contain two vertices of V_i and one in $V_{i+1 \pmod 3}$. $\mathcal{T}_3(n, 4)$ has size $\frac{5}{9}\binom{n}{3} - O(n^2)$, implying $\pi_3(4) \geq 5/9$.

$$\pi_3(4) = \frac{5}{9} \quad (?) \tag{5.1}$$

To the memory of Turán, Erdős offers \$1,000 for proving (5.1). Brown (1983) found six, Todorov (1984) eight and Kostochka (1982) $2^{(n/3)-2}$ nonisomorphic $\mathcal{K}_3(4)$-free 3-graphs of order n and sizes exactly the same as $\mathcal{T}_3(n, 4)$. This shows that if Turán's conjecture is true then there are many extremal configurations. Kalai (1985) proposed a promising new approach using the homology group of certain simplicial complexes but the conjecture is still unsolved. The upper bound for $\pi_3(4)$ was subsequently improved from 9/14 in Katona, Nemetz, Simonovits (1964), 0.6213... in de Caen (1988) to $(-1 + \sqrt{21})/6 = 0.5971\ldots$ due to Giraud (1989).

Instead of the Turán number of $\mathcal{K}_k(\ell)$, usually $T(n, \ell, k) = \binom{n}{k} - \text{ex}_k(n, \mathcal{K}_k(\ell))$ is investigated. In other words, let n, ℓ, and k be natural numbers with $n \geq \ell \geq k$. Define the function $T(n, \ell, k)$ as the minimum number of k-subsets of a set of size n needed to ensure that every ℓ-subset contains at least one of the k-subsets. This is also called the *Turán number* $T(n, \ell, k)$. The determination of $T(n, \ell, k)$ is the most widely investigated problem of this field. Obviously, $T(n, \ell, k) = C(n, n-k, n-\ell)$ where $C(n, b, a)$ is the minimum number of b-sets to cover all a-sets of an n-set. So the Turán problem and the covering problem are in fact equivalent. However, the fact that they are usually studied in different ranges of parameters, namely $T(n,\ell,k)$ is investigated for fixed ('small') values of k and ℓ and arbitrarily large n, gives them a different flavour. For example, Mills' results (1979) on covering the pairs can be reformulated as Turán numbers of the form $T(n,n-2,k)$ with $n \geq 3k/2$.

Turán (1970) conjectures that for $n > n_0$

$$T(n,5,3) = \binom{\lfloor n/2 \rfloor}{3} + \binom{\lceil n/2 \rceil}{3} \quad (?) \tag{5.2}$$

Surányi (1971) showed $T(9,5,3) = 12 < 14$, and Sidorenko (1983) proved that there is no equality for each odd $n \geq 9$, namely

$$T(n,5,3) = \binom{\lfloor n/2 \rfloor}{3} + \binom{\lceil n/2 \rceil}{3} - \Delta(n),$$

where $\Delta(n) = 3$ for $n = 13$, 2 for $n = 15$, $\lfloor n/4 \rfloor$ for $n \equiv 1 (\bmod 8)$, $\lfloor n/4 \rfloor - 1$ for $n \equiv 5 (\bmod 8)$ $(n \neq 13)$, and $\lfloor n/8 \rfloor$ for $n \equiv 3 (\bmod 4)$ $(n \neq 15)$. The construction giving the upper bound in (5.2) is the disjoint union of two complete 3 graphs on $\lfloor n/2 \rfloor$ and $\lceil n/2 \rceil$ vertices. This example implies $\pi_3(5) \geq 3/4$. Generalizing this example for all $\ell \geq k \geq 3$ (taking $r := \lfloor (\ell-1)/(k-1) \rfloor$ complete k-graphs of sizes $\sim n/r$) we have

$$T(n,\ell,k) \leq \binom{n}{k} \lfloor (\ell-1)/(k-1) \rfloor^{1-k}.$$

This upper bound is useless for $\ell \leq 2k-2$. de Caen proved first that $\lim_{n,k\to\infty} T(n,k+1,k)/\binom{n}{k} = 0$. Kim and Roush (1983) proved $T(n,k+1,k) \leq (1+o(1))(2\log k/k)\binom{n}{k}$. Their ideas were extended by Frankl and Rödl (1985a) to obtain the following:

$$T(n,k+1,k) < \frac{1}{\lceil k/\log k \rceil} \frac{\log k}{\log k - 1} \binom{n}{k} \tag{5.3}$$

$$T(n,k+a,k) \leq \frac{\log k}{\binom{k}{a}} (1 - \frac{1}{2k^a}) a(a+4) \binom{n}{k}. \tag{5.4}$$

Construction for $T(n,k+1,k)/\binom{n}{k} < 2\log k/k$ Divide V into r almost equal parts $V_1,\ldots,V_r$, $|V| = n$. Let $\mathcal{M}$ consist of the k-subsets of V lacking some of the parts V_i. If $r \sim k/(2\log k)$ then almost all of the k-subsets of V meet every V_i, yielding $|\mathcal{M}| = O(\log k/k)\binom{n}{k}$. For $0 \le \alpha < r$ define $\mathcal{E}_\alpha := \{E \in \binom{V}{k} : V_i \cap E \ne \emptyset$ for all i and $\sum_i |E \cap V_i| i \equiv \alpha (\operatorname{mod} r)\}$. Taking the smallest $\mathcal{E}_\alpha$ together with $\mathcal{M}$, we get the desired hypergraph.

There are further constructions in special cases, e.g., for $T(n, \ell, 3)$ if $n \le 9(\ell-1)/4$ by Todorov (1983), for $T(n, 6, 4)$ by Droebeske (1982), for $T(n, k+1, k)$ by Tazawa and Shirakura (1983), and for $T(n, 2k+1, 2k)$ by de Caen, Kreher, and Wiseman (1988).

The function $T(n, \ell, k)/\binom{n}{k}$ is monotone increasing in n, so the trivial fact that $T(\ell, \ell, k) = 1$ gives the lower bound $T(n, \ell, k) \ge \binom{n}{k}/\binom{\ell}{k}$. Spencer (1972) gave the almost correct order of magnitude of the coefficient of $\binom{n}{k}$, improving it to about $(k/\ell)^{k-1}$. (Actually, in Erdős, Spencer (1974) it is proved that $T(n, \ell, k)/\binom{n}{k} \ge \max_{\ell \le a \le n}(a - \ell + 1)/\binom{a}{k}$.) The current best general lower bound is due to de Caen (1983a, c):

$$T(n, \ell, k) \ge \frac{n - \ell + 1}{n - k + 1}\binom{n}{k} \Big/ \binom{\ell - 1}{k - 1}. \tag{5.5}$$

For a recent survey on this topic see de Caen (1991).

6 THE LAGRANGE FUNCTION OF HYPERGRAPHS

6.1 Two Problems Concerning Generalized Triangles

As the original Turán problem is so difficult, one of our tasks is to find solvable problems. Katona (1974) proposed the following generalization of the simplest case of Turán's graph theorem: Determine the maximum number of edges in a k-graph of order n such that the symmetric difference of two edges is not contained in a third. Let $A \triangle B := (A \setminus B) \cup (B \setminus A)$ stand for the symmetric difference, and let $\mathbf{D}_k$ be the class of k-graphs with 3 edges $\{A, B, C\}$ such that $A \triangle B \subseteq C$. ($\mathbf{D}_k$ consists of $\lfloor k^2/4 \rfloor$ hypergraphs.) Bollobás (1974) conjectured that

$$\operatorname{ex}(n, \mathbf{D}_k) = p_k(n) = \lfloor \frac{n}{k} \rfloor \lfloor \frac{n+1}{k} \rfloor \cdots \lfloor \frac{n+k-1}{k} \rfloor, \quad (?) \tag{6.1}$$

and that the complete equipartite k-graph, $\mathcal{P}_k(n)$, is the only extremal system. (6.1) was proved in Frankl, Füredi (1984b) for $n \le 2k$, and so in that case $\operatorname{ex}(n, \mathbf{D}_k) = 2^{n-k}$. By monotonicity, this implies $\operatorname{ex}(n, \mathbf{D}_k) \le \binom{n}{k} 2^k / \binom{2k}{k}$ for all $n \ge 2k$. Bollobás (1974) solved the case $k = 3$

$$\operatorname{ex}_3(n, A \triangle B \subseteq C) = \lfloor \frac{n}{3} \rfloor \lfloor \frac{n+1}{3} \rfloor \lfloor \frac{n+2}{3} \rfloor. \tag{6.2}$$

Since that time three new proofs have emerged. In the next subsection we shall present Sidorenko's proof (1987), who also verified (6.1) for $k = 4$. The other cases are still open. In Frankl, Füredi (1983) the following slightly sharper statement was proved: Let $\mathcal{D}_3$ be the hypergraph with edge set $\{123, 124, 345\}$. Then $\text{ex}(n, \mathcal{D}_3) = \lfloor n/3 \rfloor \lfloor (n+1)/3 \rfloor \lfloor (n+2)/3 \rfloor$ holds for $n > 3000$. de Caen (1985) provided another proof and posed the following simpler problem. Determine $\max |\mathcal{H}|$ where $\mathcal{H} \subseteq \binom{V}{k}$ and $\mathcal{H}$ contains no three sets A, B, C with $|A \cap B| = k-1$ and $A \triangle B \subseteq C$. Again, one can reformulate this problem in terms of $\text{ex}(n, \mathbf{F})$. Define $\mathcal{A}^i = \{\{1, 2, \ldots, k\}, \{1, 2, \ldots, k-1, k+1\}, \{i, i+1, \ldots, i+k-1\}\}$ and set $\mathbf{S}_k = \{\mathcal{A}^2, \mathcal{A}^3, \ldots, \mathcal{A}^k\}$. Then de Caen's problem asks for the determination of $\text{ex}(n, \mathbf{S}_k)$. Actually, Sidorenko (1987) proved for $k = 2, 3, 4$ that

$$\text{ex}(n, \mathbf{S}_k) = p_k(n). \tag{6.3}$$

Recently, Kleitman and Sidorenko (1991) proved (6.1) for $k = 5$ if $n > n_0$ and for all $n \equiv 0 \pmod 5$.

6.2 The Lagrange Function and Proof of (6.3)

Let $\mathcal{G}$ be a k-graph of order n, $\mathcal{G} \subseteq \binom{V}{k}$, $V = [n]$. Let us associate with $\mathcal{G}$ the homogeneous, multilinear polynomial in n variables

$$f(\mathcal{G}, x_1, \ldots, x_n) = \sum_{G \in \mathcal{G}} \prod_{i \in G} x_i.$$

The Lagrange function of $\mathcal{G}$, $\lambda(\mathcal{G})$, is defined as

$$\lambda(\mathcal{G}) = \max\{f(\mathcal{G}, \mathbf{x}) : x_1 + \cdots + x_n = 1, x_i \geq 0\}.$$

Note that $|\mathcal{G}|/n^k = f(\mathcal{G}, \frac{1}{n}, \ldots, \frac{1}{n})$, implying

$$|\mathcal{G}| \leq \lambda(\mathcal{G}) n^k. \tag{6.4}$$

The Lagrange function was introduced in Frankl and Rödl (1984), (1989) and in Sidorenko (1987). For graphs it was already used by Motzkin and Strauss (1965) to give a new proof of Turán's theorem.

Throughout the rest of this section $\mathbf{x}$ will denote a vector $(x_1, \ldots, x_n)$ with $x_1 + \cdots + x_n = 1$ and $x_i \geq 0$ for $i = 1, \ldots, n$. The *support* of a vector $\mathbf{x}$ is $\text{Supp}(\mathbf{x}) := \{i : x_i > 0\}$. We use the notation $\mathbf{x}_{-j} = (x_1, \ldots, x_{j-1}, x_{j+1}, \ldots, x_n)$. $\mathcal{G}$ is a *2-cover* if every pair $\{u, v\} \subseteq V$ is contained in some edge $G \in \mathcal{G}$. Suppose that $\mathbf{x}$ is optimal, $f(\mathcal{G}, \mathbf{x}) = \lambda(\mathcal{G})$, such that $\text{Supp}(\mathbf{x})$ is the smallest possible. Then it is easy to see (Frankl, Rödl (1984)) that the induced subhypergraph, $\mathcal{G}|\,\mathrm{Supp}(\mathbf{x})$, is a 2-cover, moreover

$$f(\mathcal{G}(j), \mathbf{x}_{-j}) = k\lambda(\mathcal{G}) \tag{6.5}$$

holds for all $j \in \mathrm{Supp}(\mathbf{x})$. Here $\mathcal{G}(j) := \{G \setminus \{j\} : j \in G \in \mathcal{G}\}$ is the *link* of the vertex j. Note that $f(\mathcal{G}(j), \mathbf{x}_{-j}) = \partial f(\mathcal{G}, \mathbf{x})/\partial x_j$ is the partial derivative.

Proof of the upper bound in (6.3) for $k \leq 4$, $n \equiv 0 \,(\mathrm{mod}\, k)$. Let $\mathcal{G}$ be an $\mathbf{S}_k$-free family. Using (6.4) it is sufficient to prove $\lambda(\mathcal{G}) \leq 1/k^k$. In view of (6.5), we may assume that $\mathcal{G}$ is a 2-cover. Since $\mathcal{G}$ is $\mathbf{S}_k$-free one has $|G \cap G'| \leq k-2$ for all distinct $G, G' \in \mathcal{G}$. Equivalently, the links $\mathcal{G}(j)$ are pairwise disjoint, $j \in V$. Thus the polynomials $f(\mathcal{G}(j), \mathbf{x}_{-j})$ have no common term. We obtain

$$\sum_j f(\mathcal{G}(j), \mathbf{x}_{-j}) \leq \sigma_{k-1}(V),$$

where σ_{k-1} is the elementary symmetric polynomial of rank $k-1$ with variables $x_{i1}, \ldots, x_{im}$ where $\mathrm{Supp}(\mathbf{x}) = \{x_{i1}, \ldots, x_{im}\}$. Since $x_i \geq 0$, we get the upper bound $\sigma_{k-1}(\frac{1}{m}, \ldots, \frac{1}{m})$ for the right hand side. Combining this with (6.5) we obtain

$$km\lambda(\mathcal{G}) \leq \binom{m}{k-1}/m^{k-1}. \tag{6.6}$$

For $k = 2, 3$ the right hand side is at most km/k^k for all $m \geq k$. The same holds for $k = 4$ and $m \geq 4$, $m \neq 5$. However, $m = 5$ is impossible, because either $|\mathcal{G}| = 1$ and then $\mathcal{G}$ is not a 2-cover, or $|\mathcal{G}| \geq 2$ and $|G \cap G'| = 3 = k-1$ holds for all $G, G' \in \mathcal{G}$, contradicting $|G \cap G'| \leq k-2$. ∎

6.3 $\mathrm{ex}(n, \mathbf{S}_k)$ and the small Witt designs

Recall that a Steiner system $S(s, k, t)$ is a family $\mathcal{F} \subseteq \binom{S}{k}$ such that every t-set $T \subseteq S$ is contained in a *unique* member of $\mathcal{F}$, $|S| = s$. For $(s, k, t) = (11, 5, 4)$ and $(12, 6, 5)$, there exist unique Steiner systems; see e.g. Cameron, van Lint (1975). They are called the Witt designs $\mathcal{W}_5$, $\mathcal{W}_6$ and consist of 66 and 132 k-tuples, respectively. In general, call a family $\mathcal{G} \subseteq \binom{S}{k}$ a ***partial*** Steiner system $PS(s, k, t)$, (or a *t-packing*), if $|G \cap G'| < t$ for all distinct $G, G' \in \mathcal{G}$. Starting with any partial $PS(s, k, k-1)$ Steiner system $\mathcal{G} \subseteq \binom{S}{k}$, $S = [s]$, one can obtain an $\mathbf{S}_k$-free family $\mathcal{H}$ over the n-element set V by the following operation, called *blow-up*. Let $V_1, \ldots, V_s$ be a partition of V and set $\mathcal{H} := \{H \in \binom{V}{k} : \{i : H \cap V_i \neq \emptyset\} \in \mathcal{G}\}$. Note that for $|V_i| = n_i$ we have $|\mathcal{H}| = f(\mathcal{G}, n_1/n, \ldots, n_s/n)n^k$. The following consequence of (6.6), due to Sidorenko (1987) and Frankl, Füredi (1988), shows that the determination of $\pi(\mathbf{S}_k)$ ($:= \lim_{n\to\infty} \mathrm{ex}(n, \mathbf{S}_k)/\binom{n}{k}$) is a finite problem although the number of cases to check increases very fast.

$$\pi(\mathbf{S}_k) = k! \max \lambda(\mathcal{G}) \tag{6.7}$$

where the maximum is taken over all partial $PS(s, k, k-1)$ Steiner systems $\mathcal{G} \subseteq \binom{S}{k}$ with $s := |S| \leq k^k/k!$. To compute or bound the value of $\lambda(\mathcal{G})$ for a specific k-graph is very difficult in general; however in Frankl, Füredi (1989) this finite process yielded

$$\pi(\mathbf{S}_k) \sim \frac{2}{e}k^{-2} + O(k^{-3})$$
$$\pi(\mathbf{S}_5) = 720/11^4 \text{ for } k = 5$$
$$\pi(\mathbf{S}_6) = 55/12^3 \text{ for } k = 6.$$

Moreover, for $n > n_0$, $k = 5, 6$ the only optimal $\mathbf{S}_k$-free k-family is obtained by blowing up the appropriate Witt design. This shows that $\mathrm{ex}(n, \mathbf{D}_k)$ is much smaller than $\mathrm{ex}(n, \mathbf{S}_k)$ for $k > 4$. In the course of calculation, the following conjecture would have been of great help. Suppose that $\mathcal{F}$ is a k-graph and $|\mathcal{F}| = x(x-1) \cdot \ldots \cdot (x-k+1)/k!$ for some real $x \geq k$. Then

$$\lambda(\mathcal{F}) \leq \binom{x}{k}/x^k. \tag{6.8}$$

Recall that $\mathcal{A}^k = \{[1, k], [1, k-1] \cup \{k+1\}, [k, 2k-1]\}$, $\mathcal{A}^k \in \mathbf{S}_k$. Replacing $\mathbf{S}_k$ by $\mathcal{A}^k$, all above cited results remain true asymptotically. Frankl, Füredi (1989) conjecture that

$$\mathrm{ex}(n, \mathbf{S}_k) = \mathrm{ex}(n, \mathcal{A}^k) \tag{6.9}$$

for all fixed k and $n > n_k$.

6.4 Families Closed Under Homomorphism

The inverse operation of a blow-up is a *homomorphism*. For $\mathcal{F}$ and $\mathcal{H}$ k-graphs, one says that the map $\varphi : V(\mathcal{F}) \to V(\mathcal{H})$ is a *homomorphism* if $\{\varphi(i) : i \in F\} \in \mathcal{E}(\mathcal{H})$ holds for all $F \in \mathcal{E}(\mathcal{F})$. For example, a graph has a homomorphism into $\mathcal{K}(s)$, the complete graph on s vertices, if and only if its chromatic number is at most s. Call $\mathbf{F}$ *closed under homomorphism* (shortly *closed*), if $\mathcal{A} \in \mathbf{F}$ implies that $\mathbf{F}$ contains (a copy of) every homomorphic image of $\mathcal{A}$. The general form of (6.7) is the following (due to Frankl, Füredi (1988), Sidorenko (1987)). For a closed family $\mathbf{F}$

$$\pi(\mathbf{F}) = \sup \lambda(\mathcal{G})k! \tag{6.10}$$

where the supremum is taken over all $\mathbf{F}$-free families $\mathcal{G}$. This equality was used in the following theorem of Sidorenko (1989a). Let $\mathcal{T} = (V, \mathcal{E}(\mathcal{T}))$ be a tree with s vertices satisfying the Erdős–Sós conjecture (3.5). Let $P^1, P^2, \ldots$ be the pairs of V not belonging to $\mathcal{E}(\mathcal{T})$ and let $B, B^1, B^2 \ldots$ be pairwise disjoint $(k-2)$-sets outside V. Define the edges of the k-graph $\mathcal{F}$ by $\mathcal{E}(\mathcal{F}) := \{E \cup B : E \in \mathcal{E}(\mathcal{T})\} \cup \{P^i \cup B^i : 1 \le i \le n(n-2)/2\}$. Suppose further that $s \ge M_k$ (where $M_k = (1/2)k^2 - (11/6)k - O(1)$, $M_2 = M_3 = 2$). Then for all n divisible by $(s+k-3)$,

$$\text{ex}(n, \mathcal{F}) = \binom{s+k-3}{k}\left(\frac{n}{s+k-3}\right)^k.$$

6.5 The Remaining Exact Results

There are only two further exact results on $\pi(\mathcal{F}) = \Theta(n^k)$. Let $\mathcal{B}$ be the 3-graph of order 9 with edges $\{123, 124, 134, 156, 257, 358, 459\}$. Using the Lagrange function method, Sidorenko (1989a) proved that $\text{ex}(n, \mathcal{B}) = n^3/16$ for $n \equiv 0 \pmod 4$. The extremal construction is a blow-up of the complete 3-graph of order 4, $\mathcal{K}_3(4)$.

Suppose that k is divisible by t and $\mathcal{F}$ is a k-graph and $\mathcal{G}$ is a t-graph. We say that $\mathcal{F}$ is a (k/t)-*expansion* of $\mathcal{G}$ if it can be obtained by replacing each vertex of $\mathcal{G}$ by a (k/t)-element set. So the expansion is a special case of the blow-up discussed in Ch. 6.3. For example, if $S^1, S^2, \ldots, S^r$ are disjoint $k/2$-element sets, then the hypergraph $\mathcal{C}_k(r)$ with $V(\mathcal{C}_k(r)) = S^1 \cup \cdots \cup S^r$, and with edge set $\{S^i \cup S^j : 1 \le i < j \le r\}$ is a $(k/2)$-expansion of the complete graph $\mathcal{K}(r)$. Considering a hypergraph with vertex set $\binom{V}{k/t}$, it is not difficult to see (Sidorenko (1989b, 1991b)) that $\pi_k(\mathcal{F}) \le \pi_t(\mathcal{G})$. Then (3.4) implies that

$$\pi(\mathcal{F}) \le \frac{r-2}{r-1}, \tag{6.11}$$

if $\mathbf{F}$ is a class of $(k/2)$-expansions of some graphs $\mathcal{G}^1, \mathcal{G}^2, \ldots$ with $r = \min\{\chi(\mathcal{G}^i)\}$. The following example of Sidorenko gives a sharp lower bound, so equality holds in (6.11) if $r-1$ is a power of 2, $r = 2^m + 1$. Let $GF(2)^m$ be the m-dimensional vector field over the 2-element field. Label the vertices of V by these vectors, $\varphi(v) \in GF(2)^m$ for all $v \in V$, such that the sizes of the parts $\varphi^{-1}(x)$ are almost equal to each other. Take all k-subsets $A \subset V$ with $\sum_{v \in A} \varphi(v) \neq 0$. (The case $\pi(\mathcal{C}_k(3)) = \frac{1}{2}$ was proved independently by Frankl (1990)).

7 SHIFTING AND THE ERDŐS, KO, RADO THEOREM

7.1 Largest t-intersecting Families

Suppose that $\mathcal{G}$ is a family of k-sets of $[n]$ such that any two members of $\mathcal{G}$ intersect in at least t elements. Erdős, Ko, and Rado (1961) proved that this condition implies $|\mathcal{G}| \leq \binom{n-t}{k-t}$ whenever $n > n_0(k,t)$. Equality holds if and only if $\mathcal{G}$ consists of all k-element subsets of $[n]$ containing a fixed t-element subset. In the case $t = 1$, they established the best possible bound $n_0(k,1) = 2k$. The exact value of $n_0(k,t)$ was determined by Frankl (1978a), for $t \geq 15$, and by Wilson (1984), for all t, proving that $n_0(k,t) = (k-t+1)(t+1)$. For smaller values of n there are larger t-intersecting families. For example, if $n \leq 2k - t$ then $\mathcal{G} = \binom{[n]}{k}$ is t-intersecting. Define

$$\mathcal{A}^r = \{G \in \binom{[n]}{k} : |G \cap [t+2r]| \geq t + r\}.$$

(To avoid trivialities, from now on it is supposed that $n > 2k - t$, $k > t \geq 2$, $k - t > r \geq 0$.) Frankl (1978a) contains the following conjecture.

(7.1) *If $\mathcal{G}$ is of maximal cardinality then $\mathcal{G} \equiv \mathcal{A}^r$ for some r.*

The case $n = 4p$, $k = 2p$, $t = 2$ (and then $r = p - 1$) was already conjectured in Erdős, Ko, Rado (1961). Of course, this is a Turán type problem since the family of forbidden k-graphs, $\mathbf{F}_k^{<t}$, consists of pairs having at most $t-1$ common elements. It is not difficult to determine the largest $\mathcal{A}^r$. $|\mathcal{A}^r|$ is the largest among the $\mathcal{A}^i$'s if

$$(7.2)\qquad n_r(k,t) = (k-t+1)(2+\frac{t-1}{r+1}) \leq n < (k-t+1)(2+\frac{t-1}{r}).$$

Moreover, if equality holds above then $|\mathcal{A}^r| = |\mathcal{A}^{r+1}|$. In Frankl, Füredi (1991) (7.1) was proved for $n > (k-t+1)c\sqrt{t/\log t}$, ($c$ is an absolute constant, $c < 50$). In other words, if n is in the range of (7.2) and $t \geq 1 + cr(r+1)/(1+\log r)$ then

$$(7.3)\qquad \mathrm{ex}_k(n, \mathbf{F}_k^{<t}) = |\mathcal{A}^r|.$$

The proof of (7.3) is elementary, it uses the so called shifting operation. Several properties of shifted families are described in subsection 7.3.

7.2 Remark on the Shannon-capacity

Wilson (1984) used Delsarte's (1973) eigenvalue method and actually obtained a stronger result for $n > n_0(k,t)$. He proved that the *Shannon capacity* of the graph $\mathrm{Kn}(n,k,t)$ is $\binom{n-t}{k-t}$. ($\mathrm{Kn}(n,k,t)$, the generalized Kneser graph, has vertex set $\binom{[n]}{k}$ and two vertices E, E' are connected by an edge if $|E \cap E'| < t$.) The case $t = 1$ was proved by Lovász (1979b) in his celebrated paper on Shannon capacity. The general case was proved by Schrijver (1981) for very large n. They both used the Johnson scheme but Wilson ingeniously utilized the properties of the Hamming scheme.

However, this method does not seem to be suitable to settle conjecture (7.1) in general, because for $n < n_0(k,t)$ the Shannon capacity of $\mathrm{Kn}(n,k,t)$ exceeds $\max|\mathcal{A}^r|$. (Private communication of R. M. Wilson.)

7.3 Shifting

The following exchange operation, or *shifting*, was defined in Erdős, Ko, Rado (1961). Let $1 \le i < j \le n$ and suppose that $\mathcal{G}$ is a t-intersecting family of k-sets over $[n]$. Define the operator $P_{ij} : \mathcal{G} \to \binom{[n]}{k}$ as follows.

$$P_{ij}(G) = \begin{cases} (G \setminus \{j\}) \cup \{i\}, \text{ if } i \notin G,\ j \in G,\ (G \setminus \{j\}) \cup \{i\} \notin \mathcal{G}, \\ \qquad\qquad G \text{ otherwise}. \end{cases}$$

Let us set $P_{ij}(\mathcal{G}) = \{P_{ij}(G) : G \in \mathcal{G}\}$. Obviously, $|P_{ij}(\mathcal{G})| = |\mathcal{G}|$ and it is easy to see that $P_{ij}(\mathcal{G})$ is t-intersecting, too. Iterating the shifting operation for all pairs $1 \le i < j \le n$, after finitely many steps one obtains a family $\mathcal{G}^*$ having the property $P_{ij}(\mathcal{G}^*) = \mathcal{G}^*$ for every pair (i,j), $i < j$. Such $\mathcal{G}^*$ is called *shifted*. This can be reformulated the following way.

$$\text{If } G \in \mathcal{G}^*, i \notin G, j \in G, i < j \text{ then } (G \setminus \{j\}) \cup \{i\} \in \mathcal{G}^* \text{ as well.}$$

From now on, we suppose that $\mathcal{G}$ is shifted. The following lemma essentially appeared first in Frankl (1978a). For all G, $G' \in \mathcal{G}$ there exists an i such that

$$(7.4) \qquad |G \cap [i]| + |G' \cap [i]| \ge t + i.$$

Note that in (7.4) $G = G'$ is allowed, hence we have the following. For all $G \in \mathcal{G}$ there exists an s such that $|G \cap [t+2s]| \ge t + s$. An important consequence of (7.4) is that $\mathcal{G}$ is t-intersecting even on the first $2k - t$ elements, i.e. for all G, $G' \in \mathcal{G}$ we have

$$(7.5) \qquad |G \cap G' \cap [2k-t]| \ge t.$$

7.4 Non-trivial Intersecting Families

An intersecting family $\mathcal{G}$ is called *non-trivial* if $\cap\mathcal{G} = \emptyset$. Define the following non-trivial families. $\mathcal{G}^1 = \{G \in \binom{[n]}{k} : 1 \in G, G \cap [2, k+1] \neq \emptyset\} \cup \{[2, k+1]\}$ and $\mathcal{G}^2 = \{G \in \binom{[n]}{k} : |[3] \cap G| \geq 2\}$. For $k = 2$ $\mathcal{G}^1 \equiv \mathcal{G}^2$, for $k = 3$ $|\mathcal{G}^1| = |\mathcal{G}^2|$ while for $k \geq 4$, $n > 2k$ $|\mathcal{G}^1| > |\mathcal{G}^2|$. Hilton and Milner (1967) proved the following generalization of the Erdős–Ko–Rado theorem. If $n > 2k$ and $\mathcal{G} \subseteq \binom{[n]}{k}$ is a non-trivial intersecting family then

$$|\mathcal{G}| \leq |\mathcal{G}^1| = \binom{n-1}{k-1} - \binom{n-k-1}{k-1} + 1.$$

Moreover, equality is possible only for $\mathcal{G} = \mathcal{G}^1$ or $\mathcal{G}^2$.

In Frankl, Füredi (1986a) a new, short proof was given based on (7.5). More exactly, first it was proved that by appropriate applications of P_{ij} (with $1 \leq i \leq 2k < j$) we can get a non-trivial family $\mathcal{G}^*$ of the same size such that $\mathcal{G}^*$ is intersecting even on the set $[2k]$. Then define $\mathcal{A}_i = \{G \cap [2k] : |G \cap [2k]| = i\}$. If $\mathcal{A}_i$ is a non-trivial intersecting family then by induction we have $|\mathcal{A}_i| \leq \binom{2k-1}{i-1} - \binom{k-i-1}{i-1} + 1$ for $2 \leq i \leq k-1$. If $\cap\mathcal{A}_i \neq \emptyset$ then $\cap\mathcal{G}^* = \emptyset$ implies $|\mathcal{A}_i| \leq \binom{2k-1}{i-1} - \binom{k-1}{i-1}$. In both cases $|\mathcal{A}_i| \leq \binom{2k-1}{i-1} - \binom{k-1}{i-1}$ holds for $2 \leq i \leq k-1$. Obviously, $|\mathcal{A}_k| \leq \binom{2k-1}{k-1}$, $\mathcal{A}_1 = \emptyset$. We obtain

$$|\mathcal{G}| \leq \sum_{i=1}^{k} |\mathcal{A}_i| \binom{n-2k}{k-i} \leq 1 + \sum_{i=1}^{k} \binom{2k-1}{i-1} - \binom{k-1}{i-1} = |\mathcal{G}^1|. \blacksquare$$

Another short proof based on the Kruskal–Katona theorem was given by Alon (1984). Other powerful applications of shifting can be found in the papers of P. Frankl. For a recent survey see Frankl (1987).

7.5 Non-trivial t-intersecting Families

A t-intersecting family $\mathcal{G}$ is called *non-trivial* if $|\cap \mathcal{G}| < t$. Define the following non-trivial families. $\mathcal{G}^1 = \{G \in \binom{[n]}{k} : [t] \subseteq G, G \cap [t+1, k+1] \neq \emptyset\} \cup \{[k+1] \setminus \{i\} : i \in [t]\}$ and $\mathcal{G}^2 = \{G \in \binom{[n]}{k} : |[t+2] \cap G| \geq t+1\}$. Frankl (1978b) proved the following generalization of the Erdős-Ko-Rado theorem. If $\mathcal{G} \subseteq \binom{n}{k}$ is a non-trivial t-intersecting family, $t \geq 2$ then

$$|\mathcal{G}| \leq \max\{|\mathcal{G}^1|, |\mathcal{G}^2|\} \tag{7.6}$$

for $n > N_k(t)$. Moreover, equality holds only if either $\mathcal{G} \equiv \mathcal{G}^1$, $k > 2t+1$ or $\mathcal{G} \equiv \mathcal{G}^2$, $k \leq 2t+1$. For the proof he used the delta system method. We shall sketch the proof in Ch. 8.2.

8 THE INTERSECTION STRUCTURE OF HYPERGRAPHS

8.1 Delta systems

A *t-star* with *center* A is a family $\mathcal{S} = \{S_1, S_2, \ldots, S_t\}$ with common intersection A, i.e. $A = S_i \cap S_j$ for all $1 \leq i < j \leq t$. The pairwise disjoint sets $S_1 \setminus A, \ldots, S_t \setminus A$ are called the *rays* of the star. This terminology was introduced by Chung. Stars are also called *delta* systems and the center is called the kernel or nucleus. Stars were first considered by Erdős and Rado (1960), who proved that every large k-graph contains a t-star. Let $\varphi_k(t)$ denote the maximum number of k-sets without a t-star. Then

$$(8.1) \qquad (t-1)^k \leq \varphi_k(t) \leq k!(t-1)^k .$$

The lower bound is given by $\mathcal{K}_k(t-1, \ldots, t-1)$, the complete equipartite k-graph. The upper bound is proved by induction on k. Erdős (1981) offers \$1,000 for deciding whether $\lim_{k\to\infty}(\varphi_k(t))^{1/k}$ is finite (for any $t \geq 3$).

The study of delta systems in Turán type problems is especially important because these are the *only* structures appearing in all hypergraphs.

8.2 The Structure of Nuclei in a t-intersecting Family

The purpose of this subsection is to give an example for the delta system method by proving theorem (7.6). We determine the next to largest families in the Erdős–Ko–Rado theorem, i.e. the largest t-intersecting families with no t-element subset contained by all members. The idea of using stars for intersection problems is due to M. Deza and the method was developed by P. Frankl. The method is based on the following two simple but very useful observations. Suppose that $\mathcal{S}$ is a t-star with center A and F is an f-set. If $f < t$ then there exists $S \in \mathcal{S}$ with

$$(8.2) \qquad F \cap S = F \cap A.$$

In fact, there are at least $t - f$ such S. The other observation is that if $\mathcal{S}$ and $\mathcal{T}$ are k-uniform t-stars with centers A and B and $t > k$ then

(8.3) there are members $S \in \mathcal{S}$, $T \in \mathcal{T}$ such that $S \cap T = A \cap B$.

Proof of (7.6) Let $\mathcal{G} \subseteq \binom{V}{k}$ be a non-trivial t-intersecting family of order n. Replace an edge $G \in \mathcal{G}$ by $G' \subset G$, $|G'| < |G|$ if $\mathcal{G} \setminus \{G\} \cup \{G'\}$ is still t-intersecting. After finitely many steps, deleting all but one of the appearing multiple edges, we get a family $\mathcal{B}$ of rank k critical to this edge-contraction. $\mathcal{B}$ contains no $k+1$-star $\mathcal{S}$, otherwise the whole star $\mathcal{S}$ could be replaced by its center. Hence $|\mathcal{B}| \le k!k^k$ by (8.1), (independently on n !). $\mathcal{B}$ is t-intersecting, too, and for every edge $G \in \mathcal{G}$ there exists $B \in \mathcal{B}$ such that $B \subseteq G$. Let $\mathcal{B}_i = \{B \in \mathcal{B} : |B| = i\}$. Obviously, $\mathcal{B}$ has no members of size less than t. If $\mathcal{B}_t \ne \emptyset$ then $\mathcal{B} = \mathcal{B}_t$ and $|\cap \mathcal{G}| = t$. We obtain

$$|\mathcal{G}| \le \sum_{i=t+1}^{k} |\mathcal{B}_i| \binom{n-i}{k-i} = |\mathcal{B}_{t+1}| \binom{n-t-1}{k-t-1} + O\left(\binom{n-t-2}{k-t-2}\right).$$

We may suppose that $|\mathcal{G}| \ge \max\{|\mathcal{G}^1|, |\mathcal{G}^2|\}$. This implies $|\mathcal{B}_{t+1}| \ge \max\{k-t+1, t+2\}$ (for $n > N_k(t)$). In particular $|\mathcal{B}_{t+1}| \ge 3$. Let us choose $B_1, B_2, B_3 \in \mathcal{B}_{t+1}$. We distinguish two cases:
Case a) $B_1 \cap B_2 = B_1 \cap B_3 := C$. Then $\mathcal{B}_{t+1}$ is a star with center C. There is a member $G^0 \in \mathcal{G}$ with $|G^0 \cap C| < t$. Then $|G^0 \cap C| = t-1$ and $|\mathcal{B}_{t+1}| \le k-t+1$. We obtain that all members of $\mathcal{G}$ contain C and intersect $G^0 \setminus C$, except those contained in $G^0 \cup C$.
Case b) $B_1 \cap B_2 \ne B_1 \cap B_3$. Then $B_1 \cup B_2 := D$ is a $(t+2)$-element set, and we obtain that $|G \cap D| \ge t+1$ holds for all $G \in \mathcal{G}$, i.e. $|\mathcal{G}| \le |\mathcal{G}^2|$. ∎

8.3 Existence of Large Subfamilies with a Semilattice Structure

For a set system $\mathcal{G}$ and a set $G \in \mathcal{G}$ define the *intersection structure* in G by $\mathcal{I}(G, \mathcal{G}) = \{G \cap H : H \in \mathcal{G} \setminus \{G\}\}$. The following theorem (Füredi (1983b)) summarizes the delta system method, asserting that every family of k-sets contains a large subfamily with a homogeneous intersection semilattice structure. Given $k, t > 1$, there exists a positive constant $c_k(t)$ such that every family $\mathcal{G}$ of k-sets contains a k-partite subfamily $\mathcal{G}^* \subseteq \mathcal{G}$ with k-partition $V = V^1 \cup \dots \cup V^k$ satisfying (8.4a-d). For $A \subseteq V$ define the *projection* $\text{Proj}(A)$ of A by $\text{Proj}(A) = \{i : A \cap V^i \ne \emptyset\}$. For a family $\mathcal{A}$ set $\text{Proj}(\mathcal{A}) = \{\text{Proj}(A) : A \in \mathcal{A}\}$.

(8.4a) $|\mathcal{G}^*| \ge c_k(t)|\mathcal{G}|$;
(8.4b) *Every pairwise intersection can be extended to a t-star in $\mathcal{G}^*$*, i.e. *for all $G^1, G^2 \in \mathcal{G}^*$ there are $G^3, \dots, G^t \in \mathcal{G}^*$ such that $\{G^1, \dots, G^t\}$ form a star;*
(8.4c) *For every two sets $G^1, G^2 \in \mathcal{G}^*$ their intersection structure with respect to $\mathcal{G}^*$ are isomorphic*, i.e. $\text{Proj}(\mathcal{I}(G^1, \mathcal{G}^*)) = \text{Proj}(\mathcal{I}(G^2, \mathcal{G}^*))$;
(8.4d) *This common intersection structure is closed under intersection*, i.e. *$B, C \in \mathcal{I}(G, \mathcal{G}^*)$ imply $B \cap C \in \mathcal{I}(G, \mathcal{G}^*)$.*

In view of (4.5), we can suppose that the original set system $\mathcal{G}$ is k-partite. For that case we shall prove a little more (due to Füredi and Komjáth). There is a decomposition $\mathcal{G} = \mathcal{G}^1 \cup \dots \mathcal{G}^s$ into at most $s \le C_k(t)$ parts such that each $\mathcal{G}^i$ fulfills (8.4b–c). Families satisfying (8.4b–c) are called *t-homogeneous*. (We do not have to deal with (8.4d) since t_1-homogenity implies t_2-homogenity for $t_1 > t_2$ and, for $t > k$, (8.4d) follows from (8.3) and (8.4b).)

8.4 Proof of the Semilattice Structure Theorem

For $G \in \mathcal{G}$ we define the *t-star center structure* $\mathcal{S}^t(G, \mathcal{G})$ of G by $\mathcal{S}^t(G, \mathcal{G}) := \{B \subset G :$ there exist $t-1$ members of $\mathcal{G}$ such that together with G they form a t-star with center $B\}$. Note that $\mathcal{S}^2(G, \mathcal{G}) \equiv \mathcal{I}(G, \mathcal{G})$. Let $\mathcal{S}^t(\mathcal{G}) = \bigcup_{G \in \mathcal{G}}(\mathcal{S}_t(G, \mathcal{G}))$, and let $\mathcal{I}(\mathcal{G}) := \mathcal{S}^2(\mathcal{G}) = \{G_1 \cap G_2 : G_1, G_2 \in \mathcal{G}, G_1 \neq G_2\}$. We need two lemmas on decompositions.

For $\mathcal{H} \subseteq \mathcal{G}$, there exists a decomposition $\mathcal{H} = \mathcal{H}_0 \cup (\bigcup_{A \in \mathrm{Proj}(\mathcal{I}(\mathcal{H}))} \mathcal{H}_A)$ such that

$$\text{(8.5)} \qquad \mathcal{H}_0 \text{ is } t\text{-homogeneous and each } A \notin \mathrm{Proj}(\mathcal{S}^t(\mathcal{H}_A)).$$

Proof of (8.5) We successively define partitions of $\mathcal{H}$ into $|\mathrm{Proj}(\mathcal{I}(\mathcal{H}))| + 1$ families. At the first step $\mathcal{H}_0^0 = \mathcal{H}$ and $\mathcal{H}_A^0 = \emptyset$ for all $A \in \mathrm{Proj}(\mathcal{I}(\mathcal{H}))$. After each step we check $\mathcal{H}_0^i$. If it satisfies (8.4b) and (8.4c) then we stop (and define $\mathcal{H}_0 := \mathcal{H}_0^i$, $\mathcal{H}_A := \mathcal{H}_A^i$). If not then we can find a set $H \in \mathcal{H}_0^i$ and a subset $B \subset H$ such that $A := \mathrm{Proj}(B) \in \mathrm{Proj}(\mathcal{I}(\mathcal{H}_0^i))$ but $B \notin \mathcal{S}^t(H, \mathcal{H}_0^i)$ holds. Modify the partition by setting $\mathcal{H}_0^{i+1} = \mathcal{H}_0^i \setminus \{H\}$, $\mathcal{H}_A^{i+1} = \mathcal{H}_A^i \cup \{H\}$, and leaving the other families unchanged. Since at each step $|\mathcal{H}_0^i|$ is decreasing, the procedure stops after finitely many steps.

Suppose $\mathcal{H} \subseteq \mathcal{G}$ and $A \notin \mathrm{Proj}(\mathcal{S}^t(\mathcal{H}))$. Then there exists a decomposition $\mathcal{H} = \mathcal{H}^1 \cup \cdots \cup \mathcal{H}^{k(t-1)}$ such that

$$\text{(8.6)} \qquad A \notin \mathrm{Proj}(\mathcal{I}(\mathcal{H}^i)) \text{ for all } i.$$

Proof of (8.6) Define $\mathcal{B} = \{B \in \binom{V}{|A|} : \mathrm{Proj}(B) = A\}$ and, for all $B \in \mathcal{B}$, let $\mathcal{H}(B) = \{H \setminus B : B \subset H \in \mathcal{H}\}$. Some of the $\mathcal{H}(B)$'s may be empty. Since $B \in \mathcal{B}$ is not the center of a t-star, $\mathcal{H}(B)$ contains no t pairwise disjoint sets. Therefore $\mathcal{H}(B)$ can be decomposed into at most $(t-1)(k-|B|)$ intersecting subfamilies $\mathcal{H}(B) = \mathcal{H}^1 \cup \cdots \cup \mathcal{H}^m(B)$. Let $\mathcal{H}^i := \bigcup_{B \in \mathcal{B}} \mathcal{H}^i[B]$.

Proof of Theorem (8.4) Applying Lemma (8.5) to $\mathcal{G}$ ($\neq \emptyset$), we obtain a homogeneous subfamily $\mathcal{G}_0$ and the families $\mathcal{G}_A$. By Lemma (8.6), each $\mathcal{G}_A$ can be decomposed into at most $(t-1)k$ families $\mathcal{G}_A^i$ with $|\mathrm{Proj}(\mathcal{I}(\mathcal{G}_A^i))| < |\mathrm{Proj}(\mathcal{I}(\mathcal{G}))|$. Now we apply the same procedure to all $\mathcal{G}_A^i$. In $|\mathcal{I}(\mathcal{G})| \leq 2^k$

steps, all of the at most $(k(t-1))^{2^k}(2^k+1)!$ obtained subfamilies will be homogeneous and we are done. ∎

9 FAMILIES WITH PRESCRIBED INTERSECTIONS

9.1 The Problem

The semilattice structure proved in (8.4) enables us to obtain a series of Turán type theorems for hypergraphs. This will be done in the next four Sections. Let $0 \le \ell_1 < \ell_2 < \cdots < \ell_s < k \le n$ be integers. We say that the family $\mathcal{G} \subseteq \binom{V}{k}$ is an $(n, k, \{\ell_1, \ldots, \ell_s\})$-system if $|G \cap G'| \in \{\ell_1, \ldots, \ell_s\}$ holds for every $G, G' \in \mathcal{G}$, $G \neq G'$. Denote $\{\ell_1, \ldots, \ell_s\}$ by L and let us denote by $m(n, k, L)$ the maximum cardinality of an (n, k, L)-system. The determination of $m(n, k, L)$ is the simplest looking Turán type problem since the family of forbidden configurations consists only of hypergraphs of size two.

$$m(n, k, L) = \text{ex}(n, \{\mathcal{F}_k^\ell : 0 \le \ell < k, \ell \notin L\})$$

where $\mathcal{F}_k^\ell$ denotes the k-graph consisting of two k-sets with intersection size ℓ. The most well-known result of this type is the Erdős–Ko–Rado theorem dealing with the case $L = \{t, t+1, \ldots, k-1\}$ (see Ch. 7.1).

Packings Another important example is the case $L = \{0, 1, \ldots, t-1\}$. An $(n, k, \{0, 1, \ldots, t-1\})$-family is called an (n, k, t)-*packing* (or, as in Ch. 6.3, a partial Steiner system $PS(n, k, t)$) and $m(n, k, \{0, 1, \ldots, t-1\})$ is denoted by $P(n, k, t)$. Obviously, $P(n, k, t) \le \binom{n}{t}/\binom{k}{t}$. Rödl (1985) proved by probabilistic methods that for every fixed pair k, t

$$P(n, k, t) = (1 - o(1))\binom{n}{t}/\binom{k}{t}. \tag{9.1}$$

This was conjectured by Erdős and Hanani (1964) and proved for $t = 2$ and for infinitely many values of k when $t = 3$.

9.2 General Bounds by Linear Algebra

Ray-Chaudhuri and Wilson (1975) proved that $m(n, k, L) \le \binom{n}{s}$ holds for all $n \ge k$ and $|L| = s$. The proof uses linear algebraic independence of some higher order incidence matrices over the reals. This was generalized for finite fields by Frankl and Wilson (1981). They obtained that if there exists an integer valued polynomial f of degree d and a prime p such that $p \mid f(\ell)$ ('p divides $f(\ell)$') for all $\ell \in L$ but $p \nmid f(k)$ then

$$m(n, k, L) \le \binom{n}{d}. \tag{9.2}$$

An important special case is (with $f(x) = \prod_{i=1,\dots,d}(x-\mu_i))$ when there exists a prime p and some distinct residues $\mu_1, \dots, \mu_d$ such that each $\ell \in L$ is congruent to some μ_i but k is not. A new proof was found by Babai (1988). Here we prove a simpler result related to (9.2) due to Babai and Frankl (1980). If the greatest common divisor of $\{\ell_1, \dots, \ell_s\}$ does not divide k then

(9.3) $$m(n,k,L) \le n.$$

Proof There exists a prime power $q = p^a$ dividing each ℓ_i but not dividing k. Let $\mathcal{G}$ be an (n,k,L)-system of size m. Let A be the $m \times n$ incidence matrix of $\mathcal{G}$ with 0-1 entries. Each entry on the main diagonal of AA^T is k and the off diagonal entries are from L. Then the product of the diagonal elements of AA^T is divisible by a lower power of p than any one of the remaining $m! - 1$ expansion terms of det AA^T. Hence this determinant is not 0 implying $m \le n$. ∎

More linear algebraic proofs and examples can be found in the book of Babai and Frankl (1988).

9.3 Necessary and Sufficient Condition for $m(n,k,L) = O(n)$

We say that the numbers $\ell_1, \dots, \ell_s$ and k satisfy property $(*)$ if

$(*)$ *There exists a family* $\mathcal{I} \subset 2^{[k]}$ *closed under intersection such that* $\cup\mathcal{I} = [k]$ *and* $|I| \in L$ *for all* $I \in \mathcal{I}$.

The following theorem was announced in Füredi (1983b)

(9.4) *If* $(*)$ *is satisfied then* $m(n,k,L) > \frac{1}{8k} n^{k/(k-1)}$.

(9.5) *If* $(*)$ *does not hold then* $m(n,k,L) \le (2^k)^{2^k} n$.

Construction for (9.4) This is a slightly modified version of a construction of Frankl (1984). Let $I_1, \dots, I_g \in \mathcal{I}$ be a subsystem covering $[k]$ (i.e. $\cup\{I_i : 1 \le i \le g\} = [k]$) such that $g \le k$. Let m be the largest integer with $km^{g-1} \le n$. We are going to define an $(\le n, k, L)$-system $\mathcal{G}$ of size m^g. The members of $\mathcal{G}$ are labelled by the vectors $(t_1, \dots, t_g) \in [m]^g$. The vertices in the edge $G(t_1, \dots, t_g)$ are integer vectors of length $g+1$ of the form $(j, x(1,j), \dots, x(g,j))$ with $1 \le j \le k$ such that

$$x(i,j) = \begin{cases} 0 & \text{if } j \in I_i, \\ t_i & \text{otherwise.} \end{cases}$$

V is k-partite with parts corresponding to the first coordinates. $|V^j| \le m^{g-1}$ for all j since there exists i_j such that $j \in I_{i_j}$ and the i_jth coordinate is 0 in all elements of V^j. Finally, it is easy to see that

$$\operatorname{Proj}(G(t_1, \dots, t_g) \cap G(t'_1, \dots, t'_g)) = \cap\{I_i : t_i \ne t'_i\}.$$

Proof of (9.5) It is an easy consequence of theorem (8.4). Let $\mathcal{G}$ be an (n,k,L)-system. Apply (8.4) (with $t = k+1$) to get a k-partite, homogeneous family $\mathcal{G}^* \subseteq \mathcal{G}$ with common intersection structure $\mathcal{I} \subset 2^{[k]}$. As $\cup\mathcal{I} \neq [k]$, there exists an element $j \in [k]$ not contained in any member of $\mathcal{I}$. Then the elements $G \cap V^j$ are pairwise distinct for $G \in \mathcal{G}^*$. Hence $|\mathcal{G}^*| \leq |V^j| \leq n$ implying $|\mathcal{G}| \leq (1/c_k(k+1))|\mathcal{G}^*| \leq (2^k)^{2^k} n$. ∎

9.4 Each Rational Occurs as an Exponent

Frankl (1986) proved the hypergraph version of conjecture (3.8). For all integers $s \geq d \geq 1$ there exists k and L such that $m(n,k,L) = \Theta(n^{s/d})$. Here we give his example. Let $a_0, a_1, \ldots, a_d$ be non-negative integers with $a_d \geq 1$ and let f be the integer valued polynomial $f(x) = \sum_{0\leq i\leq d} a_i \binom{x}{i}$. Let V be a finite set of size v. For $I \subseteq V$, $|I| = i \leq d$ let $V(I)$ be a set of size a_i such that $V(I) \cap V(J) = \emptyset$ for $I \neq J$. Define $f(X)$ for any $X \subseteq V$ as $f(X) := \cup\{V(I) : I \subseteq X, |I| \leq d\}$. We have $|f(X)| = f(|X|)$ and $f(A) \cap f(B) = f(A \cap B)$. So the family $\{f(X) : X \in \binom{V}{s}\}$ is a $(f(v), f(s), \{f(0), f(1), \ldots, f(s-1)\})$-family of size $\binom{v}{s}$ and order $f(v) \sim v^d$. This example yields the lower bound in the following theorem of Frankl (1986). (The main tool in the proof of the upper bound is (8.4).) If $a_0 + 2a_1 > f(s-1)$ then

(9.6) $$m(n, f(s), \{f(0), f(1), \ldots, f(s-1)\}) = \Theta(n^{s/d}).$$

9.5 Reductions

It is easy to see that

(9.7) $$m(n,k,\{\ell_1,\ell_2,\ldots,\ell_s\}) = \Theta(m(n,k-\ell_1,\{0,\ell_2-\ell_1,\ldots,\ell_s-\ell_1\})).$$

Hence, if we are interested only in the order of magnitude of $m(n,k,L)$ then we can always assume $\ell_1 = 0$. Of course, our most important reduction is that, by (8.4), it is enough to consider k-partite, homogeneous families. Another corollary (see Füredi (1983b)): If the greatest common divisor d of $\ell_1, \ldots, \ell_s$ divides k then

(9.8) $$m(n,k,\{\ell_1,\ell_2,\ldots,\ell_s\}) = \Theta(m(\frac{n}{d},\frac{k}{d},\{\frac{\ell_1}{d},\ldots,\frac{\ell_s}{d}\})).$$

Proof Let us consider a k-partite, homogeneous (n,k,L)-system $\mathcal{G}$ with common intersection structure $\mathcal{I} \subset 2^{[k]}$. Then the size of every *atom* in $\mathcal{I}$ is divisible by d implying (9.8). ∎

To obtain constructions one can generalize the procedure given in Ch. 9.4. Let f be the polynomial $f(x) = \sum_{0\leq i\leq d} a_i \binom{x}{i}$ with non-negative integer coefficients. Then (Frankl (1983a))

(9.9) $$m(n,k,L) \leq m(f(n), f(k), f(L)).$$

Using the above simple operations (and (8.4)) Frankl (unpublished) determined the order of magnitude of $m(n,k,L)$ for all L whenever $k \leq 10$ (for $k \leq 8$ see Frankl (1980a)).

9.6 Comparing the Algebraic and Combinatorial Bounds

The argument in the proof of (9.5) can be generalized as follows. Let $\mathcal{G}$ be a homogeneous (n,k,L)-system with common intersection structure $\mathcal{I} \subset 2^{[k]}$. Suppose that $T \subset [k]$ is not contained in any member of $\mathcal{I}$. Then

$$|\mathcal{G}| \leq \binom{n}{|T|}. \tag{9.10}$$

Using (9.10) (and the sieve method as in the proof of (9.8)), we can obtain the following version of (9.2). Let f be an integer valued polynomial of degree d and suppose that q is an integer such that $q \mid f(\ell)$ for all $\ell \in L$ but $q \nmid f(k)$. Suppose further that $\mathcal{G}$ is a k-partite, homogeneous (n,k,L)-system with common intersection structure $\mathcal{I} \subset 2^{[k]}$. Then $\mathcal{I}$ does not cover all d-subsets of $[k]$, implying

$$m(n,k,L) \leq \frac{k^k}{k!} C_k (k+1) \binom{n}{d}. \tag{9.11}$$

Proof Suppose, on the contrary, that $\mathcal{I}$ covers all d-subsets of $[k]$. Then, for all $0 \leq i \leq d$, the sieve formula gives

$$\binom{k}{i} = \sum \binom{|I|}{i} - \sum\sum \binom{|I \cap I'|}{i} + - \dots \tag{9.12}$$

As f is an integer valued polynomial, it can be written in the form $f(x) = \sum_{0 \leq i \leq d} a_i \binom{x}{i}$ with integer coefficients a_i. Multiply (9.12) by a_i and add them up for all $i \leq d$. On the left hand side we get $f(k)$ (not divisible by q) and on the right hand side we get $\sum f(|I|) - \sum\sum f(|I \cap I'|) + - \dots$, where each term is divisible by q, a contradiction. ∎

The above results illustrate properly the difference of the methods. The linear algebraic proofs are very powerful, have a larger scope, but in several cases are unable to describe the finer structure. For example, improving the original Ray-Chaudhuri–Wilson theorem (see Ch. 9.2) Deza, Erdős, and Frankl (1978) proved that $m(n,k,\{\ell_1,\dots,\ell_s\}) = O(n^{s-1})$ except in the case $(\ell_2 - \ell_1) \mid \cdots \mid (\ell_s - \ell_{s-1}) \mid (k - \ell_s)$. Even in this last case they improved the upper bound (for $n > n_k$) to

$$m(n,k,\{\ell_1,\dots,\ell_s\}) \leq \prod_{1 \leq i \leq s} \frac{n - \ell_i}{k - \ell_i}.$$

Let us close with the following, perhaps too optimistic, conjecture. If there is a family $\mathcal{I} \subset 2^{[k]}$ of sets of sizes $\ell \in L$, $\mathcal{I}$ is closed under intersection and covers all d-subsets of $[k]$ then for some positive $\varepsilon = \varepsilon(k)$

$$m(n,k,L) = \Omega(n^{d+\varepsilon}) \quad (?) \tag{9.13}$$

9.7 Remark on t-wise Intersecting Families

Let us denote by $m^t(n,k,L)$ the maximum cardinality of $\mathcal{G} \subseteq \binom{[n]}{k}$ such that $|G^1 \cap \cdots \cap G^t| \in L$ holds for every distinct $G^1, \ldots, G^t \in \mathcal{G}$. This question was posed by Sós (1976) even in a more general form. Theorem (8.4) clearly implies that

$$m^t(n,k,L) = \Theta(m(n,k,L)). \tag{9.14}$$

10 TURÁN PROBLEMS WITH NO EXPONENTS

10.1 (v,e)-free k-graphs

In this Section we interrupt the discussion of prescribed intersections and show a non-polynomial Turán number. Brown, Erdős, and Sós (1971, 1973) generalized the original Turán problem as follows. Let $g_k(n;v,e)$ denote the maximum number of edges in a k-graph of order n such that no v vertices span e or more edges. In other words, the union of any e edges has size greater than v. Let us denote the set of k-graphs of order v and size e by $\mathbf{G}_k(v,e)$. We have $\mathrm{ex}(n, \mathbf{G}_k(v,e)) = g_k(n;v,e)$. For example, a $\mathbf{G}_k(2k-t,2)$-free k-graph is an (n,k,t) packing (see (9.1)). Brown, Erdős, and Sós gave a number of examples (explicit and probabilistic ones), proofs and conjectures. Their conjecture concerning the first unsolved case was proved by Ruzsa and Szemerédi (1976).

$$g_3(n;6,3) = o(n^2), \tag{10.1}$$

$$\lim_{n\to\infty} g_3(n;6,3)/n^{2-\varepsilon} = \infty \text{ for all } \varepsilon > 0. \tag{10.2}$$

These bounds were extended to $g_k(n;3k-3,3)$ for all $k \geq 3$ by Erdős, Frankl, and Rödl (1986). The proof of (10.1) involves Szemerédi's uniformity lemma (Szemerédi (1978)).

In (10.1-2) the class of forbidden hypergraphs consists of more than one member, e.g., a triangle-like $\{123, 345, 561\}$ and a three times covered pair $\{123, 124, 125\}$. Erdős asked if a similar phenomenon can occur with only one forbidden configuration. The answer is yes, as it was shown by Frankl and Füredi (1987a) by an example based on (10.1-2). Let $\mathcal{F}_5(8,3)$ denote the 5-uniform hypergraph of order 8 with edge set $\{12346, 12357, 12458\}$. Then

$$(10.3) \qquad n^4/\exp(5\sqrt{\log n}) < \mathrm{ex}(n, \mathcal{F}_5(8,3)) = o(n^4).$$

10.2 Constructions for $\mathbf{G}_3(6,3)$ and $\mathcal{F}_5(8,3)$

Let $r_3(n)$ be the maximum cardinality of a set $A \subseteq [n]$ which contains no arithmetical progression of length 3. The following example, due to Behrend (1946), shows that

$$(10.4) \qquad n/\exp(3\sqrt{\log n}) < r_3(n).$$

Write the number $a \in [n]$ in the form $a = a_0 + a_1 m + \cdots + a_r m^r$ where $r = \lfloor\sqrt{\log n}\rfloor$, the base $m := \lfloor n^{1/(r+1)}\rfloor$ ($\sim \exp\sqrt{\log n}$), and $0 \le a_i < a$. Define $A_s := \{a : \sum a_i^2 = s\}$ for $1 \le s \le (r+1)m^2$. None of the A_s contain arithmetic progressions and the size of the largest one yields (10.4).

Define a $\mathbf{G}_3(6,3)$-free 3-graph on $[n]$ as follows. $\mathcal{E} = \{(x, 3b + x + a, 6b + x + 2a) : a \in A, x \in [b]\}$, where $b = \lfloor n/9 \rfloor$ and $A \subseteq [b]$ is a subset with no arithmetic progression of length 3. We obtained

$$(10.5) \qquad nr_3(n)/100 \le g_3(n; 6, 3).$$

Finally, let $\mathcal{G} \subseteq \binom{[2n]}{5}$ be defined as $\{E \cup P\}$ where $E \in \mathcal{E}$, $P \in \binom{[n+1,2n]}{2}$ and $\mathcal{E}$ is a $\mathbf{G}_3(6,3)$-free 3-graph over $[n]$. Then $\mathcal{G}$ is $\mathcal{F}_5(8,3)$-free giving

$$(10.6) \qquad n^2 g_3(n; 6, 3)/20 < \mathrm{ex}(n, \mathcal{F}_5(8,3)).$$

10.3 Proof of the Non-polynomial Upper Bound

To prove the upper bound in (10.3), suppose that $\mathcal{G} \subseteq \binom{[n]}{5}$ is $\mathcal{F}_5(8,3)$-free. Let us apply (8.4) to obtain a 5-partite $\mathcal{G}^* \subseteq \mathcal{G}$ with parts $V_1, \ldots, V_5$ and common intersection structure $\mathcal{I} \subset 2^{[5]}$. If there is a pair, say $\{1,2\}$, uncovered by the four-element members of $\mathcal{I}$ then, for all $x_1 \in V_1$, $x_2 \in V_2$, the 3-graph $\mathcal{G}^*(x_1, x_2) := \{G \setminus \{x_1, x_2\} : \{x_1, x_2\} \subseteq G \in \mathcal{G}^*\}$ is $\mathbf{G}_3(6,3)$-free. Hence $|\mathcal{G}^*| \le n^2 g_3(n; 6, 3)$, as desired. Thus we may suppose that all pairs $P \in \binom{[5]}{2}$ are contained in a four-element member of $\mathcal{I}$. This implies that there are at least 3 such members, say $[5] \setminus \{i\}$, $1 \le i \le 3$. Consider any edge $G = \{x_1, x_2, \ldots, x_5\} \in \mathcal{G}^*$. By (8.4c), there are sets G^i with

$\text{Proj}(G \cap G^i) = [5] \setminus \{i\}$ such that the elements $G^i \setminus G$ are distinct. Then $\{G^1, G^2, G^3\}$ is isomorphic to $\mathcal{F}_5(8,3)$, a contradiction. ∎

11 FORBIDDING JUST ONE INTERSECTION

11.1 Constructions

In 1975, Erdős (1976) raised the problem of what happens if we weaken the condition in the Erdős–Ko–Rado theorem to $|F \cap F'| \neq l$. In other words, let $\text{ex}(n, \mathcal{F}_k^\ell)$ be the maximum size of a family $\mathcal{G} \subseteq \binom{[n]}{k}$ without two edges intersecting in exactly ℓ elements. An obvious example is to take all k-sets containing $[\ell+1]$.

$$\text{(11.1)} \qquad \text{ex}(n, \mathcal{F}_k^\ell) \geq \binom{n-\ell-1}{k-\ell-1}.$$

For $k \leq 2\ell$ and $n > n_0(k)$, an (n,k,ℓ)-packing of size $\Omega(n^\ell)$ is much larger than (11.1). Erdős conjectured that one of these two families is maximal if n is large enough compared to k and ℓ. Frankl (1983b) observed that Erdős' conjecture is not completely true since one can have a slightly larger $\mathcal{F}_k^\ell$-free family as follows. Take an $(n, 2k-\ell-1, \ell)$-packing $\mathcal{P}$ and replace all $P \in \mathcal{P}$ by $\binom{P}{k}$.

$$\text{(11.2)} \quad \text{ex}(n, \mathcal{F}_k^\ell) \geq P(n, 2k-\ell-1, \ell)\binom{2k-\ell-1}{k} = (1-o(1))\binom{n}{\ell}\frac{\binom{2k-\ell-1}{k}}{\binom{2k-\ell-1}{\ell}}.$$

In the last equality we used (9.1). For $k \geq 2\ell+1$ the coefficient of $\binom{n}{\ell}$ is at least 1, so it is really larger than $P(n,k,\ell)$.

In Frankl, Füredi (1985) the following three theorems (11.3–5) were proved. For $k \geq 2\ell+2$, $n > n_k$,

$$\text{(11.3)} \qquad \text{ex}(n, \mathcal{F}_k^\ell) = \binom{n-\ell-1}{k-\ell-1}.$$

The case $\ell = 1$ was proved in Frankl (1977) and the case $k=3$, $\ell=1$ was handled by Erdős and Sós (see Sós (1976)). Larman (1978) proved $\text{ex}(n, \mathcal{F}_5^2) = O(n^2)$. Frankl (1980b) determined the order of magnitude up to $\ell \leq \lfloor k/3 \rfloor$. Moreover, if $2\ell+1 \geq k$ and $k-\ell$ is a prime power then for all n

$$\text{(11.4)} \qquad \text{ex}(n, \mathcal{F}_k^\ell) \leq \binom{n}{\ell}\frac{\binom{2k-\ell-1}{k}}{\binom{2k-\ell-1}{\ell}}.$$

(The proof uses linear algebra, namely (9.2).) Finally, for all fixed k and ℓ,

$$\text{ex}(n, \mathcal{F}_k^\ell) = \Theta(n^{\max\{k-\ell-1,\ell\}}). \tag{11.5}$$

11.2 Proof of the Order of Magnitude

Our next aim is to outline the proof of (11.3). We proceed step by step: first, we prove only the order of magnitude, then an asymptotic, and finally the exact value. The case $k = 2\ell+2$ is much more involved so we consider only the case $k \geq 2\ell + 3$. The reason for this is the following lemma. Suppose that $\mathcal{I} \subset 2^{[k]}$ is closed under intersection, $k \geq 2\ell + 3$, no ℓ-element set belongs to $\mathcal{I}$, and all $(k-\ell-2)$-element subsets of $[k]$ are contained in some $I \in \mathcal{I}$. (The last property is abbreviated as $r(\mathcal{I}) \geq k - \ell - 1$.) Then for some $A \subset [k]$, $|A| = \ell + 1$ we have

$$\{X : A \subset X \subset [k], X \neq [k]\} \subseteq \mathcal{I}. \tag{11.6}$$

The proof of (11.6) uses the following result of Frankl and Katona (1979). If $\{D_1, \ldots, D_m\}$ is a multihypergraph over the vertex set Y and $s \geq 1$ is a fixed integer such that $|D_{i1} \cap \cdots \cap D_{it}| \neq t - s$ holds for all $t \leq m$, and $1 \leq i_1 < \cdots < i_t \leq m$ then $m \leq |Y| + s - 1$. Moreover, for $s \geq 2$ (see Frankl, Füredi (1985)) $m = |Y| + s - 1$ holds only if $D_i = Y$ for all i.

Consider an $\mathcal{F}_k^\ell$-free family $\mathcal{G} \subseteq \binom{[n]}{k}$. We may suppose that it is maximal, hence $|\mathcal{G}| \geq \binom{n-\ell-1}{k-\ell-1}$. We apply (8.4) (with $t = k+1$) to obtain $\mathcal{G}^*$ and $\mathcal{I}$. As $|\mathcal{G}^*| = \Omega(n^{k-\ell-1})$, (9.10) implies that $r(\mathcal{I}) \geq k - \ell - 1$. By (11.6), there exists an $(\ell + 1)$-element set $A \subset [k]$ such that all supersets of A belong to $\mathcal{I}$. This implies that, for all $G \in \mathcal{G}^*$, the $(k - \ell - 1)$-element sets $B(G) := G \setminus \text{Proj}^{-1}(A)$ are pairwise distinct. Even more, for $G \in G^*$ and for all $H \in \mathcal{G}$ we have

$$B(G) \subset H \text{ implies } G = H. \tag{11.7}$$

Then, as we have seen in (9.10), we obtain that

$$|\mathcal{G}^*| = c|\mathcal{G}| \leq |\partial_{k-\ell-1}(\mathcal{G})| \tag{11.8}$$

where $\partial_h(\mathcal{A}) := \{H : |H| = h, H \subseteq A \text{ for some } A \in \mathcal{A}\}$ denotes the set of all h-sets covered by $\mathcal{A}$. This implies $|\mathcal{G}| \leq c^{-1}\binom{n}{k-\ell-1} = O(n^{k-\ell-1})$. In the same way we can prove (11.5), too.

11.3 Proof of the Asymptotic

We define subfamilies $\mathcal{G}^1, \ldots, \mathcal{G}^m$ of $\mathcal{G}$ recursively. Let $\mathcal{G}^1 := \mathcal{G}^*$. If $\mathcal{G}^1, \ldots, \mathcal{G}^i$

are already defined then we apply (8.4) to obtain $\mathcal{G}^{i+1} := (\mathcal{G}\setminus(\mathcal{G}^1\cup\cdots\cup\mathcal{G}^i))^*$ with intersection structure $\mathcal{I}^{i+1} \subset 2^{[k]}$. We set $m := i$ and stop when either $\mathcal{G}\setminus(\mathcal{G}^1\cup\cdots\cup\mathcal{G}^i) = \emptyset$ or $r(\mathcal{I}^{i+1}) \le k-\ell-2$. In the latter case, (9.10) implies

$$|\mathcal{G}\setminus(\mathcal{G}^1\cup\ldots\mathcal{G}^m)| \le c^{-1}\binom{n}{k-\ell-2}. \tag{11.9}$$

For $G \in (\mathcal{G}^1\cup\cdots\cup\mathcal{G}^m)$, let $A(G) := \mathrm{Proj}^{-1}(A)$ be the $(\ell+1)$-element subset of G described in (11.6). By (11.7), the sets $G\setminus A(G)$ are pairwise distinct so

$$|\mathcal{G}^1\cup\cdots\cup\mathcal{G}^m| \le \binom{n}{k-\ell-1}. \tag{11.10}$$

11.4 Proof of the Exact Value

(11.9-10) imply that $|\mathcal{G}^1\cup\cdots\cup\mathcal{G}^m| = \binom{n-\ell-1}{k-\ell-1} - O(n^{k-\ell-2})$. Instead of (11.7) we can use that for $G \in (\mathcal{G}^1\cup\cdots\cup\mathcal{G}^m)$ and for all $H\in\mathcal{G}$ we have

$$|G\cap H| \ge \ell \text{ implies } A(G)\subseteq H; \tag{11.11}$$

moreover, if $H \in (\mathcal{G}^1\cup\cdots\cup\mathcal{G}^m)$, too, then $A(G) = A(H)$.

Let $A_1,\ldots,A_s$ be the list of $(\ell+1)$-element subsets of $[n]$ and define $\mathcal{H}_i := \{G\in(\mathcal{G}^1\cup\cdots\cup\mathcal{G}^m) : A_i = A(G)\}$, $\mathcal{H}_i^- := \{G\setminus A_i : G\in\mathcal{H}_i\}$. Assume that $|\mathcal{H}_1|\ge\cdots\ge|\mathcal{H}_s|$ and let $|\mathcal{H}_i| = \binom{x_i}{k-\ell-1}$ for some reals $x_i \ge k-\ell-1$. Then, by Lovász' version (1979a) of the Kruskal (1963) – Katona (1966) theorem, $|\partial_\ell(\mathcal{H}_i^-)| \ge \binom{x_i}{\ell}$. (11.11) implies that the families $\partial_\ell(\mathcal{H}_i)$ are pairwise disjoint so $\sum\binom{x_i}{\ell}\le\binom{n}{\ell}$. Hence

$$\binom{n-\ell-1}{k-\ell-1}(1-O(\frac{1}{n})) = \sum|\mathcal{H}_i^-| = \sum_i\binom{x_i}{\ell}\frac{\binom{x_i}{k-\ell-1}}{\binom{x_i}{\ell}} \le$$

$$\sum_i\binom{x_i}{\ell}\frac{\binom{x_1}{k-\ell-1}}{\binom{x_1}{\ell}} \le \binom{n}{\ell}\frac{\binom{x_1}{k-\ell-1}}{\binom{x_1}{\ell}}.$$

This implies $x_1 \ge n - C_1$ where $C_1 = C_1(k)$ is a constant. We obtain $|\mathcal{H}_1| \ge \binom{n-\ell-1}{k-\ell-1} - O(n^{k-\ell-2})$ and using (11.9-10), $|\mathcal{G}\setminus\mathcal{H}_1| = O(n^{k-\ell-2})$. Let us define $\mathcal{K}$ as those edges G from $\mathcal{G}$ for which $A_1 \subset G$ and (11.6) holds, i.e. for all X with $A_1 \subset X \subseteq G$, $X \neq G$, X is a center of a $(k+1)$-star in $\mathcal{G}$. $\mathcal{K}$ consists of the 'regular' members of $\mathcal{G}$. We have $\mathcal{H}_1 \subseteq \mathcal{K}$. Let $\mathcal{A} = \{G : A_1 \subset G \in \mathcal{G},\ G \notin \mathcal{K}\}$ and $\mathcal{B} = (\mathcal{G} \setminus \mathcal{K}) \setminus \mathcal{A}$. Our next aim is to show that $\mathcal{K} = \mathcal{G}$. We shall derive contradiction from $\mathcal{G} \setminus \mathcal{K} \neq \emptyset$ by showing that $\partial_\ell(\mathcal{K})$ and consequently $\partial_{k-\ell-1}(\mathcal{K})$ miss too many subsets of $[n]$. We distinguish two cases according to which one of $|\mathcal{A}|$, $|\mathcal{B}|$ is larger.

$|\mathcal{A}| > |\mathcal{B}|$ Apply (4.5) to get a k-partite $\mathcal{A}^0 \subseteq \mathcal{A}$, $|\mathcal{A}^0| \geq (k!/k^k)|\mathcal{A}|$. There is a set $D \subset [k]$ such that $\mathrm{Proj}(A_1) \subsetneq D$, and a subfamily $\mathcal{A}^1 \subseteq \mathcal{A}^0$, $|\mathcal{A}^1| > |\mathcal{A}^0|/2^k$ such that $\mathrm{Proj}^{-1}(D) \cap G$ is not a center of a delta system of size $2k$ in $\mathcal{G}$ for all $G \in \mathcal{A}^1$. We get $|\mathcal{A}| + |\mathcal{B}| < 2|\mathcal{A}| = O(|\mathcal{A}^1|)$. Let $|D| = \ell+1+d$, $\mathcal{D} := \{(\mathrm{Proj}^{-1}(D) \cap G) \setminus A_1 : G \in \mathcal{A}^1\}$, $\mathcal{U} := \{G \in (\mathcal{A} \cup \mathcal{K}) : A_1 \cup E \subseteq G$ for some $E \in \mathcal{D}\}$ and let $\mathcal{V} := \mathcal{K} \setminus \mathcal{U}$. Obviously, $\mathcal{A}^1 \subseteq \mathcal{U}$. Each $A_1 \cup E$ is contained in at most $O(n^{k-\ell-1-d-1})$ members of $\mathcal{U}$, so we get

$$|\mathcal{A}|+|\mathcal{B}|+|\mathcal{U}| = O(|\mathcal{U}|) \leq |\mathcal{D}|O(n^{k-\ell-1-d-1}) \leq |\mathcal{D}|\binom{n-\ell-1}{k-\ell-1}/\binom{n-\ell-1}{d}.$$

Obviously,

$$|\mathcal{V}| = |\mathcal{V}^-| \leq |\partial_d(\mathcal{V}^-)|\binom{n-\ell-1}{k-\ell-1}/\binom{n-\ell-1}{d}$$

where $\mathcal{V}^- := \{G \setminus A_1 : G \in \mathcal{V}\}$. We have $\mathcal{D} \cap \partial_d(\mathcal{V}^-) = \emptyset$. Hence $|\mathcal{D}|+|\mathcal{V}^-| \leq \binom{n-\ell-1}{d}$. This and the above two displayed inequalities imply the desired $|\mathcal{G}| \leq |\mathcal{A}| + |\mathcal{B}| + |\mathcal{U}| + |\mathcal{V}| \leq \binom{n-\ell-1}{k-\ell-1}$.

$|\mathcal{A}| \leq |\mathcal{B}|$ First, we observe that $\partial_\ell(\mathcal{B}) \cap \partial_\ell(\mathcal{K}) = \emptyset$. Let $|\partial_{k-\ell-1}(\mathcal{B})| = \binom{x}{k-\ell-1}$. As $|\mathcal{B}| = O(n^{k-\ell-2})$ we have $x = O(n^{(k-\ell-2)/(k-\ell-1)})$ implying that $\binom{x}{\ell}/\binom{x}{k-\ell-1}$ is much larger than $\binom{n}{\ell}/\binom{n}{k-\ell-1}$. This implies

$$(11.12) \qquad |\partial_\ell(\mathcal{B})| \geq \binom{x}{\ell} > 2\frac{\binom{n}{\ell}}{\binom{n}{k-\ell-1}}|\mathcal{B}| \geq \binom{n}{\ell}/\binom{n}{k-\ell-1}|\mathcal{A} \cup \mathcal{B}|.$$

On the other hand, using the notation $|\mathcal{K}^-| = \binom{y}{k-\ell-1}$ we obtain

$$|\partial_\ell(\mathcal{K})| \geq \sum_{0 \leq i \leq \ell+1} \binom{y}{k-i}\binom{\ell+1}{i} = \binom{y+\ell+1}{\ell} \geq |\mathcal{K}|\binom{n}{\ell}/\binom{n}{k-\ell-1}.$$

This and (11.12) give the desired $|\mathcal{G}| \leq \binom{n-\ell-1}{k-\ell-1}$. ∎

12 NO TRIANGLES, NO STARS

12.1 Intersection Condensed Families

The method explained in the previous Section was further developed in Frankl, Füredi (1987a) to handle star-shaped forbidden configurations. One of the most general theorems is the following. Let $\mathcal{F}$ be a k-graph with $p = |\cap \mathcal{F}|$ such that for some set A of cardinality $|A| \leq k - 2 - 2p$ the sets $F \setminus A$, $F \in \mathcal{F}$, are pairwise disjoint. Call a set Y a *t-crosscut* if $|Y \cap F| = t$ holds for all $F \in \mathcal{F}$. Denote by $\mathcal{C}(\mathcal{F})$ the set of $(p+1)$-uniform families of the form $\{F \cap Y : F \in \mathcal{F}\}$ where Y is a $(p+1)$-crosscut. Note that all $(p+1)$-uniform delta systems of size $|\mathcal{F}|$ belong to $\mathcal{C}(\mathcal{F})$. Define $\alpha(\mathcal{F})$ as the maximum size of a $(p+1)$-graph without any member from $\mathcal{C}(\mathcal{F})$. Then

$$\text{ex}_k(n, \mathcal{F}) = (\alpha(\mathcal{F}) - o(1))\binom{n-p-1}{k-p-1}. \tag{12.1}$$

12.2 Families Without Triangles

In several cases, if the structure of $\mathcal{F}$ is more restricted than in Ch. 12.1 then we have exact results. For example, let $\mathcal{T}_k^3$ be a k-graph with edges $\{E^1, E^2, E^3\}$ such that $|E^i \cap E^j| = 1$ for all $i \neq j$ but $E^1 \cap E^2 \cap E^3 = \emptyset$. The following conjecture of Chvátal (1974) and Erdős is proved in Frankl, Füredi (1987a) (for $n > n_0(k)$).

$$\text{ex}_k(n, \mathcal{T}_k^3) = \binom{n-1}{k-1}. \tag{12.2}$$

Proof The case $k = 3$ was proved by a weight function method. Here we consider only the case $k \geq 5$. (The case $k = 4$ can be proved in a very similar way.) Instead of (11.6) we can prove the following simpler lemma. If $\mathcal{I} \subset 2^{[k]}$ is a family closed under intersection and $r(\mathcal{I}) = k - 1$ (i.e. all $(k-2)$-sets are covered) then either there exists a set $B \subset [k]$, $|B| = k - 2$ such that $2^B \subset \mathcal{I}$ or for some element $x \in [k]$

$$\{X : x \in X \subset [k], X \neq [k]\} \subseteq \mathcal{I}. \tag{12.3}$$

Consider a $\mathcal{T}_k^3$-free k-graph $\mathcal{G}$ of order n. We may suppose that $|\mathcal{G}| \geq \binom{n-1}{k-1}$. Apply (8.4) with $t = 2k$ to $\mathcal{F}$ to get $\mathcal{F}^*$ and $\mathcal{I} \subset 2^{[k]}$. Apply (12.3) to $\mathcal{I}$. In the first case of (12.3) (for $k \geq 5$), there exists a set $G \in \mathcal{G}^*$ and three pairs P^1, P^2, $P^3 \subset G$ forming a triangle, $P^i \in \mathcal{I}(G, \mathcal{G}^*)$. Then there are sets $G^i \in \mathcal{G}^*$ with $G^i \cap G = P^i$ such that they are disjoint outside G and so forming a $\mathcal{T}_k^3$, a contradiction. Finally, if the second case holds in (12.3) then we can proceed as in Ch. 11.

12.3 Families Without a Special Cover

Frankl, Füredi (1983) proved the following. Let $\mathcal{C}_k = \{[k], [k, 2k-1], [k+1, 2k]\}$. Then for $n > \Theta(k^2/\log k)$

$$\text{(12.4)} \qquad \text{ex}_k(n, \mathcal{C}_k) = \binom{n-1}{k-1}.$$

Lemma (12.3) and a variation of the argument given in Ch. 11 easily yields (12.4), although only for very large n's.

12.4 Families Without Stars

Duke and Erdős (1977) proposed the following question. Determine the maximum size of a k-graph of order n without a star of size s with a center of cardinality ℓ. Denote this maximum by $\text{ex}(n, \mathcal{S}_k^\ell(s))$. The case $s = 2$ is discussed in Ch. 11. (8.4) and (11.5) imply that $\text{ex}(n, \mathcal{S}_k^\ell(s)) = \Theta(n^{\max\{k-\ell-1, \ell\}})$ for fixed k and ℓ. The examples of Ch. 11 can be extended to the general case as follows. Let $\mathcal{A}$ be a maximal family of $(\ell+1)$-sets without any delta system of size s, so $|\mathcal{A}| = \varphi_{\ell+1}(s)$. Set $\cup\mathcal{A} = Y$, $Y \subset [n]$ and define $\mathcal{G} := \{G \in \binom{[n]}{k} : G \cap Y \in \mathcal{A}\}$. Clearly $\mathcal{G}$ contains no copy of $\mathcal{S}_k^\ell(s)$. This example gives the lower bound in the following theorem due to Frankl and Füredi (1987a). For k and s fixed, $k \geq 2\ell + 3$

$$\text{(12.5)} \qquad \text{ex}(n, \mathcal{S}_k^\ell(s)) = (1 - o(1))\varphi_{\ell+1}(s)\binom{n-\ell-1}{k-\ell-1}.$$

Proof It is a consequence of (12.1), but here we repeat the main steps. Let $\mathcal{G}$ be a maximal $\mathcal{S}_k^\ell(s)$-free family of order n. Apply (8.4) with $t = 2sk$ to get $\mathcal{G}^*$ and $\mathcal{I}$. Apply (11.6), we obtain that every set $G \in \mathcal{G}^*$ contains a $(k-\ell-1)$-element subset $B(G)$ which is not contained in any other edge $G' \in \mathcal{G}^*$. However, now $B(G)$ might be contained in edges from $\mathcal{G} \setminus \mathcal{G}^*$, (11.11) does not hold. Instead, there are at most $\varphi_{\ell+1}(s)$ edges $G \in (\mathcal{G}^1 \cup \cdots \cup \mathcal{G}^m)$ having the same $B(G)$. ∎

Let $\mathcal{P}$ be an $(n, \ell - 1 + s(k-\ell), \ell)$-packing, and replace each edge $P \in \mathcal{P}$ by the complete k-graph $\binom{P}{k}$. This example (with (9.1)) gives the lower bound in the following conjecture (Frankl and Füredi (1987a)).

$$\text{ex}(n, \mathcal{S}_k^\ell(s)) = \begin{cases} (1-o(1))\varphi_{\ell+1}(s)\binom{n-\ell-1}{k-\ell-1} & \text{if } k \geq 2\ell+1, \ (?) \\ (1-o(1))\binom{n}{\ell}\binom{\ell-1+s(k-\ell)}{k}/\binom{\ell-1+s(k-\ell)}{\ell} & \text{if } k \leq 2\ell. \quad (?) \end{cases}$$

$\text{ex}(n, \mathcal{S}_3^1(s))$ was determined for $n > 2s^3$ by Chung and Frankl (1987).

13 DESIGNS AND UNION–FREE FAMILIES

A family $\mathcal{G}$ is called *union-free* if all the $\binom{|\mathcal{G}|}{2}$ unions $A \cup B$, $A, B \in \mathcal{G}$ are distinct. Let $f_k(n)$ denote the maximum cardinality of a k-uniform union-free family of order n. We say that $\mathcal{G}$ is *weakly union-free* (*very weakly union-free*) if for any four distinct edges A, B, A', $B' \in \mathcal{G}$ $A \cup B = A' \cup B'$ implies $\{A, B\} = \{A', B'\}$ ($A \cup B = A' \cup B'$ and $A \cap B = A' \cap B'$ imply $\{A, B\} = \{A', B'\}$, respectively). That is, in a weakly union-free family $A \cup B = A \cup C$ is not excluded. Let $F_k(n)$ ($H_k(n)$) denote the maximum cardinality of a weakly union-free (very weakly union-free) k-graph of order n. As a k-partite very weakly union-free family is union-free (4.5) implies

$$\frac{k!}{k^k} H_k(n) \leq f_k(n) \leq F_k(n) \leq H_k(n). \tag{13.1}$$

In the case of graphs Erdős and Simonovits conjecture that

$$f_2(n) = \mathrm{ex}(n, \{C(3), C(4)\}) = \frac{1 + o(1)}{2\sqrt{2}} n^{3/2} \quad (?) \tag{13.2}$$

Let us mention for curiosity that Erdős and Simonovits (1982) proved that $\mathrm{ex}(n, \{C(4),\ C(5)\}) = \frac{1+o(1)}{2\sqrt{2}} n^{3/2}$. Surprisingly, the case $k = 3$ is easier (Frankl, Füredi (1984a)).

$$f_3(n) = \lfloor n(n-1)/6 \rfloor \tag{13.3}$$

$$F_3(n) = n(n-1)/3 \text{ for } n \equiv 1 (\mathrm{mod}\, 6),\ n > n_0. \tag{13.4}$$

Constructions For $n \equiv 1$ or 3 (mod 6) a Steiner triple system provides equality in (13.3). However, there are many other examples, too. For n an odd prime power, $n > 7$, $n \equiv 1$ (mod 3), $V = GF(n)$, the following $S_2(n, 3, 2)$ design $\mathcal{S}$ is weakly union-free. Let 1, g, g^2 be the solutions of $x^3 = 1$, $\mathcal{S} := \{\{a, b, c\} \in \binom{V}{3} : a + bg + cg^2 = 0\}$. We can extend this construction for all $n \equiv 1$ (mod 3) by Wilson's general existence theorem (1973), e.g., for $S_1(n, \{19, 13\}, 2)$ ($n > n_0$).

For the general case, Frankl, Füredi (1986b) proved that

$$H_k(n) = \begin{cases} \Theta(n^{2t}) & \text{for } k = 3t, \\ \Theta(n^{2t+1}) & \text{for } k = 3t+1, \\ \Theta(n^{2t+1.5}) & \text{for } k = 3t+2. \end{cases} \tag{13.5}$$

The proof of the upper bound in (13.5) is similar to that of (1.2). Let $s = \lceil (2k+1)/3 \rceil$. We have that all pairs of the form $(A \setminus S, B \setminus S)$ such that

$S \subseteq A \cap B$ are distinct. Using Jensen's inequality we have

$$\binom{\binom{n}{k-s}}{2} \geq \sum_{S \in \binom{[n]}{s}} \binom{\deg_{\mathcal{G}}(S)}{2} \geq \frac{1}{2}\binom{k}{s}|\mathcal{G}|\left(\frac{\binom{k}{s}|\mathcal{G}|}{\binom{n}{s}} - 1\right).$$

13.1 A Construction Using Symmetric Polynomials

The lower bound for $k = 3t+1$ gives the right order of magnitude for $k = 3t$ as well (by taking a family $\{G \setminus x : x \in G \in \mathcal{G}\}$). Similarly, the case $k = 6t+4$ yields a good bound for $k = 3t+2$ by defining a new family on the pairs $\binom{[n]}{2}$. So it is sufficient to deal with the case $k = 3t+1$. Note that $H_k(n)$ is monotone increasing, so it is sufficient to give constructions for n a prime. Let $GF(n)$ be the finite field of order n. The i'th elementary symmetric polynomial $\sigma_i(Y)$ for $Y \subseteq GF(n)$ is defined as

$$\sigma_i(Y) = \sum_{I \in \binom{Y}{i}} \prod_{y \in I} y.$$

In particular, $\sigma_0 = 1$. For $c_2, \ldots, c_{2t} \in GF(n)$ let

$$(13.6) \quad \mathcal{G}(c_2, \ldots, c_{2t}) := \{A \in \binom{GF(n)}{k} : \sigma_{2i}(A) = c_{2i} \text{ for } i = 1, \ldots, t\}.$$

It was shown that there exists a very weakly union-free subfamily $\mathcal{G} \subseteq \mathcal{G}(c_2, \ldots, c_{2t})$ of size $\binom{n}{k}/n^t - O(n^{2t})$ for appropriate values of c_{2i}. Here we consider the case $k = 4$ only. For $c \in GF(n)$ let

$$\mathcal{G}(c) := \{A \in \binom{GF(n)}{4} : \sigma_2(A) = c \text{ but } \sigma_1(B) \neq 0 \text{ for } B \subset A, |B| = 3\}.$$

We claim that $\mathcal{G}(c)$ is weakly union-free. First, we prove that $\mathcal{G}(c)$ is a 3-packing, i.e. $|A \cap B| \leq 2$ for all $A, B \in \mathcal{G}(c)$. Suppose, on the contrary, that $\{x, y, z, a\}, \{x, y, z, b\} \in \mathcal{G}(c)$. This means that

$$c = \sigma_2(x, y, z, a) = (xy + yz + zx) + a(x + y + z)$$
$$c = \sigma_2(x, y, z, b) = (xy + yz + zx) + b(x + y + z).$$

These imply $(a - b)(x + y + z) = 0$. As $\sigma_1(x, y, z) \neq 0$ by definition, we get the contradiction $a = b$.

Suppose, on the contrary, that for the distinct $A, B, C, D \in \mathcal{G}(c)$ we have $A \cup B = C \cup D$. (The handling of the case $A \cup B = A \cup C$ is similar.) Then $A \cap B = \emptyset$. Indeed, $A \cap B \cap C \neq \emptyset$, $C \subseteq A \cup B$ imply that either $|A \cap C| \geq 3$

or $|B \cap C| \geq 3$, a contradiction. This implies that for some disjoint pairs P, Q, U, V we have $A = P \cup Q$, $B = U \cup V$, $C = P \cup V$ and $D = U \cup Q$. By definition we have

$$c = \sigma_2(A) = \sigma_2(P \cup Q) = \sigma_2(P) + \sigma_1(P)\sigma_1(Q) + \sigma_2(Q),$$
$$c = \sigma_2(C) = \sigma_2(P \cup V) = \sigma_2(P) + \sigma_1(P)\sigma_1(V) + \sigma_2(V).$$

These imply

$$0 = \sigma_2(Q) - \sigma_2(V) + \sigma_1(P)(\sigma_1(Q) - \sigma_1(V)). \tag{13.7}$$

Starting with $c = \sigma_2(B) = \sigma_2(D)$ we obtain similarly

$$0 = \sigma_2(Q) - \sigma_2(V) + \sigma_1(U)(\sigma_1(Q) - \sigma_1(V)). \tag{13.8}$$

The difference of (13.7) and (13.8) is $(\sigma_1(P) - \sigma_1(U))(\sigma_1(Q) - \sigma_1(V)) = 0$. If $\sigma_1(Q) = \sigma_1(V)$ then (13.7) implies $\sigma_2(Q) = \sigma_2(V)$. However, if the first two symmetric polynomials coincide then $Q \equiv V$, a contradiction. The case $\sigma_1(P) = \sigma_1(U)$ is similar.

Finally, $\cup_c \mathcal{G}(c) = \{A \in \binom{GF(n)}{4} : \sigma_1(B) \neq 0 \text{ for } B \subset A, |B| = 3\}$. It is easy to see, that this union has size $\binom{n}{4} - O(n^3)$ implying the lower bound in the following equality. For n a prime power

$$F_4(n) = \frac{1}{24}n^3 - O(n^2).$$

13.2 Disjoint-union-free Families

We call the family $\mathcal{G}$ *disjoint-union-free* if for every $A, B, C, D \in \mathcal{G}$ $A \cap B = \emptyset$, $C \cap D = \emptyset$ and $A \cup B = C \cup D$ imply $\{A, B\} = \{C, D\}$. That is, all disjoint pairs have distinct unions. Erdős (1977) proposed the determination of $U_k(n)$, the maximum size of a k-uniform disjoint-union-free family of order n. In Füredi (1984) it was proved for $k \geq 3$

$$\binom{n-1}{k-1} + \lfloor \frac{n-1}{k} \rfloor \leq U_k(n) < 3.5\binom{n}{k-1}. \tag{13.9}$$

For the lower bound we take all k sets containing the element 1, and add disjoint sets in $[2, n]$. For $k = 3$ a slightly larger family (of size $\binom{n}{2}$) can be obtained by replacing each edge in a $(n, 5, 2)$-packing by a complete graph $\mathcal{K}_3(5)$. We also can prove in the case $k = 3$ that $\lim_{n\to\infty} U_3(n)/\binom{n}{2}$ exists and equals its supremum. The proof of the upper bound in (13.9) uses so-called *structure intersection* theorems. These investigations were initiated by Sós (1976).

14 PACKING BY RANDOM CHOICE

14.1 Almost Perfect Matchings in Almost Regular k-graphs

Frankl and Rödl (1985b) generalized (9.1) to obtain an almost perfect matching in an r-graph. Recall some definitions. A subhypergraph $\mathcal{M} \subseteq \mathcal{H}$ is called a *matching* if every two of its members are disjoint. The largest cardinality of a matching in $\mathcal{H}$ is the matching number $\nu(\mathcal{H})$. Finally, $\deg_{\mathcal{H}}(x, y)$ denotes the cardinality of $\mathcal{H}[\{x, y\}]$, i.e. the number of edges containing both x and y. Here we give an even more powerful version of the Frankl–Rödl theorem, which is due to Pippenger and Spencer (1989). For all fixed $K > 1$, $\varepsilon > 0$ and k there exists a $\delta > 0$ such that the following holds.

Suppose that $\mathcal{H}$ is a k-graph of order n, such that
(i) $\deg_{\mathcal{H}}(v) < Kd$ *holds for all* $v \in V$,
(ii) $d(1-\delta) < \deg_{\mathcal{H}}(v) < d(1+\delta)$ *holds for all but at most* δn *vertices,*
(iii) $\deg_{\mathcal{H}}(x, y) < \delta d$ *for all pairs. Then*

$$\nu(\mathcal{H}) > (1-\varepsilon)(n/k). \tag{14.1}$$

Of course, $\nu(\mathcal{H}) \leq n/k$.

14.2 Almost Disjoint Packing of Induced k-graphs

The following corollary of Frankl–Rödl theorem was proved in Frankl, Füredi (1987b). Let $(U, \mathcal{A})$ be a k-graph of order u size a. Whenever $n \to \infty$ we can place

$$(1-o(1))\binom{n}{k}/a \tag{14.2}$$

copies $\mathcal{A}^1, \mathcal{A}^2, \ldots$ into $\binom{[n]}{k}$ such that $|U^i \cap U^j| \leq k$ for all $i \neq j$ and $|U^i \cap U^j| = k$, $U^i \cap U^j = B$ imply $B \notin \mathcal{A}^i$, $B \notin \mathcal{A}^j$. In other words, for any given $\mathcal{A}$ there exists a hypergraph $\mathcal{H} \subseteq \binom{[n]}{k}$ of size $(1-o(1))\binom{n}{k}$ such that it can be decomposed into edge disjoint *induced* copies of $\mathcal{A}$. Moreover, these copies of $\mathcal{A}$ cannot have more than k common vertices.

Proof Let $\mathcal{P}$ be an almost optimal $(n, u, k+1)$-packing, $u := |U|$. Then $|\mathcal{P}| = (1-o(1))\binom{n}{k+1}/\binom{u}{k+1}$ holds by (9.1). Choose a positive $p < 1$ (p will be 'very close' to 1) and let $\mathcal{G}_k^n(p)$ be a random k-graph defined by $\text{Prob}(T \in \mathcal{G}_k^n(p)) = p$ for all $T \in \binom{[n]}{k}$. Define the random subhypergraph $\mathcal{P}(p) \subseteq \mathcal{P}$ as follows: $P \in \mathcal{P}(p)$ if $\mathcal{G}_k^n(p)|P \sim \mathcal{A}$. Finally, one can apply theorem (14.1) to $\mathcal{P}(p)$.

14.3 No Set Is Covered By The Union Of r Others

Corollary (14.2) was used to determine an asymptotic solution to the following problem. What is the maximum size of a k-graph $\mathcal{G}$ of order n if no edge is covered by the union of r others. Denote this maximum by $M(n,k,r)$. Clearly, $M(n,k,r) = n-k+1$ for $r \geq k$. Let $t = \lfloor k/r \rfloor$. It is easy to see, that in an r-cover-free k-uniform family $\mathcal{G}$ every edge $G \in \mathcal{G}$ contains a t-element subset which is not contained in any other edge. This implies $M(n,k,r) = O(n^t)$. In Erdős, Frankl, Füredi (1982), (1985) it was proved that for fixed k and r the limit $c_{k,r} = \lim_{n\to\infty} M(n,k,r)/\binom{n}{t}$ exists (and positive). In Frankl, Füredi (1987b) it was proved that for $k = r(t-1)+\ell+1$, $0 \leq \ell \leq r$

$$c_{k,r} = 1/\left(\binom{k}{t} - \mathrm{ex}(k, \mathcal{S}_t^0(\ell))\right), \tag{14.3}$$

where $\mathrm{ex}(k,\mathcal{S}_t^0(\ell))$ is the maximum size of a t-graph of order k not containing more than ℓ pairwise disjoint edges. This function was discussed in Ch. 12.4, (12.5) gives $\mathrm{ex}(k,\mathcal{S}_t^0(\ell)) = (1-o(1))\ell\binom{k-1}{t-1}$. Erdős conjectures

$$\mathrm{ex}(k,\mathcal{S}_t^0(\ell)) = \max\left\{\binom{t\ell+t-1}{t}, \binom{k}{t} - \binom{k-\ell}{t}\right\}. \quad (?) \tag{14.4}$$

Erdős (1965) proved (14.4) for $k \geq k_0(t,\ell)$. For a recent survey on this function see Frankl (1987).

Construction for (14.3) Let $\mathcal{B} \subseteq \binom{[k]}{t}$ be a family not containing a matching of size $\ell+1$ and suppose that $|\mathcal{B}|$ is maximal, i.e. $|\mathcal{B}| = \mathrm{ex}(k,\mathcal{S}_t^0(\ell))$. Let $\mathcal{A} := \binom{[k]}{t} \setminus \mathcal{B}$. Apply (14.2) to $\mathcal{A}$. We obtain a $\mathcal{G} \subseteq \binom{[n]}{t}$ of size $(1-o(1))\binom{n}{t}$ and $(1-o(1))\binom{n}{t}/|\mathcal{A}|$ k-subsets $U^1, U^2, \ldots$ such that U^i induces a copy of $\mathcal{A}$. Then $\mathcal{U} := \{U^1, U^2, \ldots\}$ is an r-cover-free family.

15 SOME APPLICATIONS, PROBLEMS

15.1 Three problems in Geometry

Some 45 years ago Erdős (1946) initiated the investigation of the following combinatorial geometry problem. What is the maximum number of times, $f^{(d)}(n)$, that the same distance can occur among n points in the d dimensional space $\mathbf{R}^d$? He observed that the complete bipartite graph $\mathcal{K}(2,3)$ cannot be realized on the plane hence (3.2) gives

$$f^{(2)}(n) \leq \mathrm{ex}_2(n,\mathcal{K}(2,3)) = O(n^{3/2}). \tag{15.1}$$

Erdős conjectures that the grid gives the best value,

$$f^{(2)}(n) = O(n^{1+C/\log\log n}). \quad (?) \tag{15.2}$$

The best upper bound, $O(n^{4/3})$, is due to Beck, Spencer, Szemerédi, and Trotter. A recent survey and a new proof can be found in Clarkson et al. (1990).

Concerning a convex, planar n-gon Π Erdős and Moser conjectured

$$f^{(2)}_{\text{conv}}(\Pi) = O(n) \quad (?) \tag{15.3}$$

The best upper bound, due to Füredi (1990), $f^{(2)}_{\text{conv}}(\Pi) < 7n\log n$, is achieved by an extension of Turán's problem to ordered structures. More results and problems on this, see Füredi and Hajnal (1991).

A related result, first investigated by Conway, Croft, Erdős, and Guy (1979), is the following. What is the maximum number of right angle triangles spanned by n points in the plane, $r^{(2)}(n)$. It is easy to see that (4.1) gives

$$r^{(2)}(n) = O(\mathrm{ex}_3(n, \mathcal{K}(1,2,2))) = O(n^{2.5}). \tag{15.4}$$

Pach and Sharir (1991) proved that $r^{(2)}(n)$ is $\Theta(n^2\log n)$. The $\sqrt{n}\times\sqrt{n}$ grid gives the lower bound.

Further geometric and other applications can be found in a series of papers by Erdős, Meir, Sós and Turán (1971–72). We only repeat that the most useful theorems are the density versions like (4.1) and (4.4).

15.2 Turán Numbers $T(n, \ell, k)$

There is an interesting connection between the Turán number $T(n,5,4)$ and the *crossing number* $c(\mathcal{K}(n))$, the minimum number of crossings in a planar representation of the complete graph, observed by Ringel. See de Caen, Kreher, Wiseman (1988).

Let $f_k(\chi)$ be the minimum size of a k-uniform hypergraph with chromatic number χ. (The *chromatic number* of a hypergraph is the largest integer χ, such that all $(\chi-1)$-coloration of the vertices yields a monochromatic edge.) Obviously, $f_k(\chi) \le \binom{(\chi-1)(k-1)+1}{k}$ and here equality holds for $k=2$. Alon (1985) disproved a conjecture of Erdős using (5.3–4). He showed that for $k\to\infty$, $\chi/k\to\infty$

$$f_k(\chi) = O\left(k^{5/2}\log k (\frac{3}{4})^k \binom{(\chi-1)(k-1)+1}{k}\right). \tag{15.5}$$

Lehel (1982) proved the following conjecture of Bollobás. The edge set $\mathcal{E}$ of a k-graph of order n can be covered by at most $\text{ex}(n, \mathcal{K}_k(\ell))$ copies of complete subgraphs of size ℓ and edges. The extremal graph is a maximal $\mathcal{K}_k(\ell)$-free graph. Whether $\mathcal{E}$ can be *decomposed* is unknown.

Frankl and Pach (1984) conjecture that every k-graph of size $1 + \text{ex}(k + \ell, \mathcal{K}_k(\ell))$ contains $\ell + 1$ disjointly representable edges, that is edges E^1, $\ldots$, $E^{\ell+1}$ such that $E^i \not\subseteq \cup_{j \neq i} E^j$. They prove the case $k = 2$ verifying a conjecture of Gyárfás.

Concerning the chromatic number of the generalized Kneser graph $\text{Kn}(n, k, t)$ (defined in Ch. 7.2) it was proved in Frankl, Füredi (1986c) that

$$\chi(\text{Kn}(n, k, t)) = (1 + o(1))T(n, k, t) \tag{15.6}$$

holds for k, t fixed and $n \to \infty$. We conjecture that equality holds in (15.6) for $n > n(k, t)$. This was proved by Frankl (1985) for $t = 2$. Further results can be found in Alon, Frankl, Lovász (1986).

15.3 Generalizations

Chung and Erdős (1983), (1987) determined the order of magnitude of $u_k(n, e)$, the maximum number of edges in a k-graph which is contained in every k graph of order n and size e. The maximum unavoidable k-graphs are often not stars but a combination of sunflowers of different types. (They call them *books*.)

Let $N(\mathcal{G}, \mathcal{F})$ be the number of subgraphs of $\mathcal{G}$ isomorphic to $\mathcal{F}$, and let $N(e, \mathcal{F})$ be the maximum of $N(\mathcal{G}, \mathcal{F})$ for $|E(\mathcal{G})| = e$, the maximum number of ways that $\mathcal{F}$ can be embedded as a subgraph. Alon (1981), (1986) determined the order of magnitude of $N(e, \mathcal{F})$ for all fixed $\mathcal{F}$ as $e \to \infty$. Further results can be found in Füredi (1992). However several problems remain open, for example, the cases of multigraphs, k-graphs.

As the cube is a 3-regular graph, the result of Erdős and Simonovits mentioned before (3.8) implies that every graph of order n and size at least $10n^{8/5}$ contains a 3-regular subgraph. Pyber (1985) proved that the same holds for graphs of size $50n \log n$.

Let $\mathcal{H}_3(4, 3)$ denote the 3-graph of order 4 with 3 edges. In Frankl, Füredi (1984c) and by Giraud (1983, unpublished) and in Sidorenko (1982b) an example was constructed to show the lower bound in the following conjecture.

$$\text{(15.7)} \qquad \mathrm{ex}(n, \mathcal{H}_3(4,3)) = (1+o(1))\frac{2}{7}\binom{n}{3} \quad (?)$$

We start with a $S_2(6,3,2)$ design, $\mathcal{S}$; i.e. $\mathcal{S}$ is a 3-uniform hypergraph with the vertex set $\{1,2,\ldots,6\}$ such that each pair is covered exactly twice. $\mathcal{S}$ is 5-regular and has 10 edges. Partition V into 6 almost equal parts and take all 3-element sets of $\mathcal{G}$ which intersect 3 parts corresponding to an edge of $\mathcal{S}$. We have got about $10(n/6)^3$ edges so far. Now partition each part into 6 parts and iterate this process, add all 3-tuples in a part which intersect 3 subparts and correspond to an edge, etc. The best upper bound, $(1/3)\binom{n}{3}n/(n-2)$, is established by de Caen (1983a) and Sidorenko (1982a).

Sidorenko (1991b) recently generalized his method and obtained for every k-uniform hypergraph $\mathcal{F}$

$$\text{(15.8)} \qquad \pi(\mathcal{F}) \le \frac{|\mathcal{E}(\mathcal{F})|-1}{|\mathcal{E}(\mathcal{F})|}$$

Another interesting conjecture concerning 3-graphs is due to Erdős and Sós (1982). Suppose that $\mathcal{H}$ is a 3-graph of order n such that for all $x \in V(\mathcal{H})$ $\{E \setminus \{x\} : x \in E \in \mathcal{H}\}$ is a bipartite graph. Then

$$\text{(15.9)} \qquad |\mathcal{H}| < n^3/24 \quad (?)$$

Erdős and Sós (1982) introduced a new class of extremal problems, called Ramsey–Turán type problems. Let $\mathrm{ex}(n, \mathbf{F}, \alpha) := \max\{|\mathcal{G}| : \mathcal{G} \subseteq \binom{[n]}{k}, \mathcal{G}$ is $\mathbf{F}$-free and contains no empty subset of size $\alpha n\}$. The extra condition is that $\mathcal{G}$ contains only small independent sets. The *Ramsey–Turán number* $\rho(\mathbf{F})$ is defined by

$$\sup_{\alpha(n)} \left\{ \limsup_{n\to\infty} \left(\mathrm{ex}(n,\mathbf{F},\alpha(n))\Big/\binom{n}{k}\right) : \alpha(n) \to 0 \text{ as } n \to \infty \right\},$$

that is the supremum is taken over all functions $\alpha(n) > 0$ tending to zero. Obviously, $0 \le \rho(\mathbf{F}) \le \pi(\mathbf{F})$. Frankl and Rödl (1988) proved that for all $k \ge 3$ in infinitely many cases

$$\text{(15.10)} \qquad 0 < \rho(\mathbf{F}) < \pi(\mathbf{F}).$$

Sidorenko (1991a) gave a concrete 3-graph satisfying (15.10), namely $\mathcal{F} = \{123, 145, 167, 245, 267, 345, 367, 467, 567\}$. The case of graphs, $k = 2$, was studied earlier, see, e.g., Bollobás and Erdős (1976), Erdős, Hajnal, Sós, Szemerédi (1983).

Another generalization of Turán's problem is known as the *Lotto problem.* Given $n \geq \ell, k \geq t \geq 0$, find the minimum number $L(n, \ell, k, t)$ of k-subsets of $[n]$ such that any ℓ-subset of $[n]$ meets one of these k-subsets in at least t vertices. Hanani, Ornstein, and Sós (1964) proved

$$\lim_{n\to\infty} L(n,\ell,k,2)\frac{k(k-1)(\ell-1)}{n(n-\ell+1)} = 1$$

and here the left hand side is always at most 1. A short survey can be found in Brouwer and Voorhoeve (1979).

The following problem was posed in Frankl, Füredi (1984c). Determine

$$\text{(15.11)} \qquad \max\{|\mathcal{H}| : \mathcal{H} \subseteq \binom{[n]}{k}\}, \quad (?)$$

such that every $k+1$ element subset of $[n]$ spans 0 or 2 edges. A geometric construction shows that this maximum is at least $(1-o(1))\binom{n}{k}/2^{k-1}$.

Frankl and Pach (1983) conjecture that if $\mathcal{G} \subseteq \binom{[n]}{k}$, $|\mathcal{G}| > \binom{n-1}{k-1}$ and $n > n_k$, then for some edge $G \in \mathcal{G}$ there exists an edge $G_A \in \mathcal{G}$ with $G \cap G_A = A$ for all $A \subset G$. They proved this for $|\mathcal{G}| > \binom{n}{k-1}$.

The problem of $m(n,k,L)$ discussed in Ch. 9 can be extended by dropping the size restriction. Let $m(n, L)$ be the maximum cardinality of a family $\mathcal{G} \subseteq 2^{[n]}$ such that for any two distinct $G, G' \in \mathcal{G}$ $|G \cap G'| \in L$ holds. The most important result here is due to Katona (1964): $m(n, >\ell) = \sum_{i \geq (n+\ell+1)/2} \binom{n}{i}$ for $n + \ell$ odd, and $m(n, >\ell) = 2m(n-1, \geq \ell)$ for $n + \ell$ even. Some further known results: $m(n, \{0, r\}) = \binom{\lfloor n/r \rfloor}{2} + \lfloor n/r \rfloor + (n - r\lfloor n/r \rfloor)$ for $n > 1000r^5$ (by Füredi (1982)), $m(n, \{1, 2, \ldots, r\}) = \binom{n-1}{1} + \cdots + \binom{n-1}{r}$ for $n > 6r$ (Frankl and Füredi (1981) and Pyber (1983)). In Frankl, Füredi (1984d) it was proved that $m(n, \neq \ell) = m(n, >\ell) + \sum_{j<\ell} \binom{n}{j}$ holds for $n > \ell 3^\ell$. Here there are even more unsolved problems, see the survey Frankl (1988).

A k-graph $\mathcal{H}$ is called **F**-*saturated* if it is **F**-free and whenever a new edge $H \in \binom{[n]}{k}$ is added to $\mathcal{H}$ then $\mathcal{H} \cup \{H\}$ contains a subhypergraph isomorphic to (a member of) **F**. Let $\text{sat}(n, \mathbf{F})$ be the *minimum* number of edges in an **F**-saturated k-graph of order n. For example, a celebrated theorem of Bollobás (1965) states $\text{sat}(n, \mathcal{K}_k(t)) = \binom{n}{k} - \binom{n-t+k}{k}$. The general problem was first considered by Kászonyi and Tuza (1986). Some recent results (and problems) can be found in Erdős, Füredi, Tuza (1991), e.g., $\mathrm{sat}(n, \mathcal{H}_3(4,3)) = \lfloor (n-1)^2/4 \rfloor$.

16 ACKNOWLEDGMENTS

Research supported in part by the Hungarian National Science Foundation under grant No. 1812. The author is indebted to P. Erdős and Á. Seress for continuous help and for careful reading. I am also thankful for the referees for their helpful suggestions.

17 REFERENCES

ALON, N. (1981), On the number of subgraphs of prescribed type of graphs with a given number of edges, *Israel J. Math.* **38**, 116–130.

ALON, N. (1984), Unpublished manuscript

ALON, N. (1985), Hypergraphs with high chromatic number, *Graphs and Combin.* **1**, 387–389.

ALON, N. (1986), On the number of certain subgraphs contained in graphs with a given number of edges, *Israel J. Math.* **53**, 97–120.

ALON, N., FRANKL, P., and LOVÁSZ, L. (1986), The chromatic number of Kneser hypergraphs, *Transactions Amer. Math. Soc.* **298**, 359–370.

BABAI, L. (1988), A short proof of the non-uniform Ray-Chaudhuri–Wilson inequality, *Combinatorica* **8** (1988), 133-135.

BABAI, L. and FRANKL, P. (1980), Note on set intersections, *J. Combin. Th., Ser. A* **28** (1980), 103–105.

BABAI, L. and FRANKL, P. (1988), *Linear Algebra Methods in Combinatorics*, Part 1, Dept. Comp. Sci., University of Chicago, 1988.

BEHREND, F. A. (1946), On sets of integers which contain no three elements in arithmetic progression, *Proc. Nat. Acad. Sci.* **23**, 331–332.

BENSON, C. T. (1966), Minimal regular graphs of girth eight and twelve, *Canad. J. Math.* **26**, 1091–1094.

BOLLOBÁS, B. (1965), On generalized graphs, *Acta Math. Acad. Sci. Hungar.* **16**, 447–452.

BOLLOBÁS, B. (1974), Three-graphs without two triples whose symmetric difference is contained in a third, *Discrete Math.* **8**, 21–24.

BOLLOBÁS, B. (1978), *Extremal Graph Theory*, Academic Press, London – New York.

BOLLOBÁS, B. and ERDŐS, P. (1976), On a Ramsey–Turán type problem, *J. Combin. Th., Ser. B* **21**, 166–168.

BOLLOBÁS, B., ERDŐS, P., and SIMONOVITS, M. (1976), On the structure of edge graphs II, *J. London Math. Soc.* **12** (2), 219–224.

BONDY, A. and SIMONOVITS, M. (1974), Cycles of even length in graphs, *J. Combin. Th., Ser. B* **16**, 97–105.

BROUWER, A. E., and VOORHOEVE, M. (1979), Turán theory and the lotto problem, *in* Packing and Covering in Combinatorics, (A. Schrijver, ed.), pp. 99–105. Math. Centre Tracts, **106**, Math. Centrum, Amsterdam.

BROWN, W. G. (1966), On graphs that do not contain a Thomsen graph, *Canad. Math. Bull.* **9**, 281–289.

BROWN, W. G. (1983), On an open problem of Paul Turán concerning 3-graphs, *Studies in Pure Math.*, 91–93, Birkhäuser, Basel-Boston, Mass.

BROWN, W. G., ERDŐS, P., and V. T. SÓS (1971), Some extremal problems on r-graphs, *in* New Directions in the Theory of Graphs, *Proc. 3rd Ann Arbor Conf. on Graph Th.*, Univ. Michigan, pp. 53–63, Academic Press, New York, 1973.

BROWN, W. G., ERDŐS, P., and V. T. SÓS (1973), On the existence of triangulated spheres in 3-graphs, *Periodica Math. Hungar.* **3**, 221–228.

DE CAEN, D. (1983a), Extension of a theorem of Moon and Moser on complete hypergraphs, *Ars Combin.* **16**, 5–10.

DE CAEN, D. (1983b), A note on the probabilistic approach to Turán's problem, *J. Combin. Th., Ser. B* **34**, 340–349.

DE CAEN, D. (1983c), Linear constraints related to Turán's problem, *Congressus Numerantium* **39**, 291–303.

DE CAEN, D. (1985), Uniform hypergraphs with no blocks containing the symmetric difference of any two other blocks, *in* Proc. 16th Southeastern Conf. on Combinatorics, Graph Theory and Computing, *Congressus Numerantium* **47**, 249–253.

DE CAEN, D. (1988), On upper bounds for 3-graphs without tetrahedra, *Congressus Numerantium* **62**, 193–202.

DE CAEN, D. (1991), The current status of Turán's problem on hypergraphs, manuscript

DE CAEN, D., KREHER, D. L., and WISEMAN, J. (1988), On constructive upper bounds for the Turán numbers $T(n,2r+1,2r)$, *Congressus Numerantium* **65**, 277–280.

CAMERON, P. J. and VAN LINT, J. H. (1975), *Graph Theory, Coding Theory and Block Designs* London Math. Soc. Lecture Notes **19**, Cambridge Univ. Press.

RAY-CHAUDHURI, D. K. and WILSON, R. M. (1975), On t designs, *Osaka J. Math.* **12**, 737–744.

CHUNG, F. R. K. and ERDŐS, P. (1983), On unavoidable graphs, *Combinatorica* **3**, 167–176.

CHUNG, F. R. K. and ERDŐS, P. (1987), On unavoidable hypergraphs, *J. Graph Th.* **11**, 251–263.

CHUNG, F. R. K. and FRANKL, P. (1987), The maximum number of edges in a 3-graph not containing a given star, *Graphs and Combin.* **3**, 111–126.

CHVÁTAL, V. (1974), An extremal set-intersection theorem, *J. London Math. Soc.* **9**, 355–359.

CLAPHAM, C. R. J., FLOCKART, A., and SHEEHAN, J. (1989), Graphs without four-cycles, *J. Graph Th.* **13**, 29–47.

CLARKSON, L. K., EDELSBRUNNER, H., GUIBAS, L. J. SHARIR, M., and WETZL, E. (1990), Combinatorial complexity bounds for arrangements of curves and spheres, *Discrete and Computational Geometry*, **5**, 99–160.

CONWAY, J. H., CROFT, H. T., ERDŐS, P., and GUY, M. J. T. (1979), On the distribution of values of angles determined by coplanar points, *J. London Math. Soc.* (2), **19**, 137–143.

DELSARTE, P. (1973), An algebraic approach to the association schemes in coding theory, *Philips Res. Rep. Suppl.* **10**.

DEZA, M., ERDŐS, P., and FRANKL, P. (1978), Intersection properties of systems of finite sets, *Proc. London Math. Soc. (3)* **36**, 368–384.

DROESBEKE, L. M. (1982), Determination of the values of the Turán number $T(n,4,6)$ (French), *Cahiers Centre Études Rech. Opér.* **24**, 185–191.

DUKE, R. A. and ERDŐS, P. (1977), Systems of finite sets, having a common intersection, *in:* Proc. 8th Southeastern Conf. Combinatorics, Graph Theory and Computing, *Congressus Numerantium* **19**, pp. 247–252.

ERDŐS, P. (1938), On sequences of integers no one of which divides the product of two others and some related problems, *Izvestiya Naustno-Issl. Inst. Mat. i Meh. Tomsk* **2**, 74–82. (*Mitteilungen des Forshungsinstitutes für Math. und Mechanik, Tomsk,* in Zentralblatt **20**, p. 5.)

ERDŐS, P. (1946), On sets of distances of n points, *Amer. Math. Monthly* **53**, 248–250.

ERDŐS, P. (1959), Graph theory and probability, *Canadian J. Math.* **11**, 34–38.

ERDŐS, P. (1964a), Extremal problems in graph theory, *in* Theory of Graphs and its Appl., Proc. Sympos. Smolenice, Prague, pp. 29–36.

ERDŐS, P. (1964b), On extremal problems of graphs and generalized graphs, *Israel J. Math.* **2**, 183–190.

ERDŐS, P. (1965), A problem of independent r-tuples, *Ann. Univ. Sci. Budapest. Eötvös* **8**, 93–95.

ERDŐS, P. (1967), On bipartite subgraphs of a graph, (Hungarian), *Mat. Lapok* **18**, 283–288.

ERDŐS, P. (1976), Problems and results in graph theory and combinatorial analysis, *in:* Proc. Fifth British Combinatorial Conf. (C. St. J. A. Nash-Williams and J. Sheenan, eds.), (Aberdeen 1975), *Congressus Numerantium* **15**, 169–172.

ERDŐS, P. (1977), Problems and results in combinatorial analysis, *in* Proc. 8th Southeastern Conf. on Combinatorics, Graph Theory and Computing, Baton Rouge, Lousiana State Univ., *Congressus Numerantium* **19**, 3–12.

ERDŐS, P. (1981), On the combinatorial problems which I would most like to see solved, *Combinatorica* **1**, 25–42.

ERDŐS, P., FRANKL, P., and FÜREDI, Z. (1982), Families of finite sets in which no set is covered by the union of two others, *J. Combin. Th., Ser. A* **33**, 158–166.

ERDŐS, P., FRANKL, P., and FÜREDI, Z. (1985), Families of finite sets in which no set is covered by the union of r others, *Israel Journal of Mathematics* **51**, 79–89.

ERDŐS, P., FRANKL, P., and RÖDL, V. (1986), The asymptotic number of graphs not containing a fixed subgraph and a problem for hypergraphs having no exponent, *Graphs and Combin.* **2**, 113–121.

ERDŐS, P., FÜREDI, Z., and TUZA, Zs. (1991), Saturated r-uniform hypergraphs, *Discrete Math.*, to appear

ERDŐS, P. and GALLAI, T. (1961), On the maximal number of vertices representing the edges of a graph, *MTA Mat. Kutató Int. Közl. Budapest* **6**, 181–203.

ERDŐS, P., HAJNAL, A., SÓS, V. T., and SZEMERÉDI (1983), More results on Ramsey–Turán type problems, *Combinatorica* **3**, 69–81.

ERDŐS, P. and HANANI, H. (1963), On a limit theorem in combinatorial analysis, *Publ. Math. Debrecen* **10**, 10–13.

ERDŐS, P., CHAO KO, and RADO, R. (1961), An intersecting theorem for finite sets, *Quart. J. Math. Oxford, Ser. (2)* **12**, 313–320.

ERDŐS, P. and KLEITMAN, D. J. (1968), On coloring graphs to maximize the proportion of multicolored k-edges, *J. Combin. Th.* **5**, 164–169.

ERDŐS, P., MEIR, A., SÓS, V. T., and TURÁN, P. (1971), On some applications of graph theory, II, *Studies in Pure Math.* pp. 89–99, Academic Press, London.

ERDŐS, P., MEIR, A., SÓS, V. T., and TURÁN, P. (1972a), On some applications of graph theory, I, *Discrete Math.* **2** (1972), 207–228.

ERDŐS, P., MEIR, A., SÓS, V. T., and TURÁN, P. (1972b), On some applications of graph theory, III, *Canad. Math. Bull.* **15** (1972), 27–32.

ERDŐS, P. and RADO, R. (1960), Intersection theorems for systems of sets, *J. London Math. Soc.* **35**, 85–90.

ERDŐS, P., RÉNYI, A., and SÓS, V. T. (1966), On a problem of graph theory, *Studia Sci. Math. Hungar.* **1**, 215–235.

ERDŐS, P. and SIMONOVITS, M. (1966), A limit theorem in graph theory, *Studia Sci. Math. Hungar.* **1**, 51–57.

ERDŐS, P. and SIMONOVITS, M. (1970), Some extremal problems in graph theory, *Proc. Colloq. Math. Soc. János Bolyai* **4**, Combinatorial Theory and its Appl. I, pp. 378–392.

ERDŐS, P. and SIMONOVITS, M. (1982), Compactness results in extremal graph theory, *Combinatorica* **2**, 275–288.

ERDŐS, P. and SIMONOVITS, M. (1983), Supersaturated graphs and hypergraphs, *Combinatorica* **3**, 181–192.

ERDŐS, P. and SÓS, V. T. (1982), On Ramsey-Turán type theorems for hypergraphs, *Combinatorica* **2**, 289–295.

ERDŐS, P. and SPENCER, J. (1974), *Probabilistic Methods in Combinatorics*, Academic Press, London – New York, Akadémiai Kiadó, Budapest.

ERDŐS, P. and STONE, A. H. (1946), On the structure of linear graphs, *Bull. Amer. Math. Soc.* **52** (1946), 1087-1091.

FRANKL, P. (1977), On families of finite sets no two of which intersect in a singleton, *Bull. Austral. Math. Soc.* **17**, 125–134.

FRANKL, P. (1978a), The Erdős-Ko-Rado theorem is true for n=ckt, *in Combinatorics*, Proc Fifth Hungarian Colloq. Combin., Keszthely, Hungary, 1976, (A. Hajnal et al., Eds.), *Proc. Colloq. Math. Soc. János Bolyai* **18** North-Holland, Amsterdam, pp. 365–375.

FRANKL, P. (1978b), On intersecting families of finite sets, *J. Combin. Th., Ser. A* **24**, 146–161.

FRANKL, P. (1980a), Families of finite sets with prescribed cardinalities for pairwise intersections, *Acta Math. Acad. Sci. Hungar.* **35**, 351–360.

FRANKL, P. (1980b), Extremal problems and coverings of the space, *European J. Combin.* **1**, 101–106.

FRANKL, P. (1983a), Constructing finite sets with given intersection, *Ann. Discr. Math.* **17**, 289–291.

FRANKL, P. (1983b), An extremal set theoretical characterization of some Steiner systems, *Combinatorica* **3**, 193–199.

FRANKL, P. (1984), Families of finite sets with three intersections, *Combinatorica* **4**, 141–148.

FRANKL, P. (1985), On the chromatic number of the general Kneser-graph, *J. Graph Th.* **9**, 217–220.

FRANKL, P. (1986), All rationals occur as exponents, *J. Combin. Th., Ser. A* **42**, 200–206.

FRANKL, P. (1987), The shifting technique in extremal set theory, In *Combinatorial Surveys 1987*, C. Whitehead ed., pp. 81–110. Cambridge Univ. Press.

FRANKL, P. (1988), Intersection and containment problems without size restrictions, *in* Algebraic, Extremal and Metric Combinatorics, London Math. Soc. Lecture Note Series, **131** Cambridge Univ. Press, pp. 62–111.

FRANKL, P. (1990), Asymptotic solution of a Turán-type problem, *Graphs and Combin.* **6**, 223-227.

FRANKL, P. and FÜREDI, Z. (1981), Families of finite sets with missing intersections, *Finite and Infinite Sets, Proc. Colloq. Math. Soc. János Bolyai* **37**, (Eger, Hungary), North-Holland, Amsterdam-New York, 1984. pp. 305–318.

FRANKL, P. and FÜREDI, Z. (1983), A new generalization of the Erdős-Ko-Rado theorem, *Combinatorica* **3**, 341–349.

FRANKL, P. and FÜREDI, Z. (1984a), A new extremal property of Steiner triple systems, *Discrete Math.* **48**, 205–212.

FRANKL, P. and FÜREDI, Z. (1984b), Union-free hypergraphs and probability theory, *European J. Combin.* **5**, 127–131. Erratum, *ibid* **5**, p. 395.

FRANKL, P. and FÜREDI, Z. (1984c), An exact result for 3-graphs, *Discrete Math.* **50**, 323–328.

FRANKL, P. and FÜREDI, Z. (1984d), On hypergraphs without two edges intersecting in a given number of vertices, *J. Combin. Th., Ser. A* **36**, 230–236.

FRANKL, P. and FÜREDI, Z. (1985), Forbidding just one intersection, *J. Combin. Th., Ser. A* **39**, 160–176.

FRANKL, P. and FÜREDI, Z. (1986a), Non-trivial intersecting families, *J. Combin. Th., Ser. A* **41**, 150–153.

FRANKL, P. and FÜREDI, Z. (1986b), Union-free families of sets and equations over fields, *Journal of Number Theory* **23**, 210–218.

FRANKL, P. and FÜREDI, Z. (1986c), Extremal problems concerning Kneser graphs, *J. Combin. Th., Ser. B* **40**, 270–284.

FRANKL, P. and FÜREDI, Z. (1987a), Exact solution of some Turán-type problems, *J. Combin. Th., Ser. A* **45**, 226–262.

FRANKL, P. and FÜREDI, Z. (1987b), Colored packing of sets, in *Combinatorial Design Theory, Ann. of Discrete Math.* **34**, 165–178.

FRANKL, P. and FÜREDI, Z. (1988), Extremal problems and the Lagrange function for hypergraphs, *Bull. of the Inst. of Math., Academia Sinica* **16**, 305–313.

FRANKL, P. and FÜREDI, Z. (1989), Extremal problems whose solutions are the blow-ups of the small Witt-designs, *J. Combin. Th., Ser. A* **52**, 129–147.

FRANKL, P. and FÜREDI, Z. (1991), Beyond the Erdős-Ko-Rado theorem, *J. Combin. Th., Ser. A* **56**, to appear

FRANKL, P. and KATONA, G. O. H. (1979), If the intersection of any r sets has a size $\neq r-1$, *Studia Sci. Math. Hungar.* **14** (1979), 47–49.

FRANKL, P. and PACH, J. (1983), On the number of sets in a null-t-design, *European J. Combin.* **4**, 21–23.

FRANKL, P. and PACH, J. (1984), On disjointly representable sets, *Combinatorica* **4**, 39–45.

FRANKL, P. and RÖDL, V. (1984), Hypergraphs do not jump, *Combinatorica* **4**, 149–159.

FRANKL, P. and RÖDL, V. (1985a), Lower bounds for Turán's problem, *Graphs and Combin.* **1**, 213–216.

FRANKL, P. and RÖDL, V. (1985b), Near perfect coverings in graphs and hypergraphs, *European J. Combin.* **6**, 317–326.

FRANKL, P. and RÖDL, V. (1988), Some Ramsey–Turán type results for hypergraphs, *Combinatorica* **8**, 323–332.

FRANKL P. and WILSON, R. M. (1981), Intersection theorems with geometric consequences, *Combinatorica* **1**, 357–368.

FÜREDI, Z. (1982), Set-systems with prescribed cardinalities for pairwise intersections, *Discrete Math.* **40**, 53–67.

FÜREDI, Z. (1983a), Graphs without quadrilaterals, *J. Combin. Th., Ser. B* **34**, 187–190.

FÜREDI, Z. (1983b), On finite set-systems whose every intersection is a kernel of a star, *Discrete Math.* **47**, 129–132.

FÜREDI, Z. (1984), Hypergraphs in which all disjoint pairs have distinct unions, *Combinatorica* **4**, 161–168.

FÜREDI, Z. (1985), Set-systems with three intersections, *Combinatorica* **5**, 27–31.

FÜREDI, Z. (1990), The maximum number of unit distances in a convex n-gon, *J. Combin. Th., Ser. A* **55**, 316–320.

FÜREDI, Z. (1991), Quadrilateral-free graphs with maximum number of edges, *J. Combin. Th., Ser. B*, submitted

FÜREDI, Z. (1992), Graphs with maximum number of star-forests, *Studia Sci. Math. Hungar.*

FÜREDI, Z. and HAJNAL, P. (1991), Davenport-Schinzel theory of matrices, *Discrete Math.*

GIRAUD, G., manuscript, see de Caen (1991)

HANANI, H., ORNSTEIN, D., and SÓS, V. T. (1964), On the lottery problem, *Magyar Tud. Akad. Mat. Kutató Int. Közl.* **9**, 155-158.

KALAI, G. (1985), A new approach to Turán's conjecture, *Graphs and Combin.* **1**, 107–109.

KÁSZONYI, L. and TUZA, ZS. (1986), Saturated graphs with minimal number of edges, *J. Graph Th.* **10**, 203–210.

KATONA, G. O. H. (1964), Intersection theorems for systems of finite sets, *Acta Math. Acad. Sci. Hungar.* **15**, 329–337.

KATONA, G. O. H. (1966), A theorem on finite sets, *in:* Theory of Graphs, (Proc. Colloq., Tihany, Hungary), (P. Erdős et al., eds.), Akadémiai Kiadó, Budapest and Academic Press, New York, 1968, pp. 187–207.

KATONA, G. O. H. (1974), Extremal problems for hypergraphs, *in Combinatorics*, Vol II. (M. Hall et al. eds.) Math. Centre Tracts **56**, Amsterdam, pp. 13–42.

KATONA, G., NEMETZ, T., and SIMONOVITS, M. (1964), On a problem of Turán in the theory of graphs, (Hungarian, English summary), *Mat. Lapok* **15**, 228–238.

KIM, K. H. and ROUSH, F. W. (1983), On a problem of Turán, *Studies in pure mathematics*, 423–425, Birhäuser, Basel-Boston, Mass.

KLEITMAN, D. J. and SIDORENKO, A. F. (1991), in preparation

KOSTOCHKA, A. V. (1982), A class of constructions for Turán's (3,4)-problem, *Combinatorica* **2**, 187–192.

KŐVÁRI, T., SÓS, V. T., and TURÁN, P. (1954), On a problem of K. Zarankiewicz, *Colloquium Math.* **3**, 50–57.

KRUSKAL, J. B. (1963), The number of simplices in a complex, *in* Mathematical Optimization Techniques (ed. R. Bellmann), Univ. California Press, Berkeley, 1963, pp. 251–278.

LARMAN, D. C. (1978), A note on the realization of distances within sets in Euclidean space, *Comment. Math. Helv.* **53**, 529–535.

LEHEL, J. (1982), Covers in hypergraphs, *Combinatorica* **2**, 305–309.

LOVÁSZ, L. (1979a), *Combinatorial Problems and Exercises*, Akadémiai Budapest, North-Holland Amsterdam. (Problem 13.31)

LOVÁSZ, L. (1979b), On the Shannon capacity of a graph, *IEEE Trans. Information Theory* **25**, 1–7.

MANTEL, W. (1907), Problem 28. *Wiskundige Opgaven* **10**, 60–61.

MILLS, W. H. (1979), Covering designs. I. Covering by a small number of subsets, *Ars Combin.* **8**, 199–315.

MOTZKIN, T. S. and STRAUS, E. G. (1965), Maxima for graphs and a new proof of a theorem of Turán, *Canadian Journal of Mathematics* **17**, 535–540.

PACH, J. and SHARIR, M. (1991), Repeated angles in the plane and related problems, *J. Combin. Th., Ser. A*, to appear

PIPPENGER, N. and SPENCER, J. (1989), Asymptotic behavior of the chromatic index for hypergraphs, *J. Combin. Th., Ser. A* **51**, 24–42.

PYBER, L. (1984), An extension of a Frankl-Füredi theorem, *Discrete Math.* **52**, 253–268.

PYBER, L. (1985), Regular subgraphs of dense graphs, *Combinatorica* **5**, 347–349.

RÖDL, V. (1985), On a packing and covering problem, *European J. Combin.* **6**, 69–78.

RUZSA, I. Z. and SZEMERÉDI, E. (1978), Triple systems with no six points carrying three triangles, *Combinatorics* (Keszthely, 1976), *Proc. Colloq. Math. Soc. János Bolyai* **18**, Vol. II., 939–945. North-Holland, Amsterdam-New York.

SCHRIJVER, A. (1981), Association schemes and the Shannon capacity: Eberlein polynomials and the Erdős-Ko-Rado theorem, *Proc. Colloq. Math. Soc. J. Bolyai* **25**, pp. 671–688. Szeged, Hungary, 1978, North-Holland, Amsterdam.

SIDORENKO, A. F. (1982a), Method of quadratic forms in the Turan combinatorial problem, *Moscow University Math. Bull.* **37** 1–5. MR83g:05040.

SIDORENKO, A. F. (1982b), Extremal constants and inequalities for distributions of sums of random vectors, (Russian), *PhD Thesis*, Moscow University.

SIDORENKO, A. F. (1983), The Turan problem for 3-graphs, (Russian), *in* Combinatorial Analysis, No. 6, pp. 51–57, Moscow University.

SIDORENKO, A. F. (1987), On the maximal number of edges in a uniform hypergraph that does not contain prohibited subgraphs, *Math. Notes* **41**, 247–259. Russian original: *Mat. Zametki* **41**, No.3. 433-455.

SIDORENKO, A. F. (1989a), Asymptotic solution for a new class of forbidden r-graphs, *Combinatorica* **9**, 207–215.

SIDORENKO, A. F. (1989b), Extremal combinatorial problems for spaces with continuous measure, (Russian), *Issledovanie Operatchiy i ASU*, Vol. **34**, Kiev, pp. 34–40.

SIDORENKO, A. F. (1991a), On Ramsey–Turan numbers for 3-graphs, *J. Graph Th.*, to appear

SIDORENKO, A. F. (1991b), An analytic approach to extremal problems for graphs and hypergraphs, *in* Proc. Conf. on Extremal Problems for Finite Sets, held in June 1991, Visegrád, Hungary. *Proc. Colloq. Math. Soc. János Bolyai*, to appear

SIMONOVITS, M. (1968), A method for solving extremal problems in graph theory, *in:* Theory of graphs, Proc. Colloq. Tihany, Hungary, 1966, pp. 279–319, (ed. P. Erdős et al.), Academic Press, New York.

SIMONOVITS, M. (1978), Lecture at Eötvös University, Budapest

SIMONOVITS, M. (1983), Extremal graph theory, *in:* Selected Topics in Graph Theory 2 (Beineke et al., eds.), pp. 161–200, Academic Press, New York.

SIMONOVITS, M. (1984), Extremal graph problems, degenerate extremal problems, and supersaturated graphs, *Progress in Graph Theory* (Waterloo, Ont. 1982), 419–437. Academic Press, Toronto, Ont.

SPENCER, J. (1972), Turán's theorem for k-graphs, *Discrete Math.* **2**, 183–186.

SÓS, V. T. (1976), Remarks on the connection of graph theory, finite geometry and block designs, *Teorie Combinatorie*, Proc. Colloq., Rome, 1973, Vol. II, Atti dei Convegni Lincei **17**, pp. 223–233. Accad. Naz. Lincei, Rome

SURÁNYI, J. (1971), Some combinatorial problems of geometry, (Hungarian), *Mat. Lapok* **22**, 215–230.

SZEMERÉDI, E. (1978), Regular partitions of graphs, *in* Problèmes Combinatoires et Théorie des Graphes, *Colloques Internationaux C.N.R.S.*, No. **260** Paris, pp. 399–401.

TAZAWA, S. and SHIRAKURA, T. (1983), Bounds on the cardinality of clique-free family in hypergraphs, *Math. Sem. Notes Kobe Univ.* **11**, 277–281.

TODOROV, D. T. (1983), On Turán numbers, Mathematics and mathematical education (Sl"nchev Bryag, 1983), 123–128, B'lgar. Akad. Nauk, Sofia.

TODOROV, D. T. (1984), On some Turán hypergraphs, *in:* Mathematics and mathematical education (Sunny Beach, 1984), 179–186, B'lgar. Akad. Nauk, Sofia.

TURÁN, P. (1941), On an extremal problem in graph theory, (Hungarian), *Mat. Fiz. Lapok* **48**, 436–452.

TURÁN, P. (1954), On the theory of graphs, *Colloq. Math.* **3**, 19–30.

TURÁN, P. (1961), Research problems, *MTA Mat. Kutató Int. Közl.* **6**, 417–423.

TURÁN, P. (1970), Applications of graph theory to geometry and potential theory, *in* Combinatorial Structures and their Appl., Gordon and Breach, New York, pp. 423–434.

WENGER, R. (1990), Extremal graphs with no C^4's, C^6's or C^{10}'s, preprint, Rutgers University NJ, March 1990.

WILSON, R. M. (1973), An existence theory for pairwise balanced designs I–III, *J. Combin. Th., Ser. A* **13** (1972), 220–273 and **18** (1975), pp. 71–79.

WILSON, R. M. (1984), The exact bound in the Erdős-Ko-Rado theorem, *Combinatorica* **4**, 247–257.

Printed in the United States
By Bookmasters